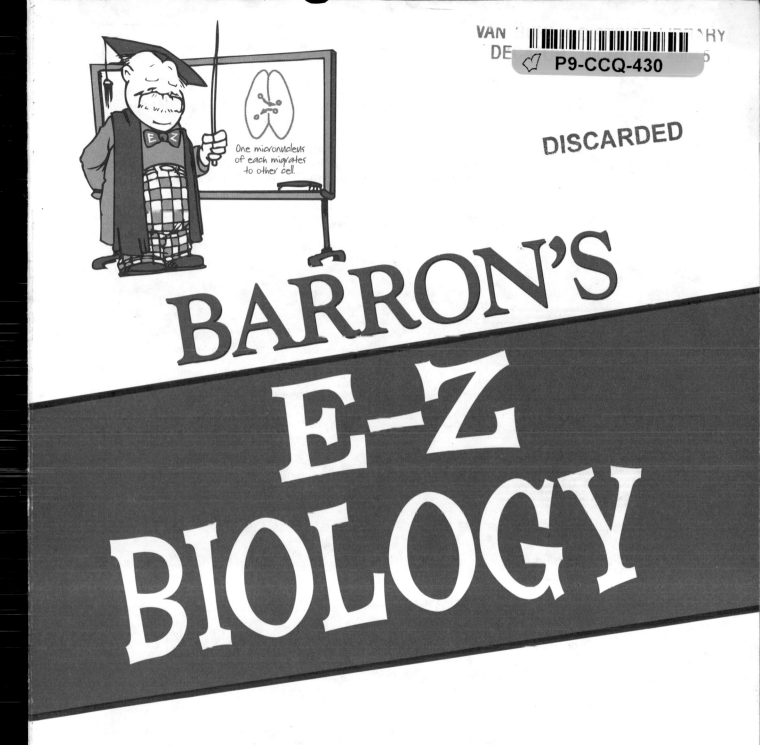

One micronucleus of each migrates to other cell.

BARRON'S E-Z BIOLOGY

Gabrielle I. Edwards
Former Assistant Principal Supervision Emerita
Science Department
Franklin D. Roosevelt High School
Brooklyn, New York

Cynthia Pfirrmann, M.S.
(revising author)
Science Department
Scotch Plains Fanwood High School
Scotch Plains, New Jersey

Better Grades or Your Money Back!

As a leader in educational publishing, Barron's has helped millions of students reach their academic goals. Our E-Z series of books is designed to help students master a variety of subjects. We are so confident that completing all the review material and exercises in this book will help you, that if your grades don't improve within 30 days, we will give you a full refund.

To qualify for a refund, simply return the book within 90 days of purchase and include your store receipt. Refunds will not include sales tax or postage. Offer available only to U.S. residents. Void where prohibited. Send books to **Barron's Educational Series, Inc., Attn: Customer Service** at the address on this page.

All inquiries should be addressed to:
Barron's Educational Series, Inc.
250 Wireless Boulevard
Hauppauge, New York 11788
www.barronseduc.com

Library of Congress Control No: 2008040996

ISBN-13: 978-0-7641-4134-8
ISBN-10: 0-7641-4134-1

Library of Congress Cataloging-in-Publication Data

Edwards, Gabrielle I.
E-Z biology / Gabrielle Edwards ; revised by Cynthia Pfirrmann.—4th ed.
p. cm.—(E-Z series)
Rev. ed. of: Biology the easy way / Gabrielle I. Edwards, 2000.
Includes index.
ISBN-13: 978-0-7641-4134-8
ISBN-10: 0-7641-4134-1
1. Biology. I. Edwards, Gabrielle I. Biology the easy way. II. Title.

QH307.2.E38 2009
570—dc22

2008040996

Printed in the United States of America
9 8 7 6 5 4 3

CONTENTS

PREFACE

E-Z Biology presents an up-to-date account of the principles and concepts of modern biology. Over the past four decades there has been an explosion of knowledge in the biological sciences, which has revolutionized our way of thinking and doing in terms of the environments of the world and the plants and animals that live therein. As a consequence of the many new discoveries in biology, some of the traditional beliefs have been discarded, others have been modified, and much new material has been added to replace the invalid theories of old. A direct result of the increase in biological knowledge has been the increase in size and weight of the modern biology textbook. Although all of the newer knowledge in biology cannot be contained under one cover, the standard high school and college biology texts have become volumes of substantial weight and cost.

E-Z Biology was prepared with the reader in mind. Although it contains the important information of modern biology, the writing is concise. The content material is presented clearly in language that is easy to read and to understand. The clarity of language serves to make difficult concepts more understandable. Technical language is avoided wherever possible. However, biology has its special vocabulary that is essential for description. The words new to the reader are presented in **boldface**, clearly defined and used appropriately in the text material. The overall quality of the writing in *E-Z Biology* is lively, modern, and interesting.

E-Z Biology is divided into 18 chapters. Each of these chapters represents an area of specialization in the field of biology. For each such area modern principles of biology are presented in an appealing way for the reader. In each chapter, the content discussion is accompanied by carefully placed line drawings of the organisms, organs, structures, or processes under discussion. Wherever possible, summary materal has been presented in tabular form, providing a quick means of study and review of a given topic.

Each chapter in *E-Z Biology* provides special study aids that are designed to enhance the learning and understanding of the biological principles or concepts under study. Because scientific discovery and invention cannot be divorced from the history of human life on Earth, a "Chronology of Famous Names in Biology" has been included at the end of each chapter. Men and women scientists representing all countries, races, and religions are included in the chronology. The contributors mentioned range in time from the ancient Greek philosopher-scientists to modern day investigators. For each name the approximate date of the discovery or invention, the country where work was done, and a brief summary of the accomplishment is given. "Connecting Through Chapter Review" is the inclusive title of the end of chapter materials that provide summary-review activities developed to connect the reader to the biology topic under study. The "Word-Study Connection" lists the vocabulary of the chapter. The reader is encouraged to review and learn the meaning of each term.

The "Self-Test Connection" includes 30 questions or more in three types of short-answer tests that are designed to assess the reader's grasp of some of the important facts and concepts in biology. Answer keys are provided to assist with self-evaluation. "Connecting to Concepts" provides open-ended questions to encourage the reader to think about and discuss concepts appearing in the chapter. "Connecting to Life/Job Skills" invites the reader to extend the biology information just learned into the living community through life skills and career information. Using reference materials in library media centers or gathering information by way of a personal computer are skills that are useful for life. Learning about careers related to biology expands one's knowledge of the kinds of opportunities available for education beyond high school and the need for science-trained persons in the everyday world of work. This section also invites you to look at the biological events taking place in your community and to assess the effects of environmental conditions.

By reading the Table of Contents you can see the range of topics that is included in this text. By design, *E-Z Biology* is versatile, having appeal and use for students in secondary school through first year college and also to adults who are not in school but who desire to learn some modern biology. A chapter may be read in its entirety or in parts. The information is presented so that one can skim or read deeply. The index provides topic-page information.

The principle aim of *E-Z Biology* has been to present the facts of biology as they are known today in such a way that the curiosity and interest of the reader are aroused.

Gabrielle I. Edwards
Cynthia Pfirrmann

How to Use This Book

E-Z Biology presents the principles and concepts of biology in easy-to-understand language using illustrations that simplify the text material further. To get the most in study help from *E-Z Biology*, however, you must use it efficiently.

Outlined below are procedures for self-study that will enable you to learn the content of biology efficiently and easily.

1. SET GOALS AND OBJECTIVES

As you begin the study of biology, you must set personal goals to chart and direct your learning objectives. A **goal** is something that you wish to achieve over a period of time. **Objectives** are short-term targets that help you reach a particular goal. For example, let us suppose that your goal is to learn the ideas of biology so well that your test scores reflect mastery. You can reach this goal by establishing objectives that will enable you to be successful. These objectives may center around the conditions for study. Therefore:

- Establish a set time for study.
- Find a place that is conducive (helpful) for study: a comfortable place, but one in which you can remain alert.
- Gather together the materials for study: *E-Z Biology*, dictionary, notebook, pens, pencils.
- Develop a study plan that will direct your course of study.
- Develop a system for note-taking.
- Learn how to use the features of this book that will help you learn.

2. FOCUS

Begin with Chapter 1. Read the chapter opener and sections in this chapter box to learn what the chapter will cover. Look through the entire chapter by scanning each page as you turn to it. Note especially terms in boldface type. Look at the illustrations and read their captions. Scan the "Word-Study Connection" list. Look over the "Self-Test Connection" questions at the end of the chapter. This procedure will help you develop a mindset for the material that you are about to learn. Avoid distractions. Turn off the radio, television, and MP3 player. Be aware also that you cannot talk on the telephone and study at the same time.

3. READ

Begin to read the chapter. Read slowly. At times you may wish to read aloud. You may also have to read a selection several times in order to grasp the meaning. Be patient. Read with deep concentration. Read only about four pages at a time, stopping at some logical place. For example, in Chapter 1 read the first three and two-thirds pages, stopping at "The Work of the Modern Biologist." Review what you have just read by skimming through the material again and by looking for familiar terms in the word list. Now complete the reading of the chapter. Review by skimming. Study the illustrations in detail.

4. THINK

Think about the chapter material. The Word-Study Connection list at the end of the chapter will help you. For each word, follow this study tip: (1) Write down the word. (2) Now try to write its meaning. Do not attempt to memorize definitions; they are easily forgotten. Learn new vocabulary in ways that are understandable to you. (3) Try to gain the meanings of words new to you from the sentences in which they are written. Think about the material you have read so that it becomes meaningful to you.

5. REVIEW EXERCISES

When you have finished reading and taking notes on the chapter material, turn your attention to the review section titled "Review Exercises for Chapter." Begin with Part A of the Self-Test Connection. Read each question carefully, and write down the answer. Now check your answers with the answer key. For wrong answers, locate the appropriate material in the chapter text and review. Repeat this procedure for Parts B and C.

6. CONNECT

Making connections between your study of biology and the world around you is important to the quality of your life. Making these connections will not only strengthen your understanding of the science but will also enable you to apply your knowledge of biology. *E-Z Biology* connects past discoveries to current information made possible by technology. The notes section "Connecting Life/Job Skills" guides you in developing skills useful in daily life and in the world of work. You can make connections to the history of biology by perusing the "Chronology of Famous Names in Biology" at the end of each chapter.

Using *E-Z Biology* efficiently will yield learning dividends.

1

Biology: The Science of Life

WHAT YOU WILL LEARN

In this chapter you will review how the science of biology began and how modern biologists gather information.

SECTIONS IN THIS CHAPTER

- The Science of Biology
- Branches of Biology
- The Work of the Modern Biologist
- Review Exercises for Chapter 1
- Connecting to Life/Job Skills
- Chronology of Famous Names in Biology

The Science of Biology

LEARNING ABOUT LIFE

Science investigates the forces that affect the Earth and its people. Through observation and experimentation, scientists strive to solve the tricky puzzles that hide the causes of events that shape the world. The work of modern science has changed the world remarkably with new and important discoveries in areas such as communication, travel, agriculture, and medicine. How scientists do their work will be discussed in this chapter and in chapters to follow. *Biology is the science that studies life and living things, including the laws that govern the events of life.*

Every type of life from the smallest microscopic living particle to the largest and most imposing plant and animal species is included in the study of biology. Biological study covers all that is known about any plant, animal, microbe, or other living thing of the past or present.

As defined above, **biology** is the science of life and living things. Biology is a *natural* science. It is the study of individual life forms within the world of life known as *nature*. It is the science of fishes and fireflies, grasses and grasshoppers, humans and mushrooms, flowers and sea stars, worms and molds. It is the study of life on top of the highest mountain and at the bottom of the deepest sea.

Biology is the accumulated knowledge about all living things and the principles and laws that govern life. Whole living things are called **organisms**. Organisms too small to be seen by the unaided eye are termed **microorganisms** or **microbes**.

Those who specialize in biology are known as *biologists* or *naturalists*, and it is through their observations of nature and natural events that the great ideas of biology have been born.

BIRTH OF BIOLOGY

The object of scientific study is to find the truth. So it was and still is that human beings have sought the truth about the nature of life. We cannot fully appreciate the present without knowing what has occurred in the past. The birth of biology as a bona fide science was slow and painful, taking place over many centuries. Study Table 1.1 below, and notice the length of time in centuries over which the ancient Greek naturalists studied and wrote about plants and animals. Now note in Table 1.2 on page 3 the number of centuries between Vesalius and Lamarck. You can see that each discovery in biology took a very long time. Learning about the historical development of the ideas of science provides a better understanding of the process and methods of inquiry. Learning about the individuals who have contributed to scientific knowledge increases our understanding of science and society.

THE STUDY OF LIFE THROUGHOUT HISTORY

The study of life is as old as humankind, dating back to ancient peoples who observed and wondered about the characteristics of the animals and plants around their limited sphere. The ancients used their observations to help them in such activities as hunting, food gathering, and crop growing.

A great deal of credit is due to the ancient Greeks for having begun a systematic study of living things, including human beings. The rise of ancient science reached a peak around A.D. 200 and then declined sharply in the European culture. During this time books were scarce, and questions were considered to be inappropriate and impious prying. These centuries of scientific regression, from A.D. 200 to A.D. 1200 were known as the Dark Ages. Religious authority was considered to be the irrefutable source of all knowledge. Individual observation and dissent were discouraged and often punished. During these centuries, though, scientific exploration and discovery were thriving in the Islamic culture. Scientific methods were refined, questions were investigated and explained, and old ideas were challenged and revised. Table 1.1 provides a brief summary of some early scientific achievements.

TABLE 1.1
SOME EARLY MILESTONES IN BIOLOGY

Hippocrates (460–370 B.C.)	Founded the first medical school, located on the Greek island of Cos.
Aristotle (382–322 B.C.)	Developed the first organized study of natural history; was a keen observer, writer, and illustrator of plants and animals.
Theophrastus (380–287 B.C.)	Specialized in the organized study of plants; is called the "ancient father of botany."
Galen (A.D. 130–200)	Became an important authority on anatomy.
Abu al-Abbas al-Abati (A.D. 1200)	Emphasized the importance of observation, experimental design, and empirical assessment of data
Ibn al-Nafis (A.D. 1213–1288)	Used experimental dissection and autopsy; corrected many of Galen's anatomical inaccuracies; was the first to correctly describe cardiovascular circulation and metabolism.

REVIVAL OF LIFE STUDIES

The 14th century ushered in a renewal of scientific thought and inquiry in the European culture. Among the reasons for the change in attitude were the invention of the printing press, the voyages of the explorers, the expansion of ideas brought on by the Crusades, and the rise of universities. All of these things contributed to a return to the study of nature and to the methods of science.

Some important contributions to the understanding of the laws of natural science between the 16th and 19th centuries are listed in Table 1.2.

TABLE 1.2
DISCOVERIES IN BIOLOGY BETWEEN THE 16TH AND 19TH CENTURIES

Andreas Vesalius (1514–1564)	Studied the human body by dissection, ignoring the authority of Galen.
Marcello Malpighi (1628–1694)	Described the metamorphosis (body change) of the silkworm.

William Harvey (1578–1667)	Demonstrated the path of the blood in the human body.
Robert Hooke (1635–1703)	Discovered and named "cells" in cork.
Anton van Leeuwenhoek (1632–1723)	Was the first person to see living cells.
Carolus Linnaeus (1707–1778)	Devised the system of *binomial nomenclature*, that is, the genus/species naming of plants and animals.
George Cuvier (1769–1832)	Founded the study of comparative anatomy.
Jean Baptiste Lamarck (1744–1829)	Devised the word *biology* by putting together two Greek words: *bios*, meaning "life," and *logus*, meaning "study."

Branches of Biology

Biology is an extensive science. It includes many life science disciplines that cover enormous areas of study and information. Biology has been divided into areas that allow for specialization of study. Several branches of biology are summarized in Table 1.3.

TABLE 1.3
BRANCHES OF BIOLOGY

Branch	Area of Study
Botany	Plants and their life cycles
Zoology	Animals and their life histories
Anatomy	Body structure visible to the naked eye
Physiology	Functions of body organs and systems
Embryology	Embryo development in plants and animals
Genetics	Inheritance and variations in living organisms
Evolution	Origins and relationships among living species
Ecology	Environmental relationships among plants and animals
Taxonomy	Naming and classification of organisms.

Biologists call each different kind of organism a **species** (the word "species" is both singular, as in 1 species, and plural, as in 10 species). Biostatistical methods indicate that there are as many as 30 million different species living on Earth, with as many as 1.8 million species identified and named. Scientists agree that many living species have not yet been discovered, and it is likely that only about 5 percent of living species have been described. Biologists who specialize in the study of ancient life believe that the number of extinct species is likely between 15 and 16 million.

Investigators of the 18th and 19th centuries were faced with the enormous task of trying to put some kind of order into the study of natural science. Thus, work during these centuries involved sorting, identifying, naming, describing, and classifying the large numbers of species that were being discovered.

The use of technology to gather information about living species has led to an explosion of data. Therefore the science of biology has become so extensive that it has been necessary to divide the branches of biology summarized in Table 1.3 into narrower areas of specialization. As you read this book, you will become familiar with a number of other divisions of the biological sciences.

The Work of the Modern Biologist

Toward the end of the 20th century, new infectious diseases began to emerge. Among these are *HIV* and *AIDS*, *Lyme disease*, *Legionnaires' disease*, and *Escherichia coli* infections. **Infectious diseases** are caused by *microorganisms* (germs) that enter the body and do it harm. In the Southwest, victims were recently attacked by the fatal hanta virus, a respiratory disease carried by rodents. Late in the summer of 1999, New York was hit with a fatal outbreak of West Nile fever, an African strain of *encephalitis* (en-sef-a-litus) spread by mosquitoes, which claimed seven lives. Now in the 21st century, public health scientists are alarmed by the rapid spread of *hepatitis C*, *malaria*, and strains of drug-resistant *tuberculosis*, which may infect 50 million people throughout the world.

Research biologists are faced with the monumental tasks of sorting out the life histories of microorganisms that can cause serious epidemics and finding ways to lessen their effects. **Epidemiologists** are scientists who track the pathways of disease outbreaks. They count and record reported cases of diseases to show patterns of occurrences. **Research physicians** and **pharmacologists** (scientists who devise plans for medicines) muster every effort to develop vaccines to prevent the spread of infectious diseases and antibiotics to cure infected persons. **Environmental scientists** study the environment, seeking ways to wipe out populations of mosquitoes, rodents, ticks, and other carriers of diseases that affect human populations.

Another kind of research biologist is the **geneticist** (jen-et-i-sist). These biologists specialize in the study of *genes*. Genes are the molecules (very small chemical particles) of heredity that may provide natural immunity against certain germs that cause disease. **Genetic engineering** uses microorganisms to make genes produce chemical substances needed to protect the body against infection. **Biochemists** and **biophysicists** devote their efforts to **technology**, devising equipment to diagnose and treat diseases and malfunctioning body organs.

The work of the modern biologist is varied. Biologists apply their knowledge in different ways. Some researchers work in private industry producing commercial food products for today's consumer. Others apply their knowledge to forestry, and some to agriculture. Still others specialize in the study of domestic animals. There are research workers who concentrate on the nonliving environment, developing ways to maintain clean water and fresh air. Research scientists try to find answers to their questions by working through the problem-solving steps of the **scientific method**.

THE SCIENTIFIC METHOD AND PROBLEM SOLVING

Research scientists strive to unravel the secrets involved in specific scientific problems. For example, a biophysicist works to devise a method of measuring particles too small to be distinguished by even the most powerful microscopes. A biologist seeks to find the cause of a newly occurring disease. A biochemist aims to identify minute (mine-ute) quantities of a strange chemical in the lungs of a deep-sea diver. A geologist attempts to determine the age of a certain rock formation. These scientists have identified problems for which they seek solutions. The solutions to the problems depend on the orderliness of experimental approach as outlined in the steps of the scientific method.

STEPS IN THE SCIENTIFIC METHOD

The first step in the scientific method is *observation*. Observation is the accurate visual noting of an event or an object. The accuracy of an observation becomes verified when a number of people who are not connected to each other agree on what they see in regard to the same object, event, or set of circumstances. For example, early in the 16th century, Vesalius presented the first accurate description of the arteries and veins in the human body. His drawings were based on firsthand observations obtained through dissection of human bodies. Over a period of 400 years succeeding generations of anatomists have attested to the accuracy of Vesalius' work. It must be noted, however, that although several trained observers may agree on the same set of observations, the conclusions they draw from these observations may differ. A *conclusion* is a judgment reached by reasoning. Therefore, problem solving depends on other activities that enable the investigator to form conclusions based on accurate observations.

An observation becomes useful only in terms of the conclusion that can be drawn from it. A tentative conclusion drawn from a set of observations is known as a hypothesis. A *hypothesis* is a very general statement made to tie together things observed prior to experimentation.

An *experiment* requires a set of procedures designed to provide data about a well-thought-out problem. The purpose of an experiment is to answer a question truthfully and accurately. A reliable experiment requires experimental conditions and control procedures. A *control* is a check on a scientific experiment and differs from the experimental procedure in one condition only. An experiment is designed to test a specific hypothesis.

A nontestable hypothesis is not useful in problem solving because it does not lend itself to experimental conditions. An example of a nontestable hypothesis follows: blowflies think that meat is an excellent site for egg laying. Because there is no way to test the thought processes of a blowfly, the hypothesis is faulty.

REDI'S BLOWFLY EXPERIMENT

A classical work illustrating the effectiveness of a well-designed experiment is *Redi's blowfly experiment*. That living things arise from lifeless matter was a popular 17th century belief known as **spontaneous generation**. In 1650, the Italian physician Francesco Redi designed and performed an excellent experiment that disproved the notion of spontaneous generation of blowflies from decaying meat:

> In each of three jars, Redi placed the same kind and quantity of meat. The first jar was left uncovered. The second jar was covered with a porous cheese cloth. The third jar was covered with a parchment thick enough to prevent the odor of the meat from diffusing to the air outside of the jar (Figure 1.1). Flies laid their eggs on the meat in the open jar and on the cheese cloth which covered the second jar. Eggs were not laid on the parchment covering the third jar. Redi watched as the eggs hatched into larvae (maggots) and observed the change of the larvae into adult flies. There were neither maggots nor flies on the meat in the third jar.
>
> On the basis of this experiment, Redi concluded that maggots or flies do not come from decayed meat, but from the fertilized eggs laid by the female fly.

FIGURE 1.1 Redi's blowfly experiment. Can you explain the results?

When an hypothesis is supported by data obtained by experimentation, the hypothesis becomes a *theory*. A theory that withstands the test of time and further experimentation becomes a *law* or *principle*. A *concept* is a blanket idea covering several related principles.

THE TOOLS OF THE BIOLOGIST

Problem solving by the modern biologist requires the use of some very special tools and laboratory techniques. The scientific method is a major idea of biology upon which all research is built. All research begins with observation, a behavior that requires training, skill, and concentration. To increase the accuracy of observation, the research biologist uses tools and techniques that will extend his or her vision and other senses. Because human sensing ability is limited and variable and not really reliable for seeing and recording fine and sometimes fleeting detail, modern scientists make use of instruments and methods that enable them to perceive, measure, and record data precisely. To *perceive* means to understand through sight.

Measurement is a basic requirement of all research. An investigator must answer questions concerning quantity, length, thickness, periods of time, mass, weight, and the like within a hair's breadth of accuracy. Modern instruments of measurement permit astounding degrees of numerical precision. *Biometrics* is the science that combines mathematics and statistics needed to deal with the facts and figures of biology. Biologists handle enormous numbers that must be organized and simplified so that they become useful in the analysis of data. Computers aid the investigator's problem-solving tasks by accepting huge quantities of numerical data and making calculations at lightning speeds. The expression **numerical data** refers to numbers, figures, and statistics used by scientists and obtained by measuring.

The research biologist gets numerical data from using all kinds of measuring instruments designed for remarkable precision. Weighing in micro quantities requires the use of an *analytical balance* that is capable of weighing accurately to infinitesimal fractions of micrograms. A **microgram** is a unit of measure indicating *one millionth* (10^{-6}) of a gram. A gram is about the size of a green pea. The *manometer* measures the uptake of gases involved in cellular respiration and photosynthesis. *Spectrophotometers* are able to measure differences in densities of fluids by color comparisons not seen by the human eye. Radioactivity is located and measured by the *Geiger counter*, a machine structured to feed pulses of electricity into an electronic counter. *Scintillation counters* measure light flashes imperceptible to the eye, while ultraviolet rays are used to measure objects thinner than onion membrane.

SEPARATING SUBSTANCES

Later in this book you will learn that all living things are built of microscopic structures called *cells* or are made of cells.

Discovery in biology involves the ability to take apart the substances of cells so that the secrets of their biochemical activities can be exposed. The technique of *chromatography* permits the biologist to separate unbelievably small quantities of a substance into its many parts. For example, 1/100,000 of a gram of protein can be separated into its basic amino acid building blocks in one hour. Before the discovery of chromatography, a scientist needed 100 grams of protein and one year's time in order to make this separation. As the name implies, chromatography is "color writing," a technique that uses paper or a column of chalk as a stationary phase. The test sample, dissolved in a suitable solvent, is allowed to travel up the paper or through the chalk column; this movement is called the mobile phase. The constituents of the sample settle out at various levels as bands or dots of color (Figure 1.2). *Electrophoresis* is a technique that uses electrical charges to separate the amino acids in proteins dotted on specially treated glass plates (Figure 1.3).

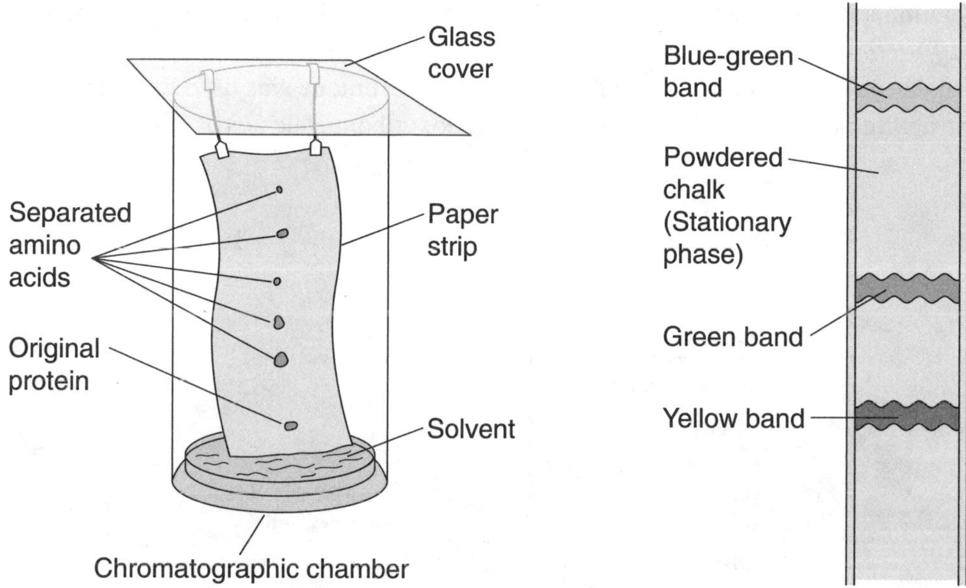

FIGURE 1.2 Paper chromatography of a protein sample

FIGURE 1.3 Column chromatography of green plant pigments

TAKING THINGS APART

Biologists learn about how living things are built (structure) and how they work (function) by observation and measurement, by separating substances and analyzing them, and also by taking things apart. The taking apart of a plant or animal body is called **dissection**. Medical students dissect cadavers (dead bodies) to learn the locations of body organs. Body organs are made up of *tissues*. Tissues are composed of cells. Cells can be seen only through a microscope.

A biologist wishing to study the processes that go on in cells must take the cells apart to separate the parts contained in the cell. (Look at the diagram of the cell on page 31.) The parts that make up cells are called *organelles*. Using two pieces of equipment, a *blender* and a *centrifuge*, the biologist is able to break open cells and obtain the cell parts for study. In Chapter 3 you will learn more about this technique.

GETTING A BETTER LOOK AT SMALL THINGS

In 1898 William Conrad Roentgen discovered X rays, that is, electromagnetic radiation of short wavelength that carries a large amount of energy. X rays are able to pass through opaque (dense) body structures, such as bone, and form an image on a photographic plate. The lengths of the various waves in the electromagnetic spectrum are shown in Figure 1.4.

A number of techniques and instruments have helped scientists see things better. *X-ray diffraction* is a method by which X rays sent through a crystal show the pattern of molecules and atoms contained in the crystal. This technique was used to determine the structure of the hereditary material DNA (deoxyribonucleic acid).

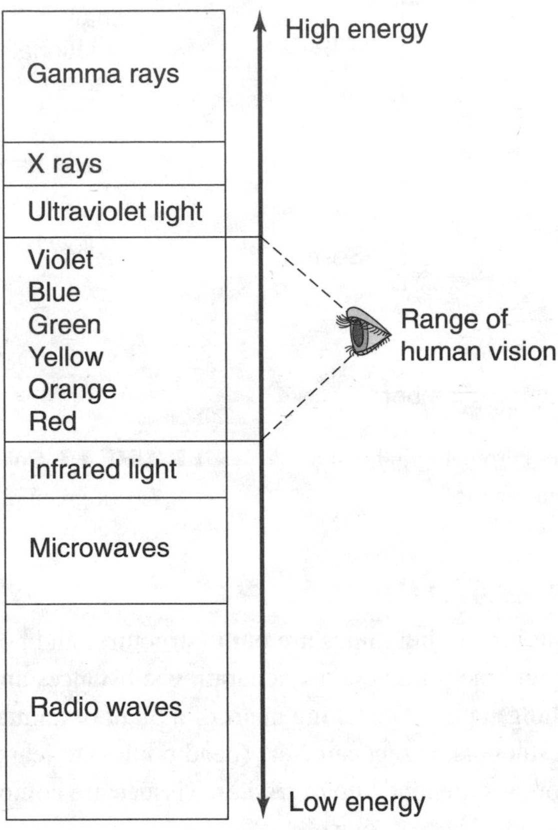

FIGURE 1.4 Range of wavelengths in the electromagnetic spectrum. Note the energy level of the X rays.

To study normal body functions and to diagnose disease conditions, physicians make use of scanning machines that produce computerized images of body organs. These scanners detect infinitesimally small tumors and other tissue abnormalities that escape imaging by standard X-ray techniques. Modern biologists make use of the knowledge of chemistry and physics as well as the technology of engineers to extend their senses and improve observation.

Magnification is the measure of how much a microscope can enlarge specimens under view. The quality of a good microscope is based on its *resolving power*, that is, its ability to distinguish two closely positioned points. The resolving power of the normal unaided human eye is 0.1 millimeter.

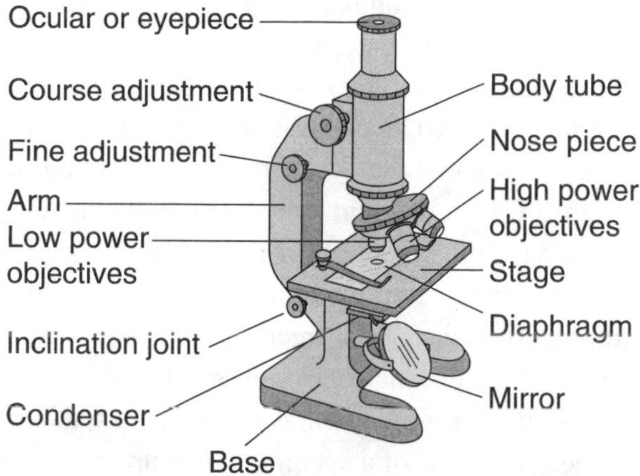

FIGURE 1.5 Parts of the compound (light) microscope

The best *light microscope* is capable of magnifying objects 2,000 times (Figure 1.5). When microscopists view a specimen through the light microscope, not only are they interested in its shape and fine detail, but also they are concerned about its size. The unit of measurement used for light microscopy is the **micrometer** (μm). In relationship to other measures, the micrometer is equal to *one thousandth* (10^{-3}) of a millimeter (mm); 1000 nanometers (nm); 10,000 Angstrom units (Å).

When using the compound light microscope, the size of a specimen may be estimated. On the low-power objective, the diameter of the field is indicated by a number, usually 1.6 mm. This means that the diameter of the low-power field is approximately 1600 μm. On the high-power objective, the diameter of the field is usually indicated as 0.4 mm, so the diameter of the high-power field of vision is 400 μm, one-fourth the length of the low-power field. Thus, if a specimen is one-half the diameter of the field under low power, it will be 1/2 of 1600 μm, that is, 800 μm. This type of measurement is an estimate. See Figure 1.6.

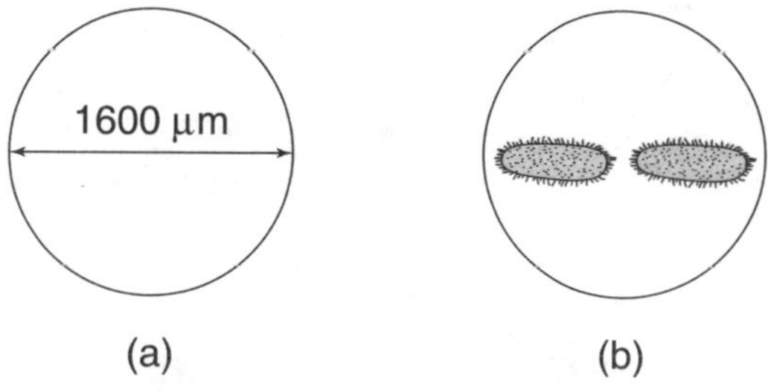

FIGURE 1.6 (a) Diameter of the field of vision under the low power of the light microscope. (b) Estimation of size.

For more exacting measurements an *ocular micrometer* is used. An ocular micrometer is a glass disk on which a micrometer scale is etched. The disk is inserted into the eyepiece (ocular) of the microscope. The scale on the ocular micrometer has 100 equal divisions, which must be calibrated against the units on a stage micrometer. This is done by superimposing the scale of the ocular micrometer on the scale of the stage micrometer. Study Figure 1.7. Notice that the 0 lines on the ocular and stage micrometers coincide. Now notice the alignment between line 20 of the ocular micrometer and the ninth line on the stage micrometer. Each division on the stage micrometer is equal to 10 μm. From 0 to line 20 on the ocular micrometer is equivalent to eight spaces on the stage micrometer. Thus the distance between 0 and 20 on the ocular micrometer is 80 μm. The ocular micrometer is now *calibrated* against a standard and can be used to make more accurate measurements of a specimen than are possible by estimation.

The *phase contrast microscope* makes living transparent specimens visible, while the *darkfield microscope* or the *ultramicroscope* gives greater clarity to transparent living organisms such as the spirochetes that cause syphilis. The *ultraviolet microscope* is used for photographing living bacteria and naturally fluorescent substances.

Most amazing of all is the electron microscope. It is capable of magnifying objects more than 200,000 times. Using electrons instead of light and magnets to direct the electrons, the electron microscope has revolutionized the study of the cell (Figure 1.8).

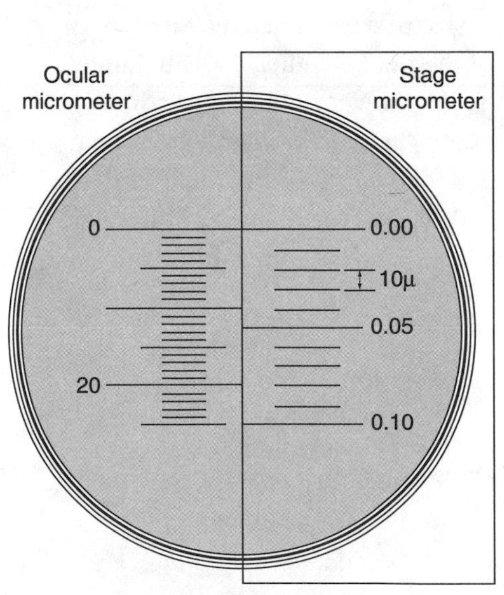

FIGURE 1.7 Calibration of the ocular micrometer

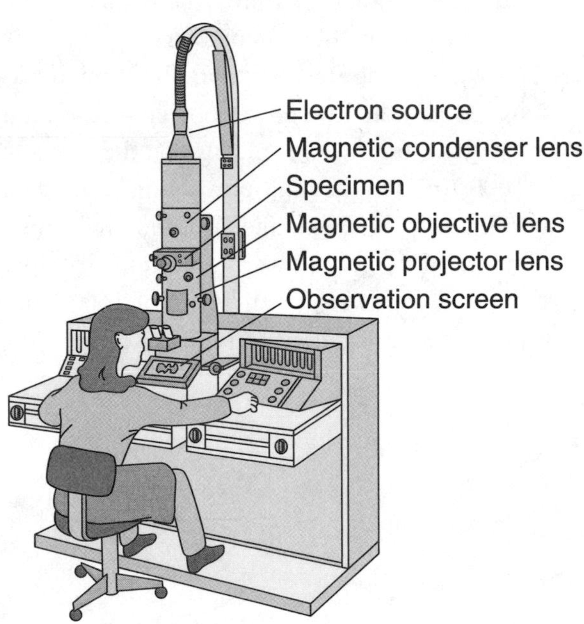

FIGURE 1.8 A biologist uses the electron microscope.

The *scanning electron microscope* (SEM) has improved upon the resolution of the electron microscope. A fine probe directs and focuses electron beams over the material being studied. This makes possible quick scanning and allows for the imaging of intact surface structures. The *transmission electron microscope* (TEM) allows us to magnify and view internal structures and cross sections of specimens. We can see cellular organelles and their internal structures using the TEM. The *high-resolution TEM* (HRTEM) allows us a view of molecules and has a resolving power up to approximately 0.78 angstroms (more than 50 million times magnification). The STEM combines scanning and transmission electron microscope technologies and allows for high resolution in viewing detailed structures of cellular anatomy.

The *high-voltage* electron microscope is a tremendous three-story structure that outputs about 3 million electron volts. This enormous amount of energy permits specimens to be viewed without the thin slicing required by other electron microscopes. In addition, the high-voltage electron microscope produces a three-dimensional micrograph (photograph) not possible with other electron microscopes.

SCIENCE AND TECHNOLOGY

Mention has been made herein of only a small sampling of the kinds of tools and techniques available to the modern biologist. There are also incubators for temperature control, refrigerators for keeping things cool, autoanalyzers for separating and classifying the elements in the blood, automatic mixers and stirrers and shakers, pH meters, autoclaves for sterilizing, timers and machines that measure time to the fraction of a second.

REVIEW EXERCISES FOR CHAPTER 1

WORD-STUDY CONNECTION

analytical balance
anatomy
biology
biometrics
botany
chromatography
concept
conclusion
controlled experiment
cytology
ecology
electron microscope
electrophoresis
embryology
epidemiologist
evolution
Geiger counter
genetics
gram
hypothesis
infectious disease
magnification
manometer
microbe
microgram
microorganism
naturalist
nature study
observation
organism
pharmacologist
physiology
resolving power
science
scientific method
scintillation counter
spectrophotometer
spontaneous generation
theory
X rays

SELF-TEST CONNECTION

PART A. Completion. *Write in the word that correctly completes each statement.*

1. The science that specializes in the study of all life is _____.
2. The ancient who founded the systemized study of natural history was _____.
3. The first person to study the dissected human body was _____.
4. The word biology was coined by _____.
5. In general, the world of life is known as _____.
6. Any individual living plant or animal is appropriately termed a (an) _____.
7. Living things too small to be seen with the naked eye are called _____.
8. The study of plants and animals in their natural environment is known as the science of _____.
9. The name *Galen* is correctly associated with the founding of the science of _____.
10. The circulation of blood in the human body was first demonstrated by _____.
11. A tentative conclusion based on observation is the _____.
12. Redi disproved the theory of _____ (2 words) of flies.
13. Laboratory specimens that cannot be seen by the naked eye can be studied by using a light _____.
14. The word *species* refers to each different kind of _____.
15. A device used to locate and measure radioactivity is the _____.
16. Figure 1.4 indicates that gamma rays, X rays, and ultraviolet light carry a great deal of _____.
17. An ocular is an essential part of the _____ microscope.
18. The word *micrograph* is best associated with the _____ microscope.
19. X rays were discovered by _____.
20. The ability of a microscope to enlarge an image is called _____.

PART B. Multiple Choice. *Circle the letter of the item that correctly completes each statement.*

1. An example of a natural science is
   (a) geology
   (b) biology
   (c) physics
   (d) chemistry

2. Theophrastus is known as the "ancient father of
   (a) biology"
   (b) botany"
   (c) bacteriology"
   (d) biometrics"

3. The "cells" in cork were named by
   (a) Hooke
   (b) Plato
   (c) Galen
   (d) Tisellius

4. The study of comparative anatomy was founded by
   (a) Malpighi
   (b) Lamarck
   (c) Harvey
   (d) Cuvier

5. The system of binomial nomenclature was developed by
   (a) Lamarck
   (b) Linnaeus
   (c) Leeuwenhoek
   (d) Lintel

6. The branch of biology that deals with function is
   (a) zoology
   (b) taxonomy
   (c) physiology
   (d) morphology

7. Included in the study of botany would be
   (a) dandelions
   (b) sparrows
   (c) guppies
   (d) skates

8. Inheritance and variation in living things are explained by the science of
   (a) ecology
   (b) genetics
   (c) paleontology
   (d) taxonomy

9. The study of animals and their life histories is known as
   (a) ichthyology
   (b) herpetology
   (c) anatomy
   (d) zoology

10. X-ray diffraction techniques were used to determine the structure of
    (a) ATP
    (b) NAD
    (c) DNA
    (d) PGA

11. The science of development is called
    (a) cytology
    (b) embryology
    (c) systematics
    (d) pteridology

12. A check on an experiment is a (an)
    (a) observation
    (b) theory
    (c) control
    (d) conclusion

13. A blanket idea covering several related principles is called a
    (a) theory
    (b) hypothesis
    (c) concept
    (d) conclusion

14. A microgram or its fraction can be weighed most accurately on a balance known as a (an)
    (a) analytical
    (b) spring
    (c) triple beam
    (d) double platform

15. Respiratory gases can be measured by using a device called a (an)
    (a) calorimeter
    (b) manometer
    (c) arometer
    (d) anemometer

16. The quality of a microscope is judged by its
    (a) volumetric capacity
    (b) light strength
    (c) magnification power
    (d) resolving power

17. A three-dimensional photograph of a specimen is made possible by the type of microscope known as
    (a) phase contrast
    (b) high-voltage electron
    (c) scanning electron
    (d) ultraviolet

18. A scientist who seeks to solve biological problems by applying the principles of the scientific method is appropriately called a
    (a) theoretical analyst
    (c) disease specialist
    (b) technological engineer
    (d) research worker

19. The types of substances used to prevent disease are
    (a) antidotes
    (b) vaccines
    (c) hormones
    (d) antibiotics

20. Scientists who specialize in tracking the pathways of disease outbreaks are
    (a) botanists
    (b) epidemiologists
    (c) chemists
    (d) pathologists

PART C. Modified True-False. *If a statement is true, write "true" for your answer. If a statement is incorrect, change the underlined expression to one that will make the statement true.*

1. The purpose of the scientific method is to devise scientific problems.
2. Experimentation is the first step in the scientific method.
3. The 17th-century notion of spontaneous generation supported the belief that decaying meat produces flies.
4. Malpighi was the first person to describe the change in body color of the silkworm.
5. Physiology is the branch of biology that studies body function.
6. A hypothesis is a tentative theory obtained from a set of observations.
7. Another term for a disease-producing microorganism is gene.
8. The control jar in Redi's blowfly experiment was covered with gauze.
9. Figures 1.5 and 1.8 show that magnetic lenses are part of the compound microscope.
10. AIDS is a respiratory disease carried by rodents in the Southwest.
11. Biochemists are scientists who develop formulas for medicines.
12. A proved hypothesis becomes a theory.
13. A combination of mathematics and statistics applied to facts of biology is known as energetics.
14. Amino acids in proteins can be separated by using electrical charges in a process known as X-ray diffraction.
15. The microscope that can magnify more than 200,000 times is the electric microscope.
16. The ability to distinguish two closely positioned points separately is known as reduction power.
17. Living, transparent microscopic organisms are best viewed by using either an electron or phase contrast microscope.

18. The human eye cannot resolve electrons.

19. Chemical engineering uses microorganisms to make genes produce protective chemical substances.

20. A cytologist studies the mind.

CONNECTING TO CONCEPTS

1. Why is it reasonable to believe that the study of life is as old as humankind?

2. Why is the science of biology divided into so many fields of study?

3. What was the control in Redi's blowfly experiment?

4. Why are numbers and mathematics so important to the modern biologist?

ANSWERS TO SELF-TEST CONNECTION

PART A

1. biology
2. Aristotle
3. Vesalius
4. Lamarck
5. nature
6. organism
7. microorganisms
8. ecology
9. anatomy
10. Harvey
11. hypothesis
12. spontaneous generation
13. microscope
14. organism
15. Geiger counter
16. energy
17. light
18. electron
19. Roentgen
20. magnification

PART B

1. **(b)**
2. **(b)**
3. **(a)**
4. **(d)**
5. **(b)**
6. **(c)**
7. **(a)**
8. **(b)**
9. **(d)**
10. **(c)**
11. **(b)**
12. **(c)**
13. **(c)**
14. **(a)**
15. **(b)**
16. **(d)**
17. **(b)**
18. **(d)**
19. **(b)**
20. **(b)**

PART C

1. solve
2. observation
3. true
4. true
5. true
6. conclusion
7. germ
8. parchment
9. electron
10. hanta virus
11. pharmacologists
12. true
13. biometrics
14. electrophoresis
15. electron
16. resolving power
17. darkfield
18. true
19. genetic
20. cell

CONNECTING TO LIFE/JOB SKILLS

You should become familiar with the metric system of measurement, which is used in all sciences. The metric system is based on multiples of 10. The unit of *length* is the **meter**. The unit of *mass* is the **gram**. The unit of *volume* is the **liter**. Turn to Appendix A and read the metric-English equivalents. Study the conversion tables. Then convert your height and weight, and perhaps the heights and weights of your family members, into metric units. Notice that milk and juice cartons provide volume information in both English and metric units. Begin to think metric!

Chronology of Famous Names in Biology

1895 **Wilhelm Konrad Roentgen** (Germany)—discovered X rays.

1903 **Mikhail Tsvett** (Russia)—discovered column chromatography.

1920 **Theodor Svedburg** (Sweden)—invented the centrifuge.

1923 **Georg von Hevesy** (England)—developed technique of using radioactive tracers.

1925 **Joseph E. Barnard** (England)—developed technique for photographing living bacteria using ultraviolet light.

1933 **Arne Tiselius** (Sweden)—developed electrophoresis.

1933 **Max Knell** and **Ernst Ruska** (Germany)—invented the single "lens" electron microscope.

1935 **Wendell Stanley** (United States)—isolated the tobacco mosaic virus by centrifugation.

1938 **James Hillier** and **Albert Prebus** (England)—invented the compound electron microscope.

1940 **A.P.J. Martin** and **R.L.M. Synge** (England)— discovered the technique of paper chromatography.

1952 **Alfred D. Hershey** and **Martha Chase** (United States)—developed the technique of using radioisotopes to study viruses.

1953 **Maurice H.F. Wilkins** (England)—developed the technique of using X-ray diffraction to decipher the structure of DNA.

2

Characteristics of Life

WHAT YOU WILL LEARN

In this chapter you will review why there can be no one best definition of life. You will also learn how living things are named and grouped.

SECTIONS IN THIS CHAPTER

- Life Functions
- How Living Things Are Named
- Five-Kingdom System of Classification
- Six-Kingdom System of Classification
- Three-Domain System of Classification
- Review Exercises for Chapter 2
- Connecting to Life/Job Skills
- Chronology of Famous Names in Biology

SIGNS OF LIFE

As you already know, biology is the science that is concerned with life and living things. Given a collection of familiar objects, you would probably experience little difficulty in separating them into a group of living and a group of nonliving things. To determine if something is living, you would look for signs of life such as movement, response to touch, patterns of growth, and ability to take in food. Let us suppose that included in the group of things you wish to classify are a mildew-covered towel, a green stain on tree bark, a large yellow turnip, and slime mold on the forest floor. What signs of life would you look for in mildew, tree stain, the turnip, and slime mold? Are these things living? What criteria are used to distinguish living organisms from nonliving objects? (See Figure 2.1.)

FIGURE 2.1 All living things share the same characteristics of life. Can you identify the living and the nonliving things in the picture?

DEFINITION OF LIFE

Scientists are unable to agree on a standard definition of life because it is impossible to define life and its qualities in a sentence or two. For every definition that might be suggested, too many exceptions are found. For example, we can say that living things grow. But so can crystals be made to grow. They do so by adding on molecules of transparent quartz. We might also say that living things move. But what of the ball that rolls down a hill or a kite that flies in the air? Are these not moving? It is examples such as growing crystals and flying kites that make life a difficult concept to define. In the descriptions of objects above, and in Figure 2.1, which things are considered living and which nonliving? The rabbit, grass, snake, birds, trees, mildew, tree stain, turnip, and slime mold are all living organisms. The water, soil, rock, sun, ball, kite, and quartz crystals are all nonliving. What is it that makes them different? Remember that life is best defined in terms of the functions performed by all living things.

Life Functions

The body of a living thing is a highly organized system. It is self-regulating, capable of adapting to changes in the environment. To maintain life, all living systems must be able to carry out biochemical and biophysical activities that together are known as **life functions**. In the paragraphs that follow, you will read about each life function.

Nutrition is the sum total of the activities through which a living organism obtains food. Food molecules are known as *nutrients*. Chemicals within the body are able to change nutrients into substances useful for fuel and for growth. Included in nutrition are the processes of **ingestion**, **digestion**, and **assimilation**. *Ingestion* is the taking in of food. *Digestion* refers to the chemical changes that take place in the body. Through these changes, nutrient molecules are made usable by the cells. From time to time in this discussion, the word **molecule** has been used. A *molecule* is the smallest part of a compound that retains the chemical characteristics of that compound. *Assimilation* involves the changing of certain nutrients into the protoplasm of cells. You will learn more about the protoplasm of cells in Chapter 3.

Transport involves the absorption of materials by living things, including the movement and distribution of materials within the body of the organism. One method of transport is **diffusion**. (See Figure 2.2.) *Diffusion* is the flow of molecules from an area where these molecules are in great concentration to an area where there are fewer of them. *Active transport* is the movement of molecules powered by energy. *Circulation* is the movement of fluid and its dissolved materials throughout the body of an organism.

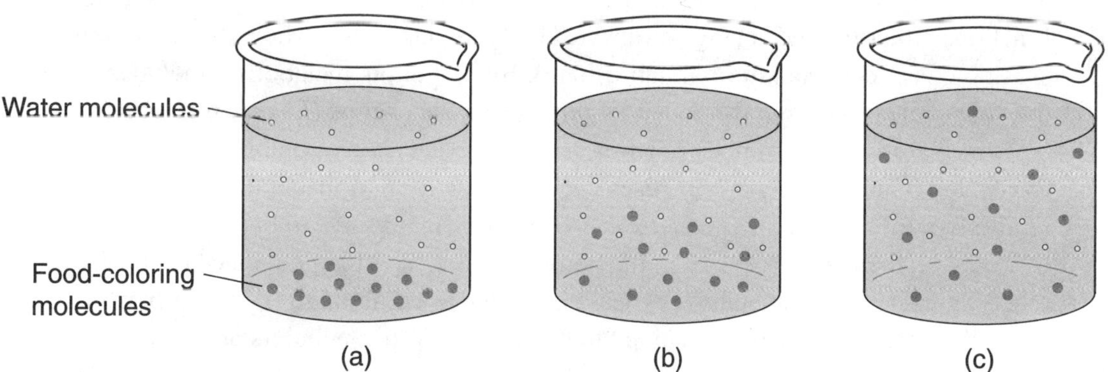

FIGURE 2.2 Diffusion. (a) Food coloring is placed at the bottom of a beaker of water. (b) Molecules of food coloring begin to move through molecules of water. (c) Molecules of food coloring diffuse throughout the water molecules.

Respiration consists of breathing and cellular respiration. *Breathing* is the intake of air (*inhaling*) and the letting out of carbon dioxide and water vapor (*exhaling*). *Cellular respiration* is a complex process through which energy is released from nutrient molecules.

Excretion removes waste products of other life functions from the body. The lungs, the skin, and the kidneys are excretory organs in humans that remove carbon dioxide, water, and urea from the blood and other body tissues.

Synthesis involves the biochemical processes by which small molecules are built into larger ones. As a result of synthesis, the building blocks of proteins are changed into more complex substances that regulate the body.

Regulation includes all processes that control and coordinate the many activities of a living thing. Regulation enables living organisms to adjust to changes in the environment. These changes may occur within the body of the organism or in the environment outside the body.

Growth describes the increase of cell size and increase of cell numbers. The orderly growth in the number of cells results in the growth of the body.

Reproduction is the process by which new individuals are produced by parent organisms. Organisms produce the same kind of individuals as themselves. There are two major kinds of reproduction: **asexual** and **sexual**. *Asexual reproduction* involves only one parent. *Sexual reproduction* requires the participation of two parents, each producing special reproductive cells known as *sex cells* or *gametes*. The continuation and survival of the species is dependent upon reproduction. Once a species has lost its *reproductive potential*, the species no longer survives and it becomes extinct.

BASIC CONCEPTS IN BIOLOGY

METABOLISM

You have just read that *life functions* are characteristic of all living organisms. Life functions either produce energy or use it. Therefore, taken as a whole, life functions are known as **metabolism** (muh-tab-o-lizm). By definition, metabolism includes all of the chemical activities in the body that produce or use energy. (*Energy* is the ability to do work.) During certain body processes, small molecules are combined into larger molecules. This building process uses energy. It is a form of metabolism called **anabolism** (an-ab-o-lizm), taken from a Greek word meaning "to build up." Conversely, there are processes in the body that break apart large molecules, thereby releasing energy. This metabolic process is known as **catabolism** (ka-tab-o-lizm), derived from a Greek word meaning "to tear down." You can understand that metabolism is ongoing in living things. Energy is released; energy is used. Metabolism is one of the basic concepts in biology.

HOMEOSTATIS

In the 19th century, the French physiologist (fizzi-ol-o-jist) Claude Bernard developed the concept of **homeostasis** (home-eo-staysis). *Homeostasis* means "staying the same." On page 25, you read that the body is self regulating and capable of adapting to changes in the environment. The body can adapt to external environmental changes because its *internal* environment is able to stay the same. For example, your normal body temperature is 98.6°F (37°C). When you go outdoors on a cold winter day, your body temperature remains the same. You will learn later in this book how the body maintains the "steady state" of its life functions in an almost unchanging internal environment. Homeostasis is another basic concept of biology.

ADAPTATION

Adaptation refers to a trait that aids the survival of an individual or a species in a given environment. An adaptation may be a structural characteristic such as the hump of a camel, a behavioral characteristic such as the mating call of a bull frog, or a physiological characteristic controlling some inner workings of tissue cells (Figure 2.3). Adaptations permit the survival of species in environments that sometimes seem forbidding. For example, some bacteria are able to live in hot springs that have temperatures up to 80°C (175°F). They have adaptations that permit the carrying out of metabolic functions at extremely high temperatures.

FIGURE 2.3 The long neck of the giraffe is an adaptation.

How Living Things Are Named

THE CONCEPT OF SPECIES

About 30 million kinds of plants and animals have been discovered. In order to identify these organisms, study them, and retrieve information about them, biologists have organized them into groups. The process of organizing living things into groups is called **classification**.

The basic unit of classification is the **species**. A species is a group of similar organisms that can mate and produce fertile offspring. The red wolf, the African elephant, the red oak, the house fly, the hair cap moss—each belongs to a separate and distinct species. For example, the red wolf belongs to the red wolf species in which the male mates with the female and produces fertile red wolf offspring. Upon maturity these red wolf offspring will reproduce just as their parents did. Species is a reproductive unit, not one defined by geography. Once brought together, a Mexican male Chihuahua can mate with a female Chihuahua born in France because they are species compatible.

In rare instances, members of closely related species—horses and donkeys, for example—can mate and produce offspring. But the products of interspecies matings are not fertile and therefore cannot reproduce. When a male donkey mates with a female horse (mare), a *mule* is produced. The mule is an infertile **hybrid** (Figure 2.4). The mating of a male horse (stallion) and a female donkey results in a *hinny*, also an infertile hybrid.

FIGURE 2.4 The mule, an infertile hybrid

BINOMIAL NOMENCLATURE

In modern biology living things are named according to the groups in which they are classified. The Swedish botanist and physician Carl von Linné (1707–1778) is credited with having devised the first orderly system of classification. Under the Latinized form of his name—*Carolus Linnaeus*, he is now known as "the father of modern taxonomy." Linnaeus based his system of classification on the anatomy and structure (morphology) of organisms. He recognized that living things fall into separate groups distinguishable by specific structural characteristics. He called each separate group of organisms a *species*. To each species he assigned a two-word Latin name or a **binomial**.

The first word in the species binomial designates the *genus*. Linnaeus assigned the genus (plural, *genera*; adjective, *generic*) as the next higher unit of classification after the species. In his system the genus is a classification group consisting of one or more related species.

The species name is the *scientific name*, a double name in Latin known as **binomial nomenclature**. The system of double naming is so orderly and practical that it is used in every country in the world. Thus the scientific name for an organism is the same worldwide no matter the language of the country.

The scientific or species name—*Quercus rubra*, for example—provides us with some information about the organism. First of all, the genus name *Quercus* is the Latin name of oak. Species that are classified as oaks are grouped into the genus *Quercus*. The name *rubra* (red) indicates that it is a separate and distinct kind of oak apart from *Quercus alba* (white oak) or *Quercus suber* (cork oak). See Table 2.1.

TABLE 2.1
THE GENUS *QUERCUS*

Scientific Name	Common Name
Quercus alba	white oak
Quercus coccinea	scarlet oak
Quercus montana	chestnut oak
Quercus rubra	red oak
Quercus suber	cork oak
Quercus virginiana	live oak

Certain rules are followed in the use of the genus and species names. The genus name is capitalized; the second word in the species name is not. The scientific name (genus and species name) is always italicized in print and underlined when written in long hand. Whenever the species name is used more than once in the same paragraph, it can be abbreviated. The first letter of the genus name is used followed by the complete name of the species: *Q. rubra*. When used alone, the genus name is always capitalized and spelled out: *Quercus*.

TAXONOMIC GROUPS

The science of **taxonomy** developed from the great need to classify the vast number of species that naturalists were discovering during the 18th and 19th centuries. The design of the classification system is a simple and practical one that lends itself easily to the addition of new names of organisms as they are discovered. The system makes use of a number of hierarchical groups each related to the other in terms of biological significance. Each group of organisms within the scheme is known as a *taxon* (plural, *taxa*). The classification groupings are as follows: *kingdom*, the largest and most inclusive group, followed by the *phylum*, *class*, *order*, *family*, *genus*, and *species*.

Linneaus based his system of classification on the morphology (structure) of organisms that appeared most closely related. The current practice is to group living things according to evolutionary relationships. **Evolution** is that branch of biology that studies changes in plants and animals that occur over long periods of time. Since the structures of plants and animals seem to reflect their evolutionary histories, the Linnaean system works well with the present-day approach to classification.

TABLE 2.2
CLASSIFICATION OF THE HUMAN

Category	Taxon	Characteristics
Kingdom	Animalia	Multicellular organisms requiring preformed organic material for food, having a motile stage at some time in their life histories.
Phylum	Chordata	Animals having a notochord (embryonic skeletal rod), dorsal hollow nerve cord, gills in the pharynx at some time during the life cycle.
Subphylum	Vertebrata	Backbone (vertebral column) enclosing the spinal cord, skull bones enclosing the brain.
Class	Mammalia	Body having hair or fur at some time in the life; female nourishes young on milk; warm-blooded; one bone in lower jaw.
Order	Primates	Tree-living mammals and their descendants, usually with flattened fingers and nails, keen vision, poor sense of smell.
Family	Hominidae	Bipedal locomotion, flat face, eyes forward, binocular and color vision, hands and feet specialized for different functions.
Genus	*Homo**	Long childhood, large brain, speech ability.
Species	*Homo sapiens*†	Body hair reduced, high forehead, prominent chin.

* *Homo* = Latin: man

† *sapiens* = Latin: wise

Five-Kingdom System of Classification

The largest and most inclusive classification category is the **kingdom**. For many decades, living things were considered to be either plants or animals and thus were grouped into one of two kingdoms, **Plantae** and **Animalia**. Improved microscopic and other research techniques have revealed many species that cannot be classified either as plant or animal. For example: the one-celled organism *Euglena*, which has both plant and animal characteristics, was commonly referred to as a "biological puzzle" because taxonomists could not classify it to meet the satisfaction of all biologists. (See Figure 7.9, page 152.) Other organisms such as bacteria, in which plant and animal characteristics can be demonstrated among the various species, also presented classifications problems.

In 1969, the ecologist Robert Whittaker proposed an updated system of classification in which every living thing is grouped into one of *five* kingdoms, based on the extent of its complexity and the methods by which nutritional needs are met.

TABLE 2.3
THE FIVE-KINGDOM SYSTEM

Kingdom	Characteristics
Monera	All monera are single-celled organisms. Unlike other cells, the monerans lack an organized nucleus, mitochondria, chloroplasts and other membrane-bound organelles. They have a circular chromosome. Examples are bacteria and blue-green algae.
Protista	Protists are one-celled organisms that have a membrane-bound nucleus. Within the nucleus are chromosomes that exhibit certain changes during the reproduction of the cell. Other cellular organelles are surrounded by membranes. Some protists take in food; others make it by means of *photosynthesis*. Some protists can move from one place to another (motile); others are nonmotile. Examples of protists are *Amoeba*, *Euglena*, *Paramecium*, *Punnularia* (diatom).
Fungi	Fungi are nonmotile, plant-like species that cannot make their own food. The fungi (fungus, singular) absorb their food from a living or nonliving organic source. Fungi differ from plants in the composition of the cell wall, in methods of reproduction and in the structure of body. Examples of fungi are mushrooms, water mold, bread mold.
Plantae	The plant kingdom includes the mosses, ferns, grasses, shrubs, flowering plants and trees. Most plants make their own food by photosynthesis and contain chloroplasts. All plant cells have a membrane-enclosed nucleus and cell walls that contain cellulose.
Animalia	All members of the animal kingdom are multicellular. The cells have a discrete nucleus that contains chromosomes. Most animals can move and depend on organic materials for food. Excluding the very simple species, most animals reproduce by means of egg and sperm cells.

Six-Kingdom System of Classification

In the five-kingdom classification system, all bacteria are grouped in the kingdom *Monera*. New evidence from molecular biology and electron microscopy, however, has indicated to biologists that all bacteria do not have the same evolutionary history. Therefore the kingdom Monera has now been divided into two kingdoms: the *Archaeobacteria* and the *Eubacteria*. (Descriptions of the species classified as Archaeobacteria and as Eubacteria are given in Chapter 6.) The current classification model exhibits six kingdoms.

TABLE 2.4
THE SIX-KINGDOM SYSTEM

Kingdom	Characteristics
Archaeobacteria	Archaeobacteria survive only where there is no oxygen. They live in harsh environments: hot springs, salt lakes and seas, anaerobic (without oxygen) marshes, sewage plants, and the sludge of lake bottoms and seas. Examples are methane-producing bacteria, salt-loving bacteria, and heat-loving bacteria.
Eubacteria	Most species of bacteria belong to the Eubacteria. Certain kinds of these bacteria are classified by their shapes: bacillus (rod-shaped), coccus (round), spirillum (spiral). This kingdom is wide ranging and consists of both helpful and disease-producing bacteria.
Protista	Same as in Table 2.3
Fungi	Same as in Table 2.3
Plantae	Same as in Table 2.3
Animalia	Same as in Table 2.3

Three-Domain System of Classification

A naming system that is gaining popularity among scientists is the three-domain system. It is based on Carl Woese's understanding of the diversity and distinctions of the bacterial kingdom. Woese recognized such extreme genetic and cellular differences between the Archaebacteria and the Eubacteria that he developed a higher, larger, more inclusive level of classification than the kingdom. Woese recognized three domains: the Bacteria (common prokaryotes/bacteria), the Archea (bacteria that live in very extreme environments), and the Eukarya (all eukaryotes and the kingdoms Protista, Fungi, Plantae, and Animalia).

REVIEW EXERCISES FOR CHAPTER 2

WORD-STUDY CONNECTION

active transport
adaptation
anabolism
Animalia
Archaeobacteria
asexual reproduction
assimilation
binomial nomenclature
breathing
catabolism
cellular respiration
circulation
class
diffusion
digestion
Eubacteria
excretion
family
Fungi
gametes
genus
growth
homeostasis
Homo sapiens
hinny
hybrid
ingestion
kingdom
life function
metabolism
mitosis
Monera
mule
nutrient
nutrition
order
phylum
Plantae
Protista
regulation
replication
reproduction
respiration
sexual reproduction
species
synthesis
taxon
taxonomy
Tetropoda
transport
vertebrate

SELF-TEST CONNECTION

PART A. Completion. *Write in the word that correctly completes each statement.*

1. The taking in of food is known as _____.

2. The type of transport requiring energy is _____ transport.

3. The pumping of air into and out of the lungs is known as _____.

4. Small molecules are built into larger ones in biochemical processes known as _____.

5. Examples of excretory organs are the kidneys, the skin, and the _____.

6. The building blocks of proteins are _____.

7. Reproduction that involves one parent is known as _____.

8. Chromosomes are duplicated in a process called _____.

9. Destructive metabolism is known as _____.

10. “To stay the same” is expressed by the biological term _____.

11. The mating of a male donkey and a mare produce the animal known as a _____.

12. Binomial nomenclature refers to _____ naming.

13. *Equus asinua* and *Equus caballus* indicate that these organisms belong to the same _____.

14. The _____ name is always capitalized.

15. The science of classification is known as _____.

16. The current classification system consists of _____ kingdoms.

17. Animals with backbones are classified as _____.

18. Linnaeus based his system of classification on the _____ of organisms.

19. Under the five-kingdom system of classification, all bacteria were grouped in the kingdom _____.

20. A bacillus is a bacterium that is shaped like a _____.

PART B. Multiple Choice. *Circle the letter of the item that correctly completes each statement.*

1. Chemical changes which change large food molecules into smaller soluble ones are known as
   (a) egestion
   (b) ingestion
   (c) mastication
   (d) digestion

2. The general term applied to the absorption and distribution of molecules within the body of an organism is
   (a) diffusion
   (b) transport
   (c) cyclosis
   (d) motion

3. Energy-releasing activities carried out by cells are known as
   (a) regulation
   (b) respiration
   (c) digestion
   (d) assimilation

4. The wastes of cellular respiration are
   (a) CO_2 and nitrogen
   (b) urea and oxygen
   (c) carbon dioxide and water vapor
   (d) urea and nitrogen

5. Auxins are most correctly associated with
   (a) bean plants
   (b) giraffes
   (c) snails
   (d) beetles

6. The movement of food-coloring molecules through water without the application of energy is known as
   (a) excretion
   (b) transportation
   (c) cremation
   (d) diffusion

7. All of the biochemical activities in the body are included in the term
   (a) anabolism
   (b) atavism
   (c) metabolism
   (d) catabolism

8. When species lose their reproductive potential, they
   (a) become extinct
   (b) increase in numbers
   (c) develop alternative systems
   (d) develop more energy

9. The long neck of a giraffe is considered to be a (an)
   (a) throw-back
   (b) affectation
   (c) effector
   (d) adaptation

10. The first word in the species binomial refers to the
    (a) kingdom
    (b) phylum
    (c) genus
    (d) class

11. The basic unit of classification is the
    (a) taxon
    (b) phylum
    (c) kingdom
    (d) species

12. A mule can properly be described as a
    (a) high breed
    (b) hybrid
    (c) mongrel
    (d) hinny

13. The name Linnaeus is best associated with
    (a) taxonomy
    (b) morphology
    (c) paleobiology
    (d) anatomy

14. The scientific name is the same as the
    (a) common name
    (b) species name
    (c) genus name
    (d) popular name

15. *Euglena* is best classified as a
    (a) plant
    (b) animal
    (c) fungus
    (d) protist

16. Tetrapoda are animals that have four
    (a) limbs
    (b) legs
    (c) fingers
    (d) toes

17. All bacteria that cause disease are placed in the kingdom
    (a) Fungi
    (b) Monera
    (c) Eubacteria
    (d) Archaeobacteria

18. Mushrooms are classified as
    (a) nongreen plants
    (b) bacteria
    (c) protists
    (d) fungi

19. "Anaerobic" means without
    (a) oxygen
    (b) carbon dioxide
    (c) methane
    (d) heat

20. All members of the animal kingdom
    (a) have legs
    (b) are multicellular
    (c) breathe in carbon dioxide
    (d) metabolize methane

PART C. Modified True-False. *If a statement is true, write "true" for your answer. If a statement is incorrect, change the* <u>*underlined*</u> *expression to one that will make the statement true.*

1. A definition of life is quite <u>easy</u> to formulate.

2. Living things are highly <u>dispersed</u> systems.

3. The process in which nutrients are changed into protoplasm is known as <u>digestion</u>.

4. The movement of materials from place to place in the body is known as <u>locomotion</u>.

5. Processes that involve control and coordination of the activities of living organisms are known collectively as <u>metabolism</u>.

6. Increase in cell size is known as growth.

7. The endocrine system is a part of the plant body.

8. Auxins help regulate growth in plants.

9. Two parents are required for the process of asexual reproduction.

10. Anabolism is a breaking down process.

11. A hinny can reproduce others like herself.

12. The genus name is binomial.

13. The largest classification group is the species.

14. The branch of biology that studies change in form of living things is classification.

15. A classification group of organisms is known as a taxon.

16. All Archaeobacteria live where there is no water.

17. Amoeba belong to the kingdom Animalia.

18. The name Carl Woese is best associated with the five-kingdom system of classification.

19. John Ray was the first to introduce the concept of the genus.

20. The fine detail of bacteria is best resolved through the light microscope.

CONNECTING TO CONCEPTS

1. Compose a definition of life. Why is life difficult to define?

2. Why must biologists classify living things?

3. Explain the meaning of homeostasis.

4. Why are bacteria not placed in the kingdom Protista?

ANSWERS TO SELF-TEST CONNECTION

PART A

1. ingestion
2. active
3. breathing
4. anabolism
5. lungs
6. amino acids
7. asexual
8. replication
9. catabolism
10. homeostasis
11. mule
12. double (scientific)
13. genus
14. genus
15. taxonomy
16. six
17. vertebrates
18. structure
19. Monera
20. rod

PART B

1. **(d)**
2. **(b)**
3. **(b)**
4. **(c)**
5. **(a)**
6. **(d)**
7. **(c)**
8. **(a)**
9. **(d)**
10. **(c)**
11. **(d)**
12. **(b)**
13. **(a)**
14. **(b)**
15. **(d)**
16. **(a)**
17. **(c)**
18. **(d)**
19. **(a)**
20. **(b)**

PART C

1. difficult
2. organized
3. assimilation
4. transport or circulation
5. regulation
6. true
7. animal
8. true
9. sexual
10. building up
11. cannot
12. species
13. kingdom
14. evolution
15. true
16. oxygen
17. Protista
18. six
19. species
20. electron

CONNECTING TO LIFE/JOB SKILLS

Museums of natural history specialize in preparing exhibits that tell the stories of the life histories of plants and/or animals. The **curator** of a natural history museum is a highly trained person who is a specialist in one of the fields displayed in the museum. The curator heads a staff of assistants and technicians who research specimens and prepare them for exhibit. The curator must be sure that each specimen included in exhibits is correctly classified. You can find information about careers in natural history museums by contacting the American Association of Museums, 1225 I Street, Washington, D.C. 20005.

Chronology of Famous Names in Biology

342 B.C. **Aristotle** (Greece) — first to try to group organisms by selecting a single outstanding feature.

1627–1705 **John Ray** (England) — introduced the word *species* to describe an organism. Catalogued about 19,000 plants of Europe.

1707–1778 **Carolus Linnaeus** (Sweden) — devised the first classification system. Wrote *Systems Natural* and *Species Plantarum*. Introduced the idea of the *type species* — typical specimen of the species.

1969 **Robert H. Whittaker** (United States) — devised the five-kingdom system of classification.

1975 **Carl R. Woese** (United States) — established a three-domain system to distinguish and classify the Bacteria, the Archea, and the Eukarya.

3

The Cell: Basic Unit of Life

WHAT YOU WILL LEARN

In this chapter you will review the structure and function of cells and the essential roles that cells play in the life of an organism.

SECTIONS IN THIS CHAPTER

- The Cell as a Basic Unit
- Parts of a Cell
- Comparison of Cells and Tissues
- Cell Reproduction
- Review Exercises for Chapter 3
- Connecting to Life/Job Skills
- Chronology of Famous Names in Biology

The Cell as a Basic Unit

CELL CONCEPT

The body of a living organism is built of units called **cells**. All living things are similar in that they are composed of one or more cells. The body of a **unicellular** organism is composed of one cell. Most plants and animals are **multicellular**, having a body made of numerous cells.

During the years 1838 and 1839, the **cell theory** was pioneered by two eminent scientists of the day. Matthias Schleiden, a botanist, and Theodor Schwann, a zoologist, combined some of their fundamental ideas about the structure of plants and animals into what has developed into a basic concept of biological thought. The work of Rudolf Virchow completed the cell theory when in 1858 he established that new cells could arise only from other living cells. The cell theory states (1) that cells are the basic units of life; (2) that all plants and animals are made of cells; and (3) that all cells arise from preexisting cells.

UNIT OF STRUCTURE

Microscopic examination of plant and animal parts indicates that the bodies of living organisms are composed of cells. Cells provide structure and form to the body. They appear in a variety of shapes: round, concave, rectangular, elongate, tapered, spherical and other. Cell shape seems to be related to specialized function (Figure 3.1).

Cells not only vary in shape, they also differ in size. Most plant and animal cells are quite small, ranging in size between 5 and 50 micrometers in diameter (Figure 3.2). Cells are measured in units that are compatible with modern microscopes. (See Table 3.1.)

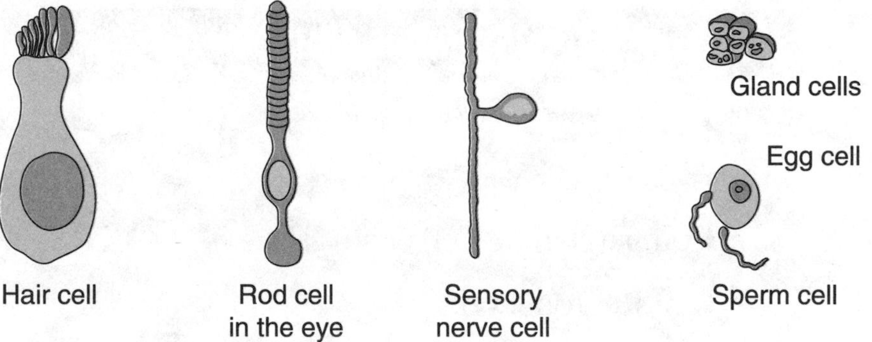

FIGURE 3.1 Types of human cells

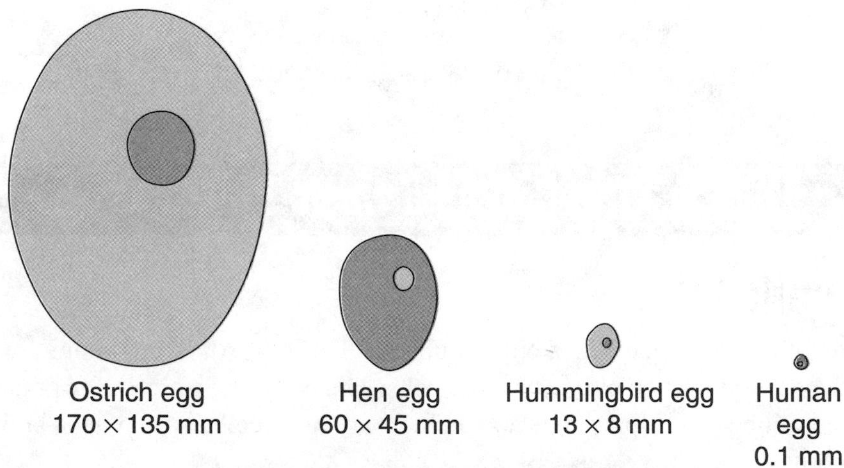

FIGURE 3.2 Comparative sizes of egg cells

TABLE 3.1
SIZE UNITS USED IN MICROSCOPY

Unit	Symbol	Equivalent	Measurement Use
Naked Eye Measurements			
Meter	m	100 cm	Naked eye measurement; standard from which microscopic measurements are derived.
Centimeter	cm	0.01 m; 0.4 in.	Naked eye measurement of giant egg cells.
Millimeter	mm	0.1 cm	Naked eye measurement of large cells.
Microscope Measurements			
Micrometer (Micron)†	μm μ*	0.001 mm	Light microscope measurement of cells and larger organelles.
Nanometer (Millimicron)†	nm mμ	0.001 μm	Electron microscopy measurement; cell fine structure; large macromolecules.
Angstrom unit†	Å	0.1 nm	Electron microscopy measurement; molecules and atoms; X-ray methods.

* Pronounced "*mew.*"

† These units of measure are being phased out but still appear in scientific literature.

UNIT OF FUNCTION

Each cell is a living unit. Whether living independently as a protist or confined in a tissue, a cell performs many metabolic functions to sustain life. Each cell is a biochemical factory using food molecules for energy, repair of tissues, growth and ultimately, reproduction. On the chemical level, the cell carries out all of the life functions that were discussed in Chapter 2. Living organisms function the way they do because their cells have the properties of life.

UNIT OF GROWTH

Each living thing begins life as a single cell. Protists remain unicellular. The protist cell body grows to a certain size and then divides into two new cells. A multicellular organism also begins life as a single cell. Its original cell grows to a certain size and then divides. Unlike the cells of protists, the cells of multicellular plants and animals hold together forming *tissues*. Later in this chapter, the role of tissues will be discussed. At this point, we are interested in cells as the unit of growth. As the number of cells increases in the body of a plant or animal, so does its size. Large organisms have more cells than small organisms. Thus the size of a living thing depends upon the increase in the number of its cells.

UNIT OF HEREDITY

New cells arise only from preexisting cells. A cell grows to optimum size and then divides, producing two other cells *identical* to itself. Paramecia produce other paramecia (Figure 3.3); onion skin cells produce new onion skin cells identical in structure and function to themselves. From single cells, multicellular organisms produce other organisms like themselves. This is so because cells carry hereditary information from one generation to the next. The information is coded in molecules of *DNA* (deoxyribonucleic acid) and is usually transmitted accurately to new cells. The reproductive machinery in the nucleus of the cell serves as a carrier of hereditary instructions.

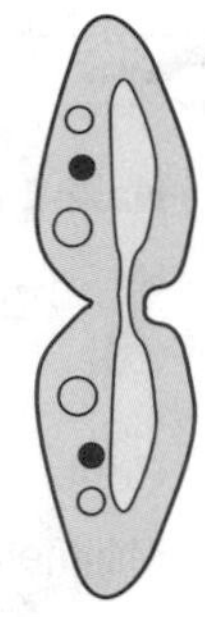

FIGURE 3.3
A dividing paramecium

Parts of a Cell

Essentially, all cells have similar parts designed to contribute to the work of the whole cell. The cell may be described as a membrane-enclosed unit consisting of **cytoplasm** and a **nucleus**. Both nucleus and cytoplasm are packed with finer structures that can be resolved or distinguished clearly only through the electron microscope.

The parts of a cell are known as **organelles**, meaning "little organs." This term is appropriate because parts of the cell have special functions, somewhat like miniature body organs. Figure 3.4 shows a typical animal cell as demonstrated through the light microscope. However, there are many organelles in the cell that cannot be seen by using the light microscope. These structures can be resolved by the electron microscope and are known collectively as the **fine structure of the cell**. Figure 3.5 presents an *electron micrograph* of a cell. You will find it helpful to refer to this diagram as you read about the structure and functions of parts of the cell.

BREAKING CELLS APART

A biologist studying cell processes wishes to remove the organelles (cell parts) from within the cell unit. Tissue and lysing (loosening) fluid are put into an electric blender or homogenizer. Here the tissue cells are separated from each other and the cell membranes are broken open. The suspension of ruptured cells and lysing fluid is then placed in a test-tube-like device and spun in an **ultracentrifuge** at 80,000 rotations per minute. The force exerted at extreme speeds of rotation is equivalent to 500,000 times the force of gravity. Such force causes the membranes enclosing the tissue cells to split open, releasing their contents into the sucrose (sugar) lysing fluid.

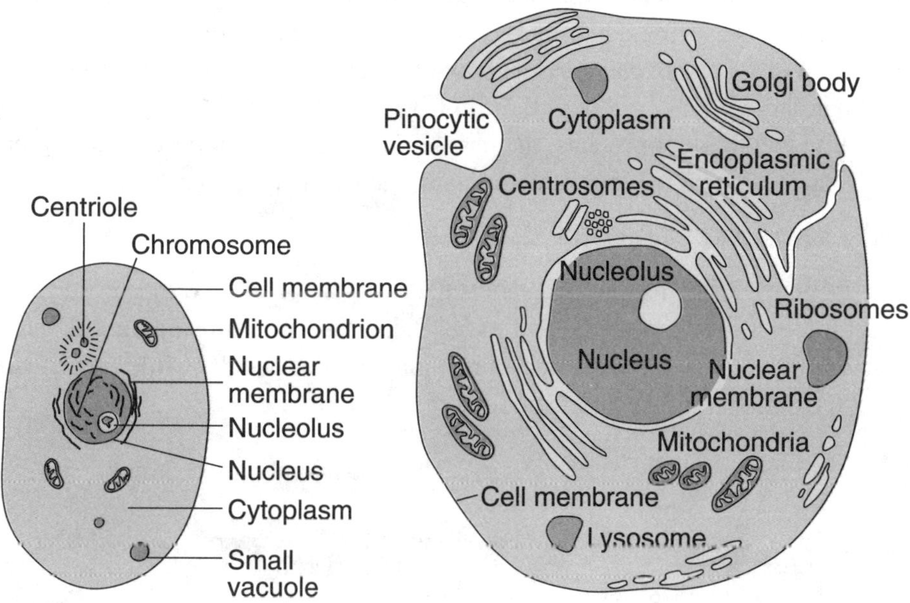

FIGURE 3.4 Typical animal cell

FIGURE 3.5 Electron micrograph of a cell

On the basis of weight differences, the organelles of each cell distribute in the lysing fluid in layers. The heavier cell structures fall to the bottom of the tube. The smaller and lighter particles remain in zones at the top. The fluid that tops a zone of heavier particles is called the **supernatant**. Organelles of intermediate weight are distributed in zones between the top and bottom layers. The technique of *ultracentrifugation* enables the research biologist to remove layered zones and to study them separately, either by biochemical means or by observation through electron microscopy. See Figure 3.6.

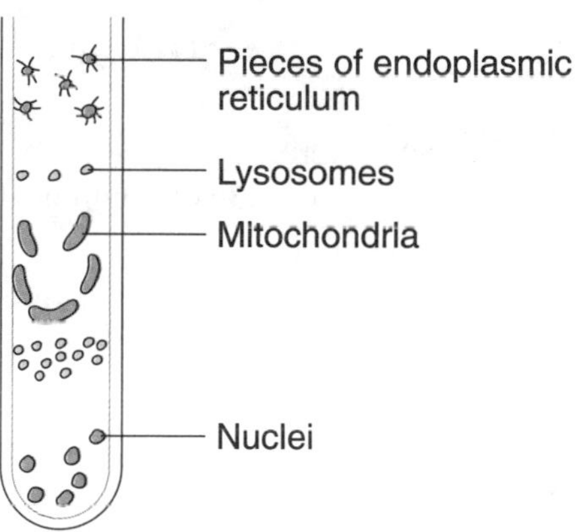

FIGURE 3.6 Results of centrifugation—the heavier particles fall to the bottom of tube

CELL MEMBRANE

The outer boundary of the cell is called the **cell** or **plasma membrane**. Although it is no thicker than 10 nanometers, the cell membrane has an intricate molecular structure. It is composed of a bilayer (double layer) of phospholipid molecules with proteins of various sizes embedded in the layers. **Pores** in the cell membrane are lined with small protein molecules.

The cell membrane controls the passage of materials into and out of the cell. It is often referred to as a living gatekeeper. The cell membrane is selectively permeable. Not every ion or molecule can cross its boundary. The movement of materials across the cell boundary and into or out of the cell is given the general term **transport**. It is controlled by the globular proteins, the phospholipids and the pores of the membrane, and the electrochemical nature of **protoplasm**, the living substance of the cell.

FLUID MOSAIC MODEL

Look at Figure 3.7. This diagram shows the structure of the cell membrane as research biologists currently believe it to be. Using this **fluid mosaic model**, biologists can explain how some molecules move into and out of cells while others cannot penetrate the cell membrane boundary.

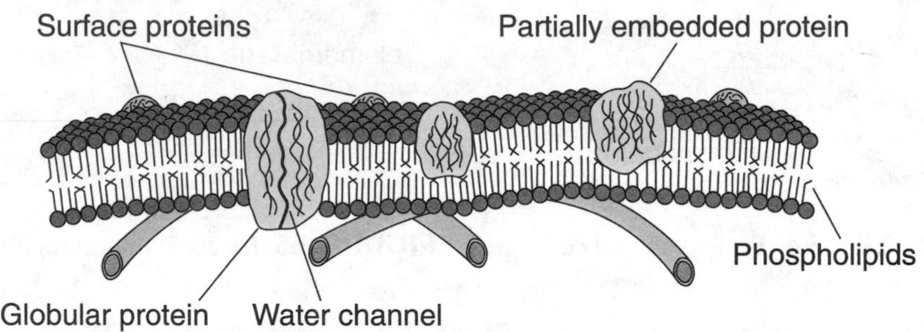

FIGURE 3.7 Fluid mosaic model of cell membrane

As you look at the fluid mosaic model, notice its structure. The core (center) of the membrane is made of compounds called *phospholipids*. Figure 3.8 shows the structure of a phospholipid. The rounded head contains a phosphate group, the tails are fatty acids, and a variable organic group is attached to the phosphate head. Notice that the tails are positioned inward, while the phosphate head points outward. The phosphate head is water-loving, but the lipid tails reject water. In other words, these fatty acid tails prevent water-soluble molecules from crossing the membrane.

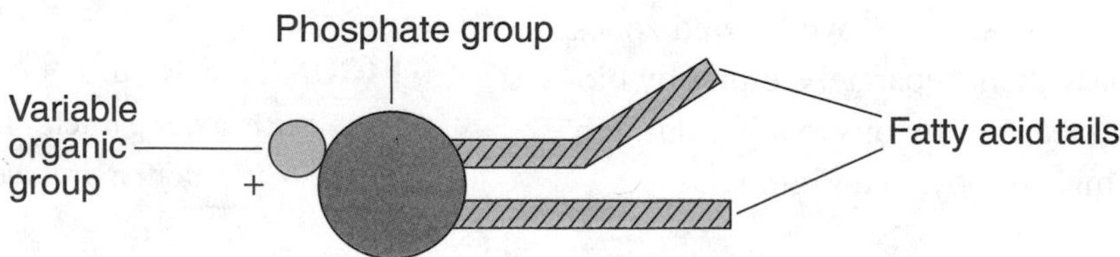

FIGURE 3.8 Phospholipid

Nevertheless, water-soluble materials do enter the cell. They do so by way of the large, circular proteins that are set into the membrane. These proteins have channels through which water and its dissolved substances can pass. Other, smaller proteins lie on the surface of the cell membrane and admit water freely. Another group of small proteins, which have only one water-rejecting end, are partly embedded in the membrane.

Each of these three groups of proteins has special functions. One group interacts with hormones and helps special substances get into cells. Another group of proteins is enzymelike and carries out chemical reactions on the surface of the membrane. The large, globular proteins are free to pass back and forth in the membrane and function as carrier molecules. These proteins actually carry certain substances across the cell membrane into the cells. The cell membrane is not regarded as rigid—thus the concept of fluid. The scattered proteins indeed form a pattern of a mosaic. The fluid mosaic model of the cell membrane is a new concept about which much remains to be learned.

TRANSPORT

Overall, there are two major types of transport into and out of the cell: passive transport and active transport. **Passive transport** does not require the cell's chemical energy to move molecules. Passive transport does depend, however, on the heat energy within the cell to increase the frequency *with which molecules move*. One type of passive transport is **diffusion**, which is the process by which molecules move from an area of greater concentration to an area of lesser concentration. **Osmosis** is the movement of water across a semipermeable membrane. **Plasmolysis** is the shrinking of cytoplasm due to the movement of water out of the cell. Osmosis and plasmolysis are forms of diffusion and thus are examples of passive transport (Figure 3.9).

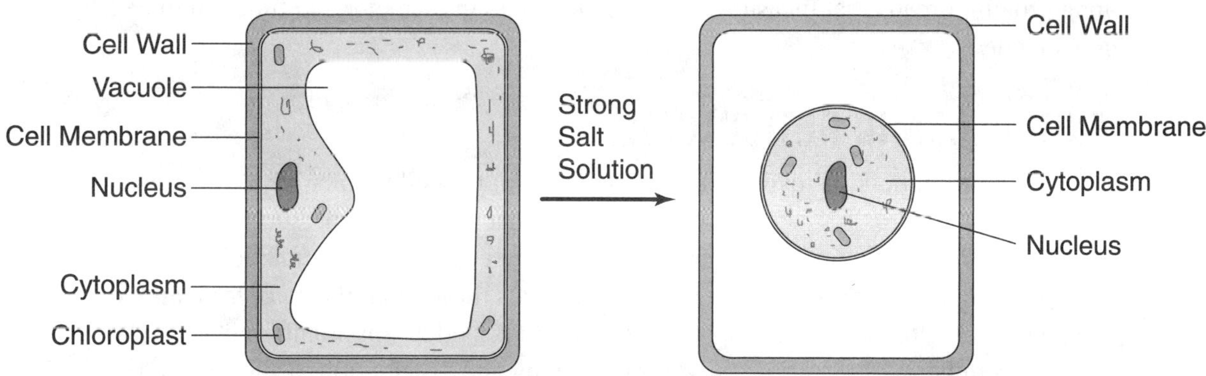

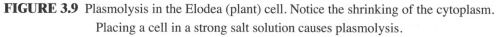

FIGURE 3.9 Plasmolysis in the Elodea (plant) cell. Notice the shrinking of the cytoplasm. Placing a cell in a strong salt solution causes plasmolysis.

Active transport requires the use of chemical energy that is stored in ATP molecules in the cell. Later you will learn how ATP molecules supply energy to the cell. **Pinocytosis** (pi-no-sigh-toe-sis), or "cell drinking," is a form of active transport by which fluid molecules are engulfed (taken in) by cells through the formation of vesicles (pockets) in the cell membrane. Solid particles are ingested by cells through a process known as **phagocytosis** (fag-o-sigh-toe-sis); white blood cells ingest bacteria by means of phagocytosis. Collectively, pinocytosis and phagocytosis are known as **endocytosis**. At times molecules are forced out of cells by **exocytosis**, a means by which they are carried to the cell surface by vacuoles or vesicles.

The sodium/potassium ion-exchange pump is an important mechanism in active transport. Sodium/potassium ion-exchange pumps push sodium ions out of the cell and force potassium ions into the cell. This process involves a protein carrier that is lodged in the cell membrane. Three sodium ions become attached to a special site (place) on the protein. Interaction with ATP causes the protein to change shape and release two of the sodium ions to the outside of the cell. Potassium ions take the place of the released sodium ions and are deposited within the cell. Sodium/potassium ion-exchange pumps are the means by which sodium ions are forced out of nerve cells, while potassium ions are pulled into the cells.

CYTOPLASM

All of the living material of the cell that lies outside the nucleus is the **cytoplasm**. It is packed full of organelles and is highly structured by a network of fine tubes and fibers that are spread throughout the cell. These supporting tubes (microtubules) and fibers (microfilaments) are known collectively as the **cytoskeleton**. **Microtubules** are long, thin, hollow tubules (little tubes), measuring 25 nanometers in diameter. They are composed of globular proteins and act as the framework of the submicroscopic cytoskeleton. Microtubules also seem to direct the flow of circulating materials within the cytoplasm and to create pathways for organelle movement. **Microfilaments** are long protein threads that measure about 6 nanometers in diameter and function in cell movement.

ENDOPLASMIC RETICULUM

Spreading throughout the cytoplasm, extending from the cell membrane to the membranes of the nucleus is a network of membranes that form channels, tubes and flattened sacs; this network is named the **endoplasmic reticulum**. One function of the endoplasmic reticulum is the movement of materials throughout the cytoplasm and to the plasma membrane. The endoplasmic reticulum has other important functions related to the synthesis of materials and their packaging and distribution to sites needed. Some membranes of the endoplasmic reticulum are dotted with thousands of organelles known as *ribosomes*; others are smooth.

RIBOSOMES

Ribosomes are small, circular organelles measuring 25 nanometers in diameter. They are by far the most numerous organelles in a cell; in one bacterial cell ribosomes may number upward of 15,000. Some ribosomes are attached to the membranes of the endoplasmic reticulum. These are engaged in the synthesis of proteins that will be exported from the cell to be used by various organs of the body. Digestive enzymes and hormones are examples of the kinds of protein molecules that are secreted by cells to be used elsewhere in the body. Ribosomes that lie free in the cytoplasm synthesize proteins to be used in the cell in activities such as cellular respiration. In general, ribosomes are the sites of protein synthesis.

GOLGI BODY

The **Golgi body**, sometimes referred to as the Golgi apparatus, was discovered by the Italian biologist Camillio Golgi in 1898. It consists of a series of membranes loosely applied to one another and forming *vesicles* (fluid-filled pouches) that are surrounded by microtubules. The Golgi body receives vesicles and their fluids from membranes of the endoplasmic reticulum. The vesicles are rewrapped in membranes by the Golgi body and transported to the cell membrane where they leave the cell. In general, the function of the Golgi body is to temporarily store, package and transport materials synthesized by other parts of the cell. A number of compounds formed in the endoplasmic reticulum are funneled into the Golgi apparatus, where they are modified and concentrated. These compounds are then rewrapped and transported to other parts of the cell. Among these compounds are many types of proteins, glycoproteins, proteins that become part of the plasma membrane, and glycolipids. Plant cells have several hundred Golgi bodies; animal cells, usually 10 to 20.

LYSOSOMES

The **lysosomes** are vacuoles bounded by double membranes that keep them separate from other cellular organelles. The lysosomes contain hydrolytic enzymes that are capable of destroying the cell. By a well-controlled mechanism, lysosomes become attached to food vacuoles and release some of the digestive enzymes into the food vacuoles. Bacteria engulfed by white blood cells are digested in this way.

PEROXISOMES

Peroxisomes are membrane-bound sacs that resemble lysosomes. Unlike lysosomes, however, peroxisomes contain oxidizing enzymes. These enzymes make certain substances that are toxic to the cells harmless by adding oxygen to them. Membranes that form peroxisomes are derived from the endoplasmic reticulum.

GLYOXYSOMES

Glyoxysomes are membrane-bound sacs in the cytoplasm that resemble peroxisomes. However, glyoxysomes contain enzymes that convert fatty acids to carbohydrates. Glyoxysomes are usually found in plant seedlings. The enzymes in glyoxysomes change the fats stored in the seed into carbohydrates that are used to build cell wall structures.

MITOCHONDRION

Aside from the nucleus, the **mitochondrion** is the largest organelle in the cell. Because there are so many of these organelles, the plural *mitochondria* is used more frequently than the singular *mitochondrion*. In diameter, mitochondria measure between 0.5 and 1 micrometer; in length they measure up to 7 micrometers. Mitochondria, known as the "power houses of the cell," exhibit a variety of shapes: round, ovoid, elongated, and the like. Each mitochondrion is surrounded by a double membrane and has a membrane of many folds fitted into its internal structure. The many folds are covered with enzymes necessary for the chemical reactions that release energy. The degree of folding of the inner membrane is related to the energy requirements of the cell. The internal membranes, *cristae*, actually increase the surface area, permitting a great amount of biochemical activity to take place. In addition to increasing surface area, the cristae form compartments that provide additional work and storage areas for the complex task of *cellular respiration*. Depending on their energy needs, cells may have more than 2,500 mitochondria (Figure 3.10).

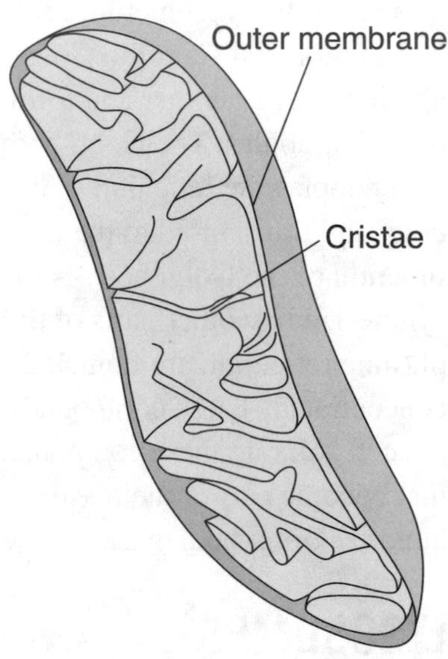

FIGURE 3.10 Mitochondrion

NUCLEUS

The largest and most prominent organelle in the cell is the **nucleus**. A few primitive species such as bacteria and blue-green algae do not have an organized nucleus. These types of cells are known as **prokaryotes** (pro-kary-ots). Most cells do, however, have a double membrane-bound nucleus and are classified as **eukaryotes** (you-kary-ots).

The double membrane of the nucleus fuses at certain places forming openings called *pores*. The pores measure about 65 nanometers in diameter. Inside the nuclear membrane a clear semi-solid material seems to fill up the nucleus. Embedded in this material are one or two small spherical bodies called *nucleoli* (sing., nucleolus). The nucleolus is the site of the synthesis and storage of the nucleic acid RNA (ribonucleic acid). In the nucleus of the nondividing cell is a tangle of very fine threads which absorb stain quite readily. In the granular stage these threads are known as *chromatin*. The chromatin threads come together, shorten and thicken, forming *chromosomes* that can be seen quite prominently in the dividing cell.

Comparison of Plant and Animal Cells

PLANT CELLS

A typical plant cell is shown in Figure 3.11; a typical animal cell is also shown for comparison. Notice that a plant cell has as its outer boundary a *cell wall* which surrounds the plasma membrane. The cell wall is a nonliving structure composed of *cellulose*, a complex starch molecule. Cellulose molecules are bound together in an intricate pattern and held in place by gluey carbohydrates known as *pectins*. The carbohydrate molecules that compose the cell wall are synthesized by the cytoplasm of the cell and secreted through the cell membrane to form a rigid boundary around the cell.

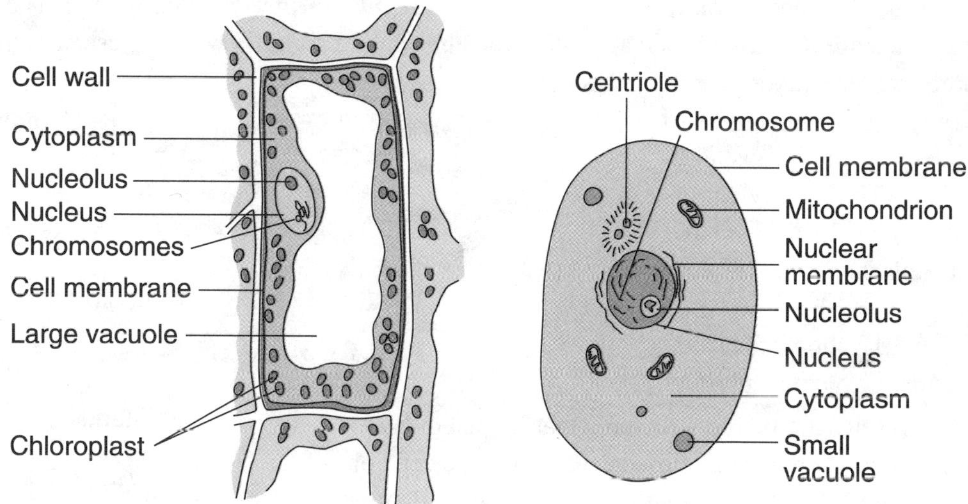

FIGURE 3.11 A typical plant cell (left); a typical animal cell (right)

The cell wall serves to support the cell, to protect it from drying out and to inhibit bacterial invasion. If the cytoplasm inside the cell loses water and shrinks, the cell wall still retains its shape and remains fairly rigid. Animal cells do not have a structure that can be compared to the cellulose cell wall.

Figure 3.11 shows a plant cell with a very large *vacuole*. A vacuole is a space in the cytoplasm filled with water and dissolved substances such as salts, sugars, minerals and other materials. The *cell sap* in the vacuoles of some cells contains pigments that color the flowers, fruits, leaves and stems of plants. Like other organelles, the vacuole is surrounded by a membrane which separates it from the cytoplasm of the cell.

ANIMAL CELLS

Storage areas in animal cells are quite small, occupying very little space in the cytoplasm. These storage areas in animal cells are known as *vesicles*. They are neither fixed in number nor as lasting as organelles, but rather appear and disappear as they move materials from the endoplasmic reticulum to the Golgi body to the cell membrane.

Chloroplasts belong to a group of structures which have the general name *plastids*. Plastids are membrane-bound organelles found only in plant cells. Usually plastids are spherical bodies that float freely in the cytoplasm, holding pigment molecules or starch. Storage plastids include leucoplasts and chromoplasts. Chloroplasts contain the green pigment *chlorophyll*, a substance that gives plants their green color. Chlorophyll is a special molecule that has the ability to trap light and to convert it to a form of energy that plants can use in carrying out the chemical steps of the food-making process known as **photosynthesis**.

Each chloroplast is surrounded by a double membrane. Inside the chloroplast are numerous flattened membranous sacs called *thylakoids* (formerly called *grana*). The thylakoids are the structures that contain the chlorophyll and it is within these sacs that photosynthesis takes place. *Stroma* is the name given to the dense ground substance that cushions the thylakoids.

Animal cells do not have chloroplasts and therefore cannot make their own food. Figure 3.12 shows the fine structure of a chloroplast.

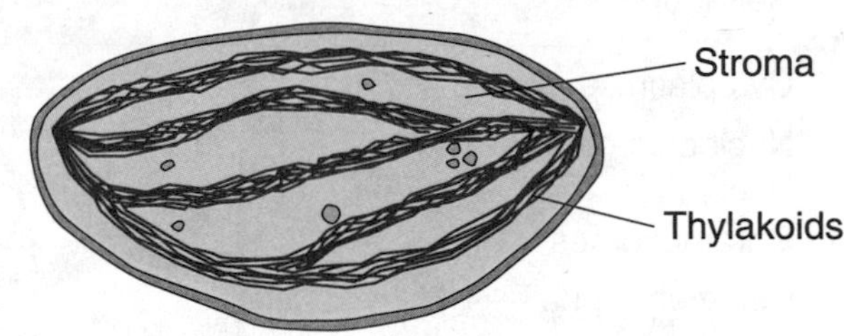

FIGURE 3.12 A chloroplast

Centrioles are paired structures that lie just outside the nucleus of nearly all animal cells and some cells of lower plants. They are absent in cells of higher plants. Under the light microscope, the centrioles look like two insignificant granules, but the electron microscope demonstrates that they have a very intricate structure (Figure 3.13).

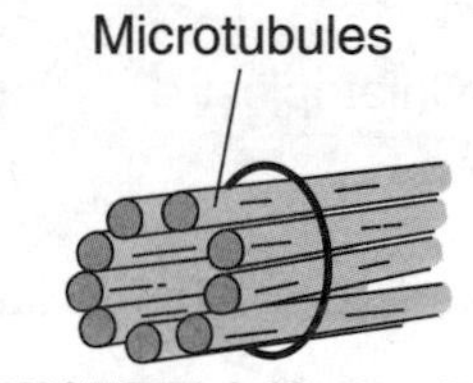

FIGURE 3.13 The fine structure of a centriole

Flagella and **cilia** are fine threads of cytoskeleton that extend from the surfaces of some cells. Both of these structures are involved in the locomotion of some protist species. Cilia are relatively short extensions but appear in great numbers, usually surrounding the body of the protist. Flagella are much longer than cilia and appear in fewer numbers.

In addition to serving the locomotive needs of one-celled organisms, flagella and cilia help functions of other types of cells. Sperm cells of animals and plants are propelled through fluid media by the whip-like actions of their flagella. Tissue cells of the human windpipe are lined with cilia which wave back and forth catching dust particles and pushing them away from the lungs. The microstructure of the flagella and cilia resembles that of the centrioles.

Organization of Cells and Tissues

Cells in the body of the multicellular organism are arranged in structural and functional groups called **tissues**. A tissue is a group of similar cells that work together to perform a particular function. Tissues that are grouped together and work for a common cause form **organs**. Groups of organs that contribute to a particular set of functions are called *systems*. The ability of cells to carry out special functions in addition to the usual work of cells exemplifies *specialization*. When different jobs are accomplished by the various tissues in an organ, we call this *division of labor*.

Tables 3.2 and 3.3 provide a summary of major animal and plant tissues.

TABLE 3.2
ANIMAL TISSUES

Tissue Type	Function of Tissue
Epithelial	Made of closely packed cells specialized for covering and lining organs and protecting underlying tissues from drying out, mechanical injury and bacterial invasion (Figure 3.14).
squamous cells	Flattened epithelial cells; specialized for lining body cavities such as the mouth, esophagus, eardrum and vagina.
cuboidal cells	Cube-like cells; specialized for gland tissue and the lining of the kidney tubules.
columnar cells	Tall and column-like cells; specialized for lining the alimentary canal in such organs as the stomach and intestines.
ciliated columnar cells	Ciliated columnar epithelial cells line the tubes that serve as air passageways in the respiratory system; the cilia push dust particles away from the lungs.
goblet cells	Specialized for secreting and storing fluid products, such as milk, hormones, enzymes, and oils, of cells.

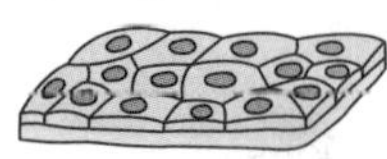

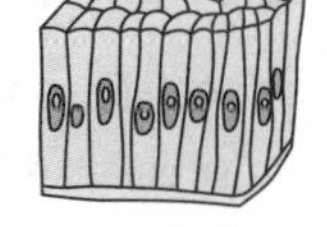

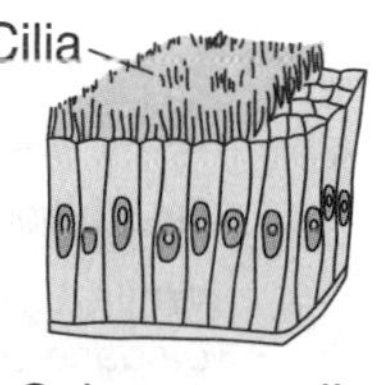

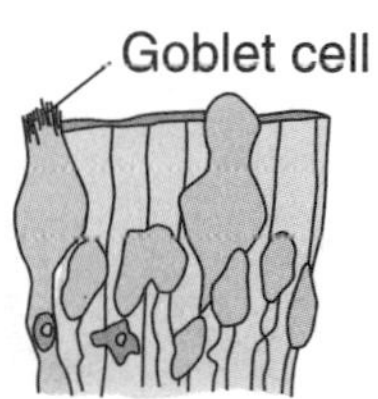

FIGURE 3.14 Types of epithelial tissue

Nervous	Made of cells called neurons that are specialized for carrying impulses; the nervous system coordinates all of the body's activities (Figure 3.15).
sensory neuron	Carries information from sense organs to other neurons in the brain and spinal cord.
motor neurons	Carries impulses to muscles and glands.
interneurons	Transmits impulses from sensory neurons to motor neurons.

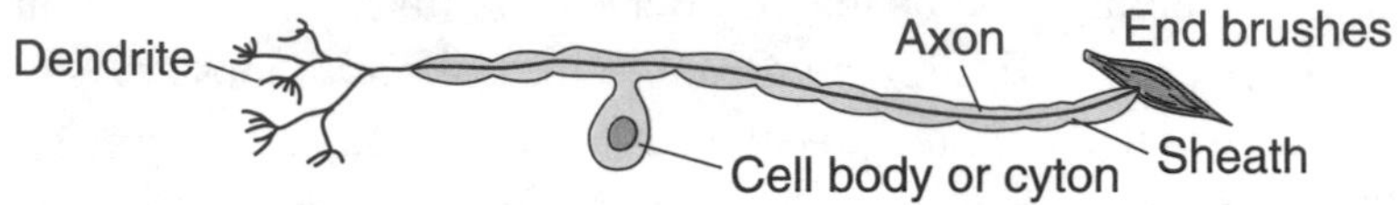

FIGURE 3.15 Nerve cell

Muscle	Specialized to respond to stimuli transmitted by motor neurons; characterized by electrical excitability and the ability to contract (Figure 13.16).
smooth	Involuntary muscle; composes organs not under the control of the individual—e.g., small intestine.
cardiac	Specialized for the heart.
striated	Voluntary muscle; composes organs that are controlled by the will of the individual—e.g., arm and leg muscles.

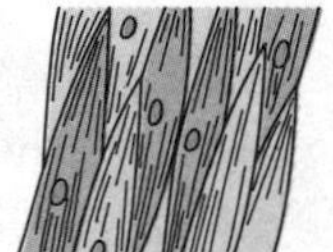

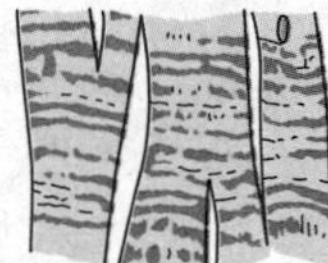

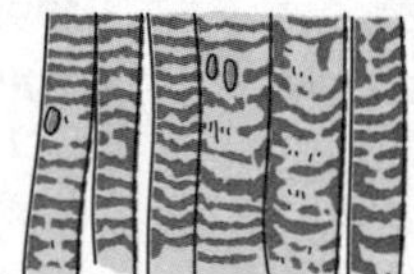

FIGURE 3.16 Types of muscle tissue

Blood	Composed of three types of cells suspended in plasma (Figure 3.17).
erythrocytes	Red blood cells specialized for carrying oxygen.
leucocytes	White blood cells specialized for counteracting invasions of disease organisms.
platelets	Cell fragments that function in blood clotting.

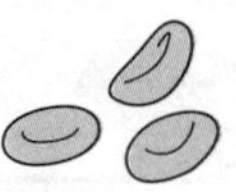

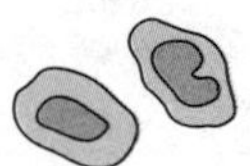

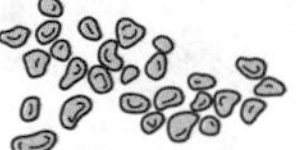

FIGURE 3.17 Types of blood cells

Connective	Characterized by large deposits of nonliving material that surrounds living cells; the nonliving *matrix* supports and binds tissues to the body skeleton (Figure 3.18).
cartilage	Firm and elastic matrix found at tip of nose, end of long bones, ear lobes and wherever strength and flexibility are needed.
fibrous connective tissue	Parallel protein fibers give strength to the matrix; ligaments—elastic fibers—connect bone to bone. *Tendons*—nonelastic fibers—connect muscle to bone.

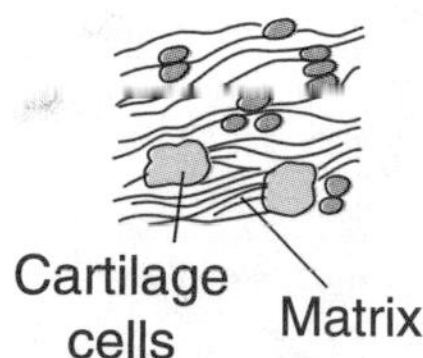

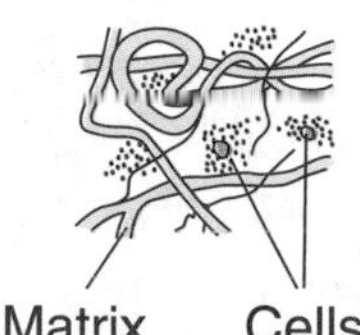

FIGURE 3.18 Types of connective tissue

Bone	Calcium and phosphorus compounds make up the matrix; living cells arranged in rings around a canal through which nerve fibers and blood vessels extend. The skeleton of vertebrates is made of bone (Figure 3.19).

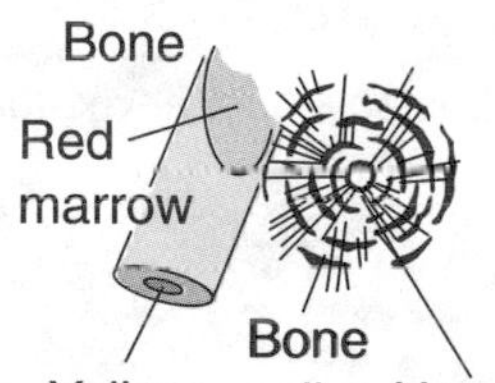

FIGURE 3.19 Bone tissue

TABLE 3.3
PLANT TISSUES

Tissue Type	Function of Tissue
Epidermis	One cell-layer thick; covering surfaces of leaves, stems and roots, protecting inner tissues (Figure 3.20).

Leaf epidermis

Cork cells

FIGURE 3.20 Epidermal plant tissue

Vascular	Conducting tissues; transport materials throughout plant (Figure 3.21).
xylem	Composed of elongated cells known as *tracheids* and *vessels*, or conducting tubes; conducts water and its dissolved minerals upward from roots of plant.
phloem	Transports food materials to all parts of the plant; composed of *sieve tubes* and *companion cells*.

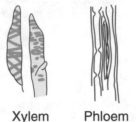

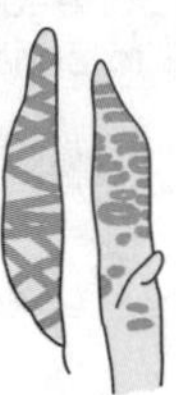

FIGURE 3.21 Vascular plant tissues

Fundamental	Basic tissues that make up most of the plant body (Figure 3.22).
parenchyma	Thin-walled cells containing chloroplasts and other plastids; tissues where photosynthesis takes place.
sclerenchyma	Tough supporting tissues composed of sturdy cell walls and dead cells; gives mechanical support to plant stems and forms tough coverings of seeds.
ollenchyma	Living cells in stems and leaves that provide support.

FIGURE 3.22 Fundamental plant tissue

Cell Reproduction

CELL CYCLE

Most cells go through an endless cycle of growth, replication of chromosomes, mitosis, and cytokinesis. You will learn about each of these cell events in the sections that follow.

The *cell cycle* consists of four repeating phases known as the *M phase, G-1 phase, S phase*, and *G-2 phase* (M stands for mitosis, G for gap, and S for synthesis). Mitosis and cell division take place during the M phase. During the G-1 phase, chromosomes are single stranded. Proteins necessary for the growth of the cell are synthesized in this phase. During the S phase DNA molecules are replicated. In the G-2 stage, the cell is getting ready for mitosis. The G-1, S, and G-2 phases are known collectively as *interphase*.

Cells that do not cycle have a much shortened life span. Human red blood cells do not have nuclei and die without undergoing cell division. The nuclei in the red blood cells of fish, birds, and reptiles are not active and cannot function in the process of cell division. These cells also die without reproducing themselves. Most cells, however, cycle continuously. Red blood cells are examples of the exceptions.

CELL DIVISION

When a cell reaches a certain size, it divides into two new cells, identical to each other and very similar to the original *parent* cell. The new cells are known as **daughter cells**. The events marking cell division differ in prokaryotes and eukaryotes.

As you recall, a prokaryotic cell does not have an organized nucleus. The nuclear material is in the form of a single circular chromosome attached to the *mesosome* on the cell membrane. Before cell division, the chromosome replicates, forming another chromosome exactly like itself. The new chromosome is attached to its own mesosome on the cell membrane. As the cell elongates, the chromosomes move further apart. Just as the cell doubles in length, the cell membrane pinches inward. New cell wall material surrounds the pinched membrane and separates the two cells (Figure 3.23).

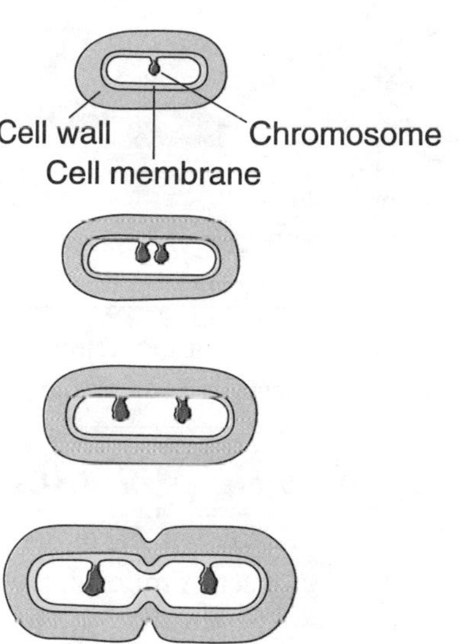

FIGURE 3.23 Division of a bacterium

Cell division in the eukaryotic cell includes two separate events: events in the nucleus through which the chromosomes are distributed to the daughter cells and *cytokinesis*, the division of the cytoplasm and its organelles.

Each species has a characteristic number of chromosomes in the nuclei of its cells. It is referred to as the *species number*, or simply as the *chromosome number*. As indicated in Table 3.4, the number of chromosomes in the cells of a given species is not an indicator of size or complexity. The somatic, or body, cells of an organism contain the full complement of chromosomes, referred to as the *diploid number* and designated by the symbol 2*N*. The nucleus of a somatic cell divides through *mitosis*, and the chromosomes are distributed equally to the daughter cells.

The organism's sex cells, or gametes—eggs and sperm, contain half the species number of chromosomes. This number is called the *haploid number* and symbolized as *N*. *Meiosis* is the kind of nuclear division that leads to the formation of sperm and egg cells.

TABLE 3.4
CHROMOSOME NUMBERS OF SOME COMMON SPECIES

Organism	Haploid No.	Diploid No.
mosquito	3	6
fruit fly	4	8
gall midge	20*	8*
evening primrose	7	14
onion	8	16
corn	10	20
grasshopper (female)	11	22
grasshopper (male)	10	21†
frog	13	26
sunflower	17	34
cat	19	38
human	23	46
plum	24	48
dog	39	78
sugar cane	40	80
goldfish	47	94

* In the fertilized egg of the gall midge, 32 chromosomes become nonfunctional leaving 8 functional chromosomes.

† The male grasshopper has only one sex chromosome.

PHASES OF MITOSIS

Mitosis (also known as karyokinesis) concerns the cell nucleus and its chromosomes. Before the onset of mitosis, the cell is in a stage known as *interphase*. During interphase, the chromosomes are exceptionally long and very thin, appearing as fine granules through the light microscope. It is during this stage that DNA molecules in the nucleus *replicate*. The result of replication is that each chromosome now has an exact copy of itself. When interphase comes to an end, the cell has enough nuclear material for two cells. The orderly process that divides the chromosomes equally between the two daughter cells is known as **mitosis**.

There are four stages of mitosis: prophase, metaphase, anaphase, and telophase. The significant events which mark each of these stages and interphase are outlined below and shown in Figure 3.24.

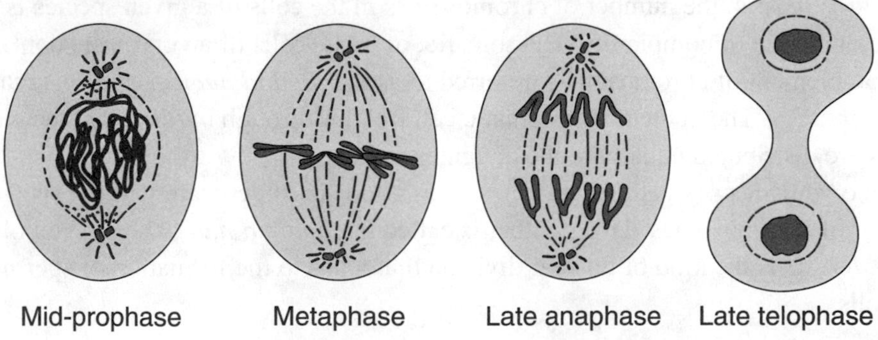

FIGURE 3.24 Stages of mitosis

Prophase

1. Chromosomes become shorter and thicker. Each chromosome consists of two *chromatids* attached at the centromere (Figure 3.25).
2. The nuclear membrane begins to disintegrate.
3. Spindle fibers form extending from the centromeres to the poles.
4. The centrioles in animals cells, fungi, algae, and some ferns replicate and a pair migrates toward each pole.
5. Chromosomes begin to move toward the equator of the cell.

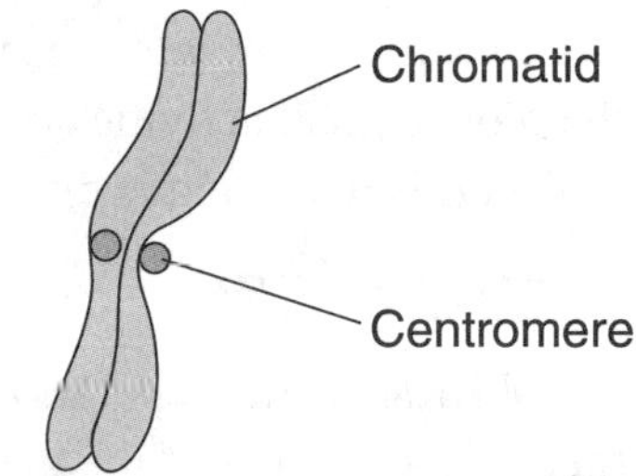

FIGURE 3.35 Chromatids hold together at centromere

Metaphase

1. The centrioles have migrated to the poles.
2. The chromosomes are lined up at the equator of the spindle.
3. Spindle fibers are attached to the centromeres connecting them to the poles of the spindle.
4. Both the nuclear membrane and the nucleolus have disappeared.

Anaphase

1. The centromeres split apart.
2. The chromatid pairs of each chromosome separate from each other. They move quickly in opposite directions, one toward each pole.

Telophase

1. The recently separated chromosomes reach the poles. A pole is the place where the new nucleus of each daughter cell will be located.
2. The spindle fibers extending from the poles to the centromeres disappear. Those fibers that lie in the plane between the opposing rows of chromosomes remain for a longer time.
3. A nuclear membrane reforms around each bundle of chromosomes at the poles. At this time, all remnants of the spindle fibers have disappeared.
4. At the equator of animal cells, the cytoplasm turns inward, pinching the old cell into two new ones. In plant cells, a cell plate of rigid cellulose separates the two new cells.

Interphase

Two new daughter cells are formed identical in genetic material to the parent and to each other. In a relatively short span of time, the daughter cells will accumulate more cytoplasm and grow to the size of the original parent cell. The cell nucleus controls the biochemical activities of the cell during interphase. The only work that the "resting" cell is not doing is dividing.

CYTOKINESIS

Cytokinesis is the separation of the cytoplasm following nuclear division. At first there is a doubling of the molecules that make up the cytoplasm. Cell structures that are composed of protein subunits, such as microtubules, microfilaments, and ribosomes, are synthesized from molecules within the cytoplasm. The membranous organelles, such as the Golgi apparatus, lysosomes, vacuoles and vesicles, are assembled by the membranes of the endoplasmic reticulum, which is restored and enlarged by self-assembling molecules. Chloroplasts and mitochondria replicate themselves from existing chloroplasts and mitochondria, respectively; it is theorized that these organelles have their own chromosomes much as in a prokaryotic cell. (Some scientists believe that the chloroplast and the mitochondrion were once independently living organisms that now live symbiotically in cells.)

The cytokinetic activities discussed above cannot be seen. The visible part of cytokinesis commences in late anaphase and reaches completion in telophase. The first sign of cytokinesis is the formation of a *cleavage furrow* which cuts across the equator of the cell through the spindle. In animal cells, microfilaments appear at the furrow which separates the daughter cells. In plants a cell plate separates the daughter cells. The cell plate is produced from a series of vesicles that are provided by the Golgi apparatus.

In some algae and fungi, mitosis without cytokinesis is common. When cytokinesis does not occur, cells with many nuclei but without separating membranes or walls result. Plant bodies made up of multinucleated cells without membrane or cell wall separations are described as being *coenocytic*.

PHASES OF MEIOSIS

Meiosis, or reduction division, is a kind of cell division that occurs in the primary sex cells leading to the formation of viable egg and sperm cells. The purpose of meiosis is to reduce the number of chromosomes to one half in each gamete so that upon fertilization (the fusing of sperm and egg nuclei) the species chromosome number is kept constant. If the gametes contained the full complement of chromosomes, on fertilization the number would be doubled. Instead, a gamete contains the haploid or monoploid number of chromosomes, one of each type of chromosome.

Meiosis occurs in primary sex cells—oocytes in the female, spermatocytes in the male—which have the diploid chromosome number. The process includes two cell division events—Meiosis I and Meiosis II—coming one after the other (Figure 3.26).

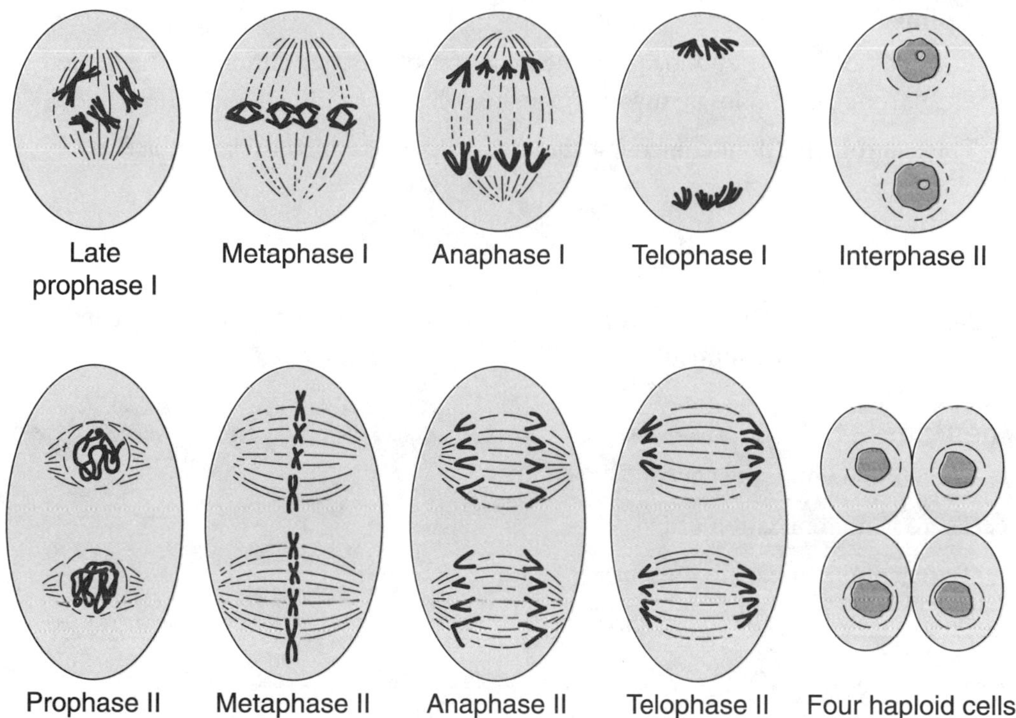

FIGURE 3.26 Stages of meiosis

MEIOSIS I

The stages of Meiosis I result in the reduction of the number of chromosomes.

Prophase I

1. The chromosomes become shorter and thicker.
2. The nucleolus disappears.
3. Chromosomes pair with their homologues (mates) forming a group of four chromatids referred to as a *tetrad*.
4. The tetrads wrap around each other (synapse) and may exchange like parts.
5. The centrioles migrate and the spindle fibers appear.
6. The nuclear membrane disappears.

Metaphase I

1. The tetrads move as a unit to the equator.
2. The centromeres (kinetochores) of each of the homologous pairs of chromosomes become attached to spindle fibers extending from opposite poles.

Anaphase I

1. Each pair of double-stranded chromosomes (a set of sister chromatids) is pulled away from its homologue toward opposite poles.
2. The centromeres do not uncouple and the sister chromatids remain attached.

Telophase I

1. The chromosomes are double-stranded.
2. In some organisms the nuclear membrane reappears; in others, it does not and Metaphase II starts immediately.

Interkinesis

1. The chromosomes disappear.
2. There is no replication of DNA.
3. Two haploid nuclei are present.
4. Interkinesis lasts for a very short time.

MEIOSIS II

The stage of Meiosis II results in the separation of the chromatids, terminating in four haploid cells.

Prophase II

1. The chromosomes reappear as do the spindle fibers.
2. The centrioles migrate to opposite poles.

Metaphase II

1. Spindles form.
2. The double-stranded chromosomes migrate to the equator. Their centromeres become attached to the spindle fibers.
3. The centromeres uncouple as they did in mitosis.

Anaphase II

The chromosomes pull apart to opposite poles.

Telophase II

1. Four haploid nuclei are formed. Each nucleus has one member of each pair of chromosomes that began the original meiosis.
2. The nuclear membrane reforms.
3. Cytokinesis comes to completion.

Mechanically, Meiosis II is primarily a mitotic division. As in mitosis, the chromosomes do not synapse. Since the nucleus is now haploid, there are no homologous chromosome pairs. Each double-stranded chromosome moves to the equator independently, not being attached to the spindle with a homologue. Each haploid cell produced during Meiosis I divides again during Meiosis II, producing four new haploid cells.

REVIEW EXERCISES FOR CHAPTER 3

WORD-STUDY CONNECTION

acid
active site
active transport
cell
cell cycle
cell membrane
cell plate
cell theory
cellulose
centriole
centromere
chlorophyll
chloroplast
chromatin
chromosome
cilia
cleavage furrow
cristae
cytokinesis
cytoplasm
cytoskeleton
diffusion
diploid number
disaccharide
DNA
endocytosis
endoplasmic
enzyme
eukaryote
exocytosis
flagella
fluid mosaic model
Golgi body
glycolysis
glyoxysome
haploid number
induced fit
inorganic
interphase
lysosome
meiosis
mesosome
metaphase
microfilament
microtubule
mitochondrion
mitosis
monosaccharide
multicellular
nanometer
nucleic acid
nucleolus
nucleus
organ
organelle
organic
osmosis
passive transport
pectin
peroxisome
phagocytosis
pinocytosis
plasma membrane
plasmolysis
plastid
prokaryote
prophase
protoplasm
replicate
resolving power
reticulum
ribosome
selectively permeable
specificity
sodium/potassium ion exchange pump
stroma
substrate
system
telophase
thylakoid
tissue
unicellular
vacuole
vesicle

SELF-TEST CONNECTION

PART A. Completion. *Write in the word that correctly completes each statement.*

1. The largest organelle in the cell is the _____.
2. The general term given to the movement of materials into and out of cells is _____.
3. The movement of water across a cell membrane is known as _____.
4. Solid particles are ingested by cells through a process known as _____.
5. The flow of circulating materials inside of a cell is directed by protein structures known as _____.
6. The network of membranous canals that are spread through the cytoplasm are known collectively as the _____.
7. Globular proteins are embedded in the _____ bilayer in the cell membrane.
8. Small storage areas in the cytoplasm of animal cells are known as _____.
9. Proteins for export are synthesized by ._____ that are attached to membranes of the endoplasmic reticulum.
10. The granular stage of the fine threads that are contained in the nucleus during interphase is known as _____.
11. Cell walls of green plants are for the most part made of the compound _____.
12. Plant cells are colored green by the pigment named _____.
13. Protists use for locomotion fine cytoplasmic hairs known as flagella and _____.
14. The type of tissue that covers and lines organs is _____ tissue.
15. Cells specialized for the carrying of nervous impulses are nerve cells, also known as ._____.
16. The nanometer is the unit used to measure objects resolved by the _____ microscope.
17. One micrometer is equal to 0.001 of a _____.
18. All cells arise from pre-existing _____.

19. The symbol $2N$ refers to the _____ number of chromosomes.

20. Meiosis takes place in egg and sperm cells which are known collectively as _____.

21. Mitosis is a process in cells that involves the _____ and its chromosomes.

22. Mitosis takes place during the _____ phase of the cell cycle.

23. The G-1, S, and G-2 phases of the cell cycle are known collectively as _____.

24. As a direct result of mitosis, the parent and daughter cells have _____ genetic material.

25. The formation of the cleavage furrow signals the onset of _____.

PART B. Multiple Choice*. Circle the letter of the item that correctly completes each statement.*

1. Multicellular organisms are produced from
   (a) one cell
   (b) two cells
   (c) three cells
   (d) four cells

2. The fine structures of the cell can be seen with the
   (a) naked eye
   (b) light microscope
   (c) phase contrast microscope
   (d) electron microscope

3. The cell's energy is used in the process of
   (a) passive transport
   (b) active transport
   (c) diffusion
   (d) osmosis

4. Because the cell membrane is highly selective in regard to the materials that can cross its boundary, it is described as being
   (a) selectively porous
   (b) selectively permeable
   (c) selectively fluid
   (d) selectively solid

5. The shrinking of cytoplasm due to the loss of water molecules is known as
   (a) evaporation
   (b) osmosis
   (c) diffusion
   (d) plasmolysis

6. Molecules carried to the cell surface by vesicles are forced out of the cell by the process of
   (a) exocytosis
   (b) endocytosis
   (c) facilitated transport
   (d) adsorption

7. The most numerous organelles in the cell are the
   (a) mitochondria
   (b) lysosomes
   (c) ribosomes
   (d) microtubules

8. Fluid substances for export outside of the cell are stored temporarily in the membrane of the
   (a) ribosomes
   (b) lysosomes
   (c) endoplasmic reticulum
   (d) Golgi apparatus

9. Proteins are synthesized and temporarily stored in the
   (a) ribosomes
   (b) mitochondria
   (c) nucleoli
   (d) lysosomes

10. The internal membranes of the mitochondria are known collectively as
   (a) grana
   (b) thylakoids
   (c) mesosoma
   (d) cristae

11. Cell sap is stored in areas of plant cells known as
   (a) lysosomes
   (b) centrosomes
   (c) vesicles
   (d) vacuoles

12. Thylakoids are the same as
    (a) cristae
    (b) nucleoli
    (c) grana
    (d) stroma

13. The type of tissue that is specialized to respond to stimuli transmitted by motor nerve cell is
    (a) blood
    (b) muscle
    (c) connective
    (d) bone

14. A nonliving matrix is most correctly associated with the type of tissue classified as
    (a) epithelial
    (b) muscle
    (c) connective
    (d) nerve

15. A type of plant tissue specialized for conducting water and its dissolved materials is
    (a) vascular
    (b) epidermal
    (c) fundamental
    (d) collenchyma

16. The two strands that make up a chromosome are the
    (a) centromeres
    (b) chromatids
    (c) centrioles
    (d) spindle fibers

17. The red blood cells of birds
    (a) lack nuclei
    (b) carry out mitosis
    (c) divide asexually
    (d) do not cycle

18. Passive transport requires
    (a) a semipermeable membrane
    (b) chemical energy from ATP
    (c) heat energy
    (d) an ion transfer pump

19. Examples of passive transport are
    (a) diffusion and osmosis
    (b) osmosis and endocytosis
    (c) diffusion and exocytosis
    (d) endocytosis and exocytosis

20. Solid particles are engulfed by cells during a process called
    (a) phagocytosis
    (b) plasmolysis
    (c) exocytosis
    (d) pinocytosis

PART C. Modified True-False. *If a statement is true, write "true" for your answer. If a statement is incorrect, change the underlined expression to one that will make the statement true.*

1. Scientists believe that cell shape seems to be related to the age of the cell.
2. Groups of similar cells that work together to do a particular job are designated as systems.
3. Large organisms have fewer cells than small organisms.
4. The parts of a cell are known as organs.
5. Passage of materials into and out of the cell is controlled by the cell wall.
6. Another name for "cell drinking" is imbibing.
7. All of the material that is outside of the nucleus and inside of the cell membrane is called protoplasm.
8. Aside from the nucleus, the centriole is the largest organelle in the cell.
9. Areas in plant cells that store starch and pigment molecules are given the general name of plastics.
10. The food-making process of green plants is known as phototropism.

11. The names Schleiden and Schwann are correctly associated with the <u>chromosome</u> theory.

12. A type of white blood cell specialized to counteract disease organisms that enter the body is the <u>erythrocyte</u>.

13. Another name for involuntary muscle is <u>striated</u> muscle.

14. Tracheids and vessels are best associated with <u>phloem</u> tissues.

15. As a result of mitosis, the daughter cells receive the <u>haploid</u> number of chromosomes.

16. The number of chromosomes in a human skin cell is <u>23</u>.

17. Chromosome replication takes place during <u>prophase</u>.

18. Ultracentrifugation of tissue cells separates cellular organelles according to <u>function</u>.

19. Plasmolysis is the shrinking of <u>peroxisomes</u> due to water loss.

20. The sodium/potassium pump forces sodium ions <u>out of</u> the cell.

CONNECTING TO CONCEPTS

1. Why is the cell the unit of life for all living things?

2. Why is the selectivity of the cell membrane important to the life of the cell?

3. What is the relationship between the endoplasmic reticulum and transport?

4. Why do cytologists (cell biologists) take apart cells and separate the organelles?

5. How does the cell cycle differ from cell division?

ANSWERS TO SELF-TEST CONNECTION

PART A

1. nucleus
2. transport
3. osmosis
4. phagocytosis
5. microtubules
6. endoplasmic reticulum
7. phospholipid
8. vesicles
9. ribosomes
10. chromatin
11. cellulose
12. chlorophyll
13. cilia
14. epithelial
15. neurons
16. electron
17. millimeter
18. cells
19. diploid
20. gametes or sex cells
21. nucleus
22. M
23. interphase
24. identical
25. cytokinesis

PART B

1. **(a)**
2. **(d)**
3. **(b)**
4. **(b)**
5. **(d)**
6. **(a)**
7. **(c)**
8. **(d)**
9. **(a)**
10. **(d)**
11. **(d)**
12. **(c)**
13. **(b)**
14. **(c)**
15. **(a)**
16. **(b)**
17. **(d)**
18. **(c)**
19. **(a)**
20. **(a)**

PART C

1. function
2. tissues
3. more
4. organelles
5. cell membrane
6. pinocytosis
7. cytoplasm
8. mitochondrion
9. plastids
10. photosynthesis
11. cell
12. leucocyte
13. smooth
14. xylem
15. diploid
16. 46
17. interphase
18. weight
19. cytoplasm
20. true

CONNECTING TO LIFE/JOB SKILLS

There is an increasing need for **laboratory technicians**. The work of the laboratory technician involves a kind of scientific problem-solving, using the knowledge of science and laboratory investigative tools. Medical laboratory technicians work under the direction of a physician in a hospital laboratory; they perform tests and keep records of the results. In a commercial laboratory a technician works under the direction of a scientist, testing products to determine compliance with government standards. The basic education for a laboratory technician is usually two years of study in a community college, with a concentration on courses in biology, chemistry, and physics. Many technicians go on to earn four-year college degrees. On-the-job training in the specialized equipment needed in a given laboratory is usually provided. You may wish to search the Internet for more information.

Chronology of Famous Names in Biology

1665 **Robert Hooke** (England)—was the first to view the pores in cork through the microscope and applied the word *cell* to these box-like structures.

1702 **Antonie van Leeuwenhoek** (Netherlands)—described red blood cells and protists.

1828 **Robert Brown** (England)—discovered and described the nucleus in orchid plants.

1830 **Johann Evangelista Purkinje** (Bohemia)—described the nucleus of the hen's egg.

1835 **Felix Dujardin** (France)—described cell sap as the essential substance of life.

1838 **Matthias Schleiden and Theodor Schwann** (Germany)—formulated the cell theory: all living organisms are made of cells.

1839 **Johann Evangelista Purkinje** (Bohemia)—invented the term *protoplasm*.

1858 **Rudolf Virchow** (Germany)—founded the study of cellular pathology and completed the cell theory when he established that new cells could arise only from existing cells.

1869 **Fredrick Miescher** (Switzerland)—was the first to isolate nucleic acid, which he did from pus cells.

1880 **Walther Flemming** (Austria)—discovered and described mitosis in the cells of amphibious larvae. He called mitosis "the dance of the chromosomes."

1898 **Camillio Golgi** (Italy)—discovered the membranes known as the Golgi apparatus by staining cells with silver nitrate and osmium tetroxide.

1938 **James Danielli** (United States)—accurately described the phospholipid layers of the cell membrane.

1945 **Albert Claude** (Belgium) and **Keith Porter** (United States)—discovered the endoplasmic reticulum.

1949 **Christian de Duve** (Belgium)—discovered lysosomes.

1952 **Fritiof Sjostrand** (Sweden) and **George Palade** (United States)—discovered cristae in the mitochondria.

1953 **Keith Porter** (United States)—gave the name endoplasmic reticulum to the network of membranes in the cytoplasm.

1954 **George Palade** (United States)—discovered ribosomes. In 1956 he discovered the function of rough endoplasmic reticulum.

1956 **Philip Siekevitz** and **George Palade** (United States)—isolated the ribosomes.

1960 **Hans Moor** (Switzerland)—developed the freeze-fracture technique used to study the interior of membranes.

1962 **Earl W. Sutherland** (United States)—discovered the role of cyclic AMP in the transport of certain hormones across the cell membranes. He was awarded the Nobel Prize for this work in 1971.

1962 **Marshall Nirenberg**, **Severo Ochoa**, and **Har Gobind Khorana** (United States)—deciphered the genetic code.

1966 **S.J. Singer** (United States)—proposed the "fluid mosaic" model of the cell membrane.

1967 **Edwin Taylor** (United States)—discovered the role of microtubules in mitosis.

1974 **Christian de Duve** (Belgium), **Albert Claude** and **George Palade** (United States)—won the Nobel Prize for discovering the submicroscopic structures of the cell and determining their functions.

1977 **Hagan Bayley** (United States)—devised a method of building artificial pores in the cell membrane.

2001 **Leland H. Hartwell, R. Timothy Hunt,** and **Paul M. Nurse** (United States)—discovered key regulators of the cell cycle.

4

The Chemistry of Life

WHAT YOU WILL LEARN

In this chapter you will review some basic principles of chemistry and the ways in which atoms, elements, molecules, and compounds are involved in chemical changes.

SECTIONS IN THIS CHAPTER

- Some Basic Principles of Chemistry
- Chemical Bonding
- Chemical Reactions
- Review Exercises for Chapter 4
- Connecting to Life/Job Skills
- Chronology of Famous Names in Biology

Some Basic Principles of Chemistry

MATTER AND ENERGY

Matter is anything that occupies (takes up) space. Matter may be solid or liquid or gas. It may be visible, as an iron nail, or invisible, as oxygen gas. The amount of matter in a substance is called its **mass**. The pull of gravity on the mass of a substance is **weight**. The greater the mass, the greater the weight. Previously, you read that **energy** is the ability to do work. Expressed another way, energy is also the capacity to move matter. Matter and energy cannot be separated from the processes of life.

ATOMS, ELEMENTS, AND COMPOUNDS

All matter is made up of **atoms**, particles invisible to the eye and to enlarging instruments. There are currently more than 110 different kinds of atoms that combine in different ways to form a variety of substances. The number of substances in the world is counted in the hundreds of thousands. Therefore the kind of matter in the world is almost infinite. A substance (matter) made from the same type of atom is an **element**. Iron, sulfur, and calcium are elements. A substance made of two or more kinds of atoms held together in chemical combination is a **compound**.

By looking at a substance you cannot tell whether it is an element, made of the same kind of atom, or a compound, made of different atoms. For example, through chemical analysis it can be shown that iron is an element, made of iron atoms only, while water is a compound, made of two different atoms, hydrogen and oxygen. Examples of other compounds are sugar, starch, and salt.

TABLE 4.1
TWELVE ELEMENTS IMPORTANT IN CELL BIOLOGY

Name of Element Symbol	Atomic Number	Name of Element Symbol	Atomic Number
Hydrogen H	1	phosphorus P	15
carbon C	6	sulfur S	16
nitrogen N	7	chlorine Cl	17
oxygen O	8	potassium P	19
sodium Na (natrium)	11	calcium Ca	20
magnesium Mg	12	iron Fe (ferrum)	26

Table 4.1 lists twelve elements that are important in the life processes of cells. Notice that each element has a symbol that begins with a capital letter, while the name of the element begins with a small letter. In most cases, the symbol consists of the first letter or the first two letters of the name of the element. For example, H is the symbol for hydrogen; Ca, for calcium. In some cases, to avoid confusion, the symbol is taken from the element's Latin name; K for potassium is derived from kalium. You will also notice that an atomic number is given for each element. In the next section you will learn about the significance of this number.

GENERAL STRUCTURE OF ATOMS

All chemical activity depends on atoms, the building blocks of elements and compounds. Although invisible, atoms are made up of still smaller particles. In the center part of the atom is the **nucleus**. The nucleus contains **protons** and **neutrons**, which are smaller structures. In orbits outside the nucleus are other small structures known as **electrons**.

Figure 4.1 shows the structure of the atom. The nucleus contains one or more protons and one or more neutrons. Protons have a *positive* electric charge of one unit. Neutrons have no electric charge, but do have mass (weight) that almost equals the mass of a proton. Revolving around the nucleus are electrically charged particles called electrons. Electrons have a negative charge. (Electric charges are either positive or negative.) The number of protons in the nucleus equals the number of electrons in the atomic orbitals.

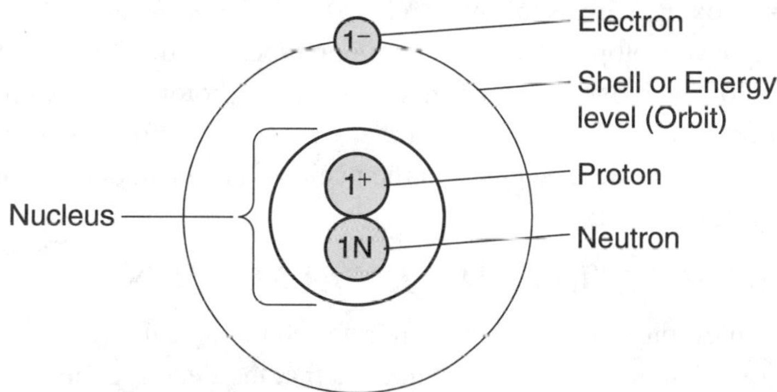

FIGURE 4.1 Structure of the atom

ELEMENTS: ATOMIC NUMBER AND ATOMIC MASS

The number of protons in the nucleus of an atom is known as its **atomic number**. The element hydrogen has one proton in its nucleus, making its atomic number 1. Calcium has 20 protons in its nucleus, and therefore its atomic number is 20. Each element has its own atomic number. Since the number of protons in a neutral atom equals the number of electrons, there is no electric charge on the atom. The number of positive charges cancels out the number of negative charges, so that the atom is electrically neutral.

The **mass number** is the sum of the protons plus the neutrons in the nucleus. In the nucleus of calcium, for example, there are 20 protons and 20 neutrons. Adding the number of protons and the number of neutrons results in a mass number of 40. Protons and neutrons are measured in units called *daltons*. Each proton and neutron weighs slightly more than one dalton. Therefore, the **atomic mass** of an element is slightly greater than the mass number. Turn to Table 4.2, Modern Periodic Table. Find calcium (Ca) in the Group 2 elements. Note that the atomic mass is recorded as 40.078, indicating that the combined mass of the protons and neutrons in calcium is slightly greater than the sum of these particles. In some textbooks, you may find the terms *atomic mass* and *atomic weight* used interchangeably.

ISOTOPES

Atoms of the same element have the same atomic number, that is, number of protons, but may differ in atomic mass. Such atoms are called **isotopes**. The difference in atomic mass is caused by a variance in the number of neutrons in the nucleus. For example, most oxygen atoms have 8 neutrons in the nucleus, resulting in an atomic mass of 16 (8 protons and 8 neutrons). There are, however, two other kinds of oxygen atoms. One of these has 9 neutrons in the nucleus and therefore an atomic mass of 17. Another type of oxygen atom has 10 neutrons in the nucleus and consequently an atomic mass of 18. These three kinds of oxygen atoms are isotopes. The symbols for these isotopes of oxygen are as follows: ^{16}O; ^{17}O, ^{18}O.

All elements have isotopes. The most common type of hydrogen atom has 1 proton in its nucleus. However, one isotope of hydrogen has 1 proton and 1 neutron in its nucleus and thus an atomic mass of 2. Another isotope of hydrogen has an atomic mass of 3 because there are 2 neutrons in the nucleus. The symbols for the isotopes of hydrogen are as follows: H; ^{2}H, ^{3}H.

THE PERIODIC TABLE OF THE ELEMENTS

In the discussion of mass number, you were referred to Table 4.2 to find the atomic mass number of calcium. You may have noticed that the **Periodic Table** has 112 elements arranged according to their atomic numbers. (Of the 112 currently known elements, 92 are natural; the others are products of nuclear research.) Remember that the atomic number is the number of protons in the nucleus of an atom, and these particles equal the number of electrons that orbit the atomic nucleus. The *Periodic Law* states that the properties of elements are periodic functions of their atomic numbers. The horizontal rows of the table are called *periods* or *rows*. Each period begins with an element that has one valence electron and ends with an element that is inactive. The vertical columns in the Periodic Table are called *groups* or *families*. The elements in a family have related properties. Note that each element is placed in a single box. The information given about each element is its symbol, name, atomic number, atomic mass (weight), oxidation state, and electron configuration. At times, a kind of chemical shorthand is used to give information about an element. For example: the element helium may be written as $^{4}_{2}He$. The *subscript* number 2 indicates the atomic number of helium. The *superscript* number 4 stands for the mass number. Basic information about the behavior of elements is necessary to the understanding of many biological processes.

KEY

1 +1 -1 H Hydrogen 1.00794 $1s^1$

- Common oxidation states
- Atomic number
- Element symbol
- Element name
- Atomic mass (or number of longest-lived isotope)
- Electron configuration

Note: Atomic masses are based on carbon-12=12.000...u

s-block | *d*-block (Transition Elements) | *p*-block

GROUP

Period	1	2	3	4	5	6	7	8	9	10	11	12	13	14	15	16	17	18
1	1 +1 -1 H Hydrogen 1.00794 $1s^1$																	2 0 He Helium 4.00260 $1s^2$
2	3 +1 Li Lithium 6.941 $[He]2s^1$	4 +2 Be Beryllium 9.01218 $[He]2s^2$											5 +3 B Boron 10.81 $[He]2s^22s^1$	6 -4 +2 +4 C Carbon 12.011 $[He]2s^22s^2$	5 -3 -2 -1 +1 +2 +3 +4 +5 N Nitrogen 14.0067 $[He]2s^22s^3$	6 -2 O Oxygen 15.9994 $[He]2s^22s^4$	5 +3 F Flourine 18.998403 $[He]2s^22s^5$	6 -4 +2 +4 Ne Neon 20.1797 $[He]2s^22s^6$
3	11 +1 Na Sodium 22.98977 $[Ne]3s^1$	12 +2 Mg Magnesium 24.305 $[Ne]3s^2$											13 +3 Al Aluminum 26.98154 $[Ne]3s^23p^1$	14 -4 +2 +4 Si Silicon 28.0855 $[Ne]3s^23p^2$	15 -3 +3 +5 P Phosphorus 30.97376 $[Ne]3s^23p^3$	16 -2 +4 +6 S Sulfur 32.066 $[Ne]3s^23p^4$	17 -1 +1 +3 +5 +7 Cl Chlorine 35.453 $[Ne]3s^23p^5$	18 0 Ar Argon 39.948 $[Ne]3s^23p^6$
4	19 +1 K Potassium 39.0983 $[Ar]4s^1$	20 +2 Ca Calcium 40.078 $[Ar]4s^3$	21 +3 Sc Scandium 44.9559 $[Ar]3d^14s^2$	22 +2 +3 +4 Ti Titanium 40.078 $[Ar]3d^24s^2$	23 +2 +3 +6 V Vanadium 50.9415 $[Ar]3d^34s^2$	24 -2 -3 -6 Cr Chromium 51.996 $[Ar]3d^54s$	25 +2 +3 +4 +7 Mn Manganese 54.9380 $[Ar]3d^54s^2$	26 +2 +3 Fe Vanadium 55.847 $[Ar]3d^64s^2$	27 +2 +3 Co Chromium 58.9332 $[Ar]3d^74s^1$	28 +2 +3 Ni Nickel 58.69 $[Ar]3d^34s^2$	29 +2 +3 Cu Copper 63.546 $[Ar]3d^{10}4s^1$	30 +2 Zn Zinc 65.93 $[Ar]3d^{10}4s^2$	31 +2 +3 Ga Gallium 69.72 $[Ar]3d^{10}4s^24p^1$	32 +2 +3 Ge Germanium 72.61 $[Ar]3d^{10}4s^24p^2$	33 +2 +4 +5 As Arsenic 74.9216 $[Ar]3d^{10}4s^24p^3$	34 -2 +4 +6 Se Selenium 78.96 $[Kr]3d^{10}4s^24p^4$	35 -1 +1 +5 Br Bromine 121.757 $[Kr]3d^{10}4s^25p^5$	36 0 +2 Kr Krypton 121.757 $[Kr]3d^{10}4s^25p^6$
5	37 +1 Rb Rubidium 85.4678 $[Kr]5s^1$	38 +2 Sr Strontium 87.62 $[Kr]5s^2$	39 +3 Y Yttrium 88.9059 $[Kr]4d^15s^2$	40 +4 Zr Zirconium 87.62 $[Kr]4d^15s^2$	41 +3 Nb Nobelium 92.9064 $[Kr]4d^45s^1$	42 -3 -6 Mo Molybdenum 95.94 $[Kr]4d^55s$	43 +4 +6 +7 Tc Technetium (98) $[Kr]4d^65s^2$	44 +3 Ru Ruthenium 101.07 $[Kr]4d^75s^1$	45 +3 Rh Rhodium 102.9055 $[Kr]4d^85s^1$	46 +2 +4 Pd Palladium 106.42 $[Kr]4d^{10}$	47 +1 Ag Siver 107.8682 $[Kr]4d^{10}5s^1$	48 +2 Cd Cadnium 112.41 $[Kr]4d^{10}5s^2$	49 +3 In Indium 114.82 $[Kr]4d^{10}5s^25p^1$	50 -4 +2 +4 Sn Tin 118.710 $[Kr]4d^{10}5s^25p^2$	51 -3 +3 +5 Sb Antimony 121.757 $[Kr]4d^{10}5s^25p^3$	52 -2 +4 +6 Te Tellurium 121.757 $[Kr]4d^{10}5s^25p^4$	53 -1 +1 +5 +7 I Iodine 121.757 $[Kr]4d^{10}5s^25p^5$	54 0 +2 +4 +6 Xe Xenon 121.757 $[Kr]4d^{10}5s^25p^6$
6	55 +1 Cs Cesium 132.9054 $[Xe]6s^1$	56 +2 Ba Barium 87.62 $[Xe]6s^2$	57 +3 La Lanthanum 138.9055 $[Xe]5d^16s^2$	72 +4 Hf Hafnium 178.49 $[Xe]4f^{14}5d^26s^2$	73 +5 Ta Tantalum 180.9479 $[Xe]4f^{14}5d^36s^2$	74 +6 W Tungsten 183.85 $[Xe]4f^{14}5d^46s^2$	75 +4 +6 +7 Re Rhenium 186.207 $[Xe]4f^{14}5d^56s^2$	76 +3 +4 Os Osmium 190.2 $[Xe]4f^{14}5d^66s^2$	77 +3 +4 Ir Iricium 192.22 $[Xe]4f^{14}5d^76s^2$	78 +2 +4 Pt Platinum 195.08 $[Xe]4f^{14}5d^96s^1$	79 +1 +3 Au Gold 196.9665 $[Xe]4f^{14}5d^{10}6s^1$	80 +1 +2 Hg Mercury 200.59 $[Xe]4f^{14}5d^{10}6s^2$	81 +1 +3 Ti Thallium 204.383 $[Xe]4f^{14}5d^{10}6s^26p^1$	82 +2 +4 Pb Lead 207.2 $[Xe]4f^{14}5d^{10}6s^26p^2$	83 +3 +5 Bi Bismuth 208.9804 $[Xe]4f^{14}5d^{10}6s^26p^3$	84 -2 +4 Po Potassium (209) $[Xe]4f^{14}5d^16s^26p^4$	85 At Astatine (210) $[Xe]4f^{14}5d^{10}6s^26p^5$	86 0 Rn Radon (222) $[Xe]4f^{14}5d^{10}6s^26p^6$
7	87 +1 Fr Francium (223) $[Rn]7s^1$	88 +2 Ra Radium 87.62 $[Rn]7s^2$	89 +1 Ac Actinium 227.0278 $[Rn]6d^17s^2$	104 Rf Rutherfordium (261) $[Rn]5f^{14}6d^27s^2$	105 Db Dubnium (262) $[Rn]5f^{14}6d^37s^2$	106 Sg Seaborgnium (263) $[Rn]5f^{14}6d^47s^2$	107 Bh Bohrium (262) $[Rn]5f^{14}6d^57s^2$	108 Hs Hassium (265) $[Rn]5f^{14}6d^67s^2$	106 Mt Meitenerium (266) $[Rn]5f^{14}6d^77s^2$	110 Uun Ununnilium (269) $[Rn]4f^{14}6d^87s^2$	111 Uuu Unununium (272) $[Rn]5f^{14}6d^97s^2$	112 Uub Ununbium (277) $[Rn]5f^{14}6d^{10}7s^2$						

f-block

58 +3 +4 Ce Cerium 140.12 $[Xe]4f^15d16s^2$	59 +3 Pr Praseodymium 140.0359 $[Xe]4f^36s^2$	60 +3 Nd Neodymium 144.24 $[Xe]4f^46s^2$	61 +3 Pm Promethium (145) $[Xe]4f^56s^2$	62 +2 +3 Sm Samarium 150.36 $[Xe]4f^66s^2$	63 +2 +3 Eu Europium 151.96 $[Xe]4f^76s^2$	64 +3 Gd Gadolinium 157.25 $[Xe]4f^75d^16s^2$	65 +3 Tb Terbium 158.9254 $[Xe]4f^96s^2$	66 +3 Dy Dysprosium 162.50 $[Xe]4f^{10}6s^2$	67 +3 Ho Holmium 164.9304 $[Xe]4f^{11}6s^2$	68 +3 Er Erbium 167.26 $[Xe]4f^{12}6s^2$	69 +3 Tm Thulium 168.9342 $[Xe]4f^{13}6s^2$	70 -2 -3 Yb Ytterbium 173.04 $[Xe]4f^{14}6s^2$	71 +3 Lu Lutetium 174.967 $[Xe]4f^{14}5d^16s^2$
90 +4 Th Thorium 232.0381 $[Rn]6d^27s^2$	91 +4 +5 Pa Protactinum 231.0359 $[Rn]5f^26d^17s^2$	92 +3 U Uranium 238.0289 $[Rn]5f^36d^17s^2$	93 +3 Np Promethium 237.048 $[Rn]5f^46d^17s^2$	94 +2 +3 Pu Samarium (244) $[Rn]5f^56d^17s^2$	95 +4 Am Americium (243) $[Rn]5f^77s^2$	96 +3 Cm Curium (247) $[Rn]5f^26d^17s^2$	97 +3 +4 Bk Berkelium (247) $[Rn]5f^97s^2$	98 +3 Cf Californium (251) $[Fn]5f^17s^2$	99 Es Einsteinium (252) $[Rn]5f^{11}7s^2$	100 Fm Fermium (257) $[Rn]5f^{12}7s^2$	101 Md Mendelevium (258) $[Rn]5f^{13}7s^2$	102 No Nobelium (259) $[Rn]5f^{14}7s^2$	103 Lr Lawrenncium (262) $[Rn]5f^{14}6d^17s^2$

Chemical Bonding

COVALENT BONDS

Electrons occupy shells, or energy levels, outside the nucleus of the atom. The first shell can hold a maximum of 2 electrons. The second and third shells are completed with 8 electrons.

Look at Figure 4.2. Notice the single electron in shell 1. This shell can hold 2 electrons. In order to complete this shell, 1 electron must be added.

Now look at Figure 4.3. You can see that shell 1 is complete with 2 electrons. Shell 2 has 6 electrons. However, this shell can hold 8 electrons. In order to complete the shell, two electrons must be added.

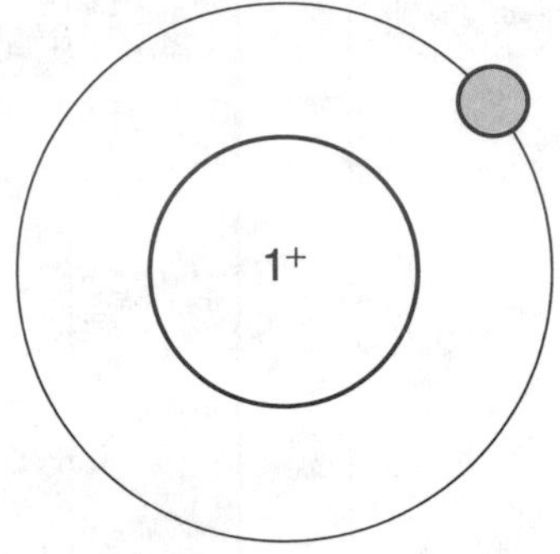

FIGURE 4.2 Structure of the hydrogen atom

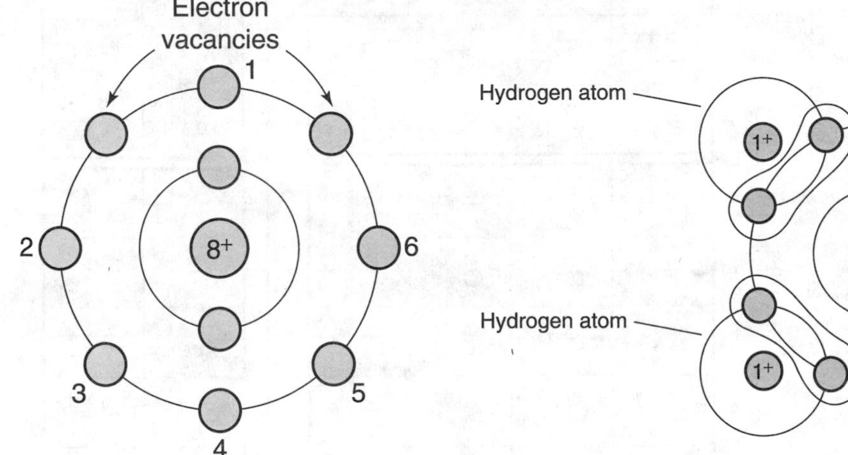

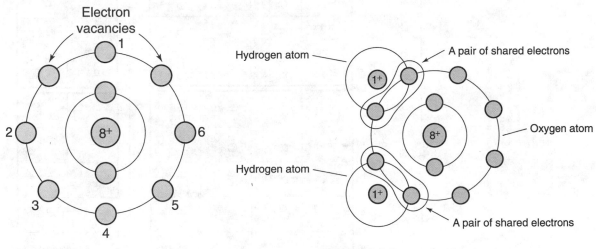

FIGURE 4.3 Structure of the oxygen atom. The outer shell needs 2 more electrons for completion.

FIGURE 4.4 A molecule of water

Study Figure 4.4 closely. Two atoms of hydrogen have each shared 1 electron with an atom of oxygen. Now, the second shell of oxygen has been filled by 2 electrons from 2 atoms of hydrogen. Atoms of hydrogen and oxygen have formed a **molecule** of water. Water is a compound because it is made up of two different kinds of atoms. The atoms are held together by a force called a **chemical bond**. Chemical bonds are the result of attraction of electrons in the outer energy levels of atoms. A chemical bond that is formed by the sharing of electrons is called a **covalent bond**.

Carbon has an atomic number of 6. This means that carbon has 6 protons and therefore 6 electrons. The first shell of carbon holds 2 electrons. The second shell holds 4 electrons and can be completed by gaining 4 more electrons. Figure 4.5 provides a graphic example of covalent bonding. Notice how 4 atoms of hydrogen share their electrons with an atom of carbon, thus completing its outer shell. The compound known as methane gas is formed. Water and methane are compounds with *single bonds*. Each bond consists of one pair of shared electrons. During covalent bonding, atoms may share two pairs of electrons. A bond that consists of two pairs of shared electrons is known as a *double bond*.

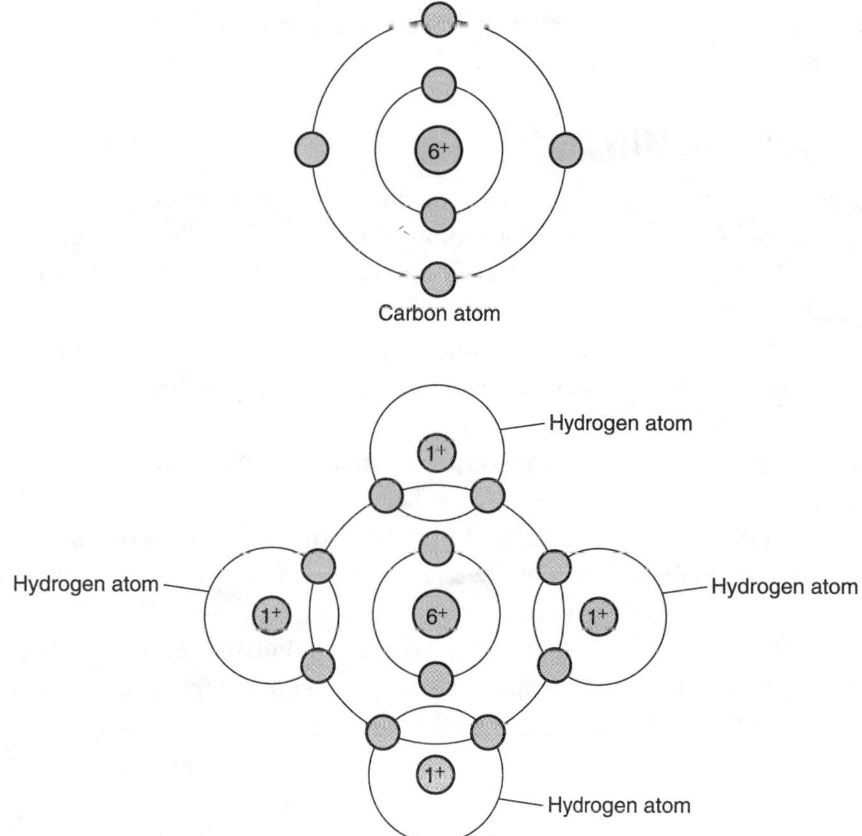

FIGURE 4.5 A molecule of methane gas. Notice the single carbon atom above. It has 4 electrons in its outer shell.

Chemical Reactions

ACTIVATION ENERGY

Chemical bonds are formed by completion of the outer energy levels of atoms. As the result of bonding between atoms, stable compounds are produced. Breaking bonds between atoms requires an input of energy. Released atoms form new bonds with other atoms to make new compounds. The process in which existing chemical bonds are broken and new bonds are formed between atoms is called a **chemical change** or a **chemical reaction**.

To start a chemical reaction, sufficient energy to break existing bonds is necessary. The energy needed to begin a chemical reaction is termed the **activation energy**. The substances that were present before the reaction started are the **reactants**. The new substances formed as the result of the reaction are the **products**. When you learn about the chemical reactions that take place in living cells and organisms, you will understand why a knowledge of chemical bonding and chemical change is required to comprehend the processes necessary for life.

CHEMICAL FORMULAS

In chemistry a chemical compound is represented by a **chemical formula**. H_2O is the chemical formula for water; NaCl, for sodium chloride. A formula provides specific information about a compound. First, each element is represented by a chemical symbol. Second, the formula provides information about the proportions in which atoms combine. The formula for water shows that 2 hydrogen atoms combine with 1 atom of oxygen. In like manner, the formula for sodium chloride shows that 1 atom of sodium combines with 1 atom of chlorine.

These examples indicate that atoms combine in definite proportions to form compounds. Notice H_2 in the formula for water. The number 2 is called a **subscript**, a small number written after a symbol and slightly below it. The subscript shows the proportions in which the atoms combine to form water. When no subscript is written, as in O, it is understood that the subscript is 1.

The **empirical formula** shows the simplest proportion (smallest number) of atoms in a compound. A **molecular formula** shows the makeup of a molecule of a compound. The formula H_2O is both the empirical and the molecular formula. It shows, using the simplest proportion of atoms, the makeup of a molecule of water: 2 atoms of hydrogen and 1 atom of oxygen. The molecular formula for glucose is $C_6H_{12}O_6$, showing the kinds and numbers of atoms present in one molecule. The empirical formula for glucose, CH_2O, shows the atoms in their simplest proportion.

A **structural formula** shows not only the kinds and numbers of atoms in a compound, but also the arrangement of the atoms. As you have learned, a covalent bond is a pair of shared electrons. In a structural formula a covalent bond is shown by a short line that joins atoms.

```
  H
  |
H—O
```

Water: H2O

H—C—C—H

Acetylene: C2H2

In the next chapter structural formulas will be discussed in more detail in relationship to the chemistry of cells.

DIATOMIC MOLECULES

An atom of hydrogen is highly reactive and cannot exist alone. An atom of hydrogen forms a covalent bond with another hydrogen atom; the result is a hydrogen molecule of 2 atoms, called a **diatomic molecule**. A molecule of hydrogen is written as H_2. Most elements that form diatomic molecules are gases. Examples of other atoms that form diatomic molecules are oxygen, chlorine, and nitrogen.

CHEMICAL EQUATIONS

An equation is a statement of equality. A **chemical equation** represents a chemical reaction between *reactants* that results in the formation of *products*. Let us consider the formation of water from atoms of hydrogen and oxygen. The reactants are hydrogen (H_2) and oxygen (O_2). Remember that hydrogen and oxygen are diatomic molecules and are written with subscripts indicating 2 atoms of hydrogen and 2 atoms of oxygen. Let us put this information in the form of an equation:

$$H_2 + O_2 \rightarrow H_2O$$

Since an equation represents an equality, the number of hydrogen atoms and the number of oxygen atoms on each side of the yield sign (_) should be equal. On the left side of the equation above, there are 2 atoms of hydrogen and 2 atoms of oxygen. On the right side, however, there are 2 atoms of hydrogen but only 1 atom of oxygen. As written, therefore, this equation does not show equality. To make the number of atoms on the left side of the equation equal to the number of atoms on the right side, the equation must be **balanced**:

$$2H_2 + O_2 \rightarrow 2H_2O$$

This equation shows that 2 molecules of hydrogen combine with 1 molecule of oxygen to produce 2 molecules of water. Notice that the number 2 was placed in front of hydrogen on the left side of the equation and in front of water on the right side. Numbers placed in front of formulas to *balance* an equation are called **coefficients**. The coefficient is a *multiplier* for the entire formula. Therefore on the equation left, there are 4 hydrogen atoms and 2 oxygen atoms. On the equation right, there are 4 hydrogen atoms and 2 oxygen atoms.

The chemical reactions that take place within cells follow the same rules for the making and breaking of chemical bonds that are discussed in this chapter. You will learn about a number of chemical reactions that are necessary for producing, storing, and using energy.

IONIC BONDING

An atom has a neutral electrical charge because its number of protons equals its number of electrons. An atom becomes a charged particle if its outer shell loses or gains electrons. Charged atoms are called **ions**. Table 4.1 on page 74 indicates that the atomic number of sodium is 11, so it has 11 protons in the nucleus and 11 electrons in shells outside the nucleus.

Figure 4.6 shows that there is a single electron in the outer shell of the sodium atom. To complete the outer shell, sodium must either gain 7 electrons or lose 1 electron. Which of these situations is the most reasonable? Of course, it is easier for the sodium atom to lose 1 electron, thus becoming a positively charged ion represented by the symbol Na^+. The number of protons in the nucleus remains 11, whereas the number of electrons is reduced to 10. Therefore a sodium ion has an excess of 1 positive charge.

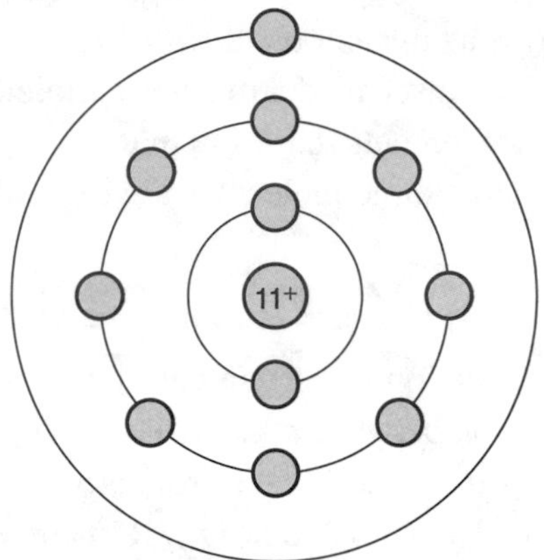

FIGURE 4.6 A sodium atom

Table 4.1 indicates that the atomic number of a chlorine atom is 17. How many protons are in the nucleus, and how many electrons are in the energy levels that surround the nucleus? Figure 4.7 indicates that the outer shell of chlorine contains 7 electrons. You know that 8 electrons complete shell number 3. Is it easier for chlorine to gain 1 electron or to lose 7 electrons? The answer is obvious. If chlorine gains an electron, there will be 17 protons in the nucleus and 18 electron shells outside the nucleus. The result is an excess of 1 unit of negative charge. The charged chlorine atom is now an ion with a negative charge and is represented by the symbol Cl^-.

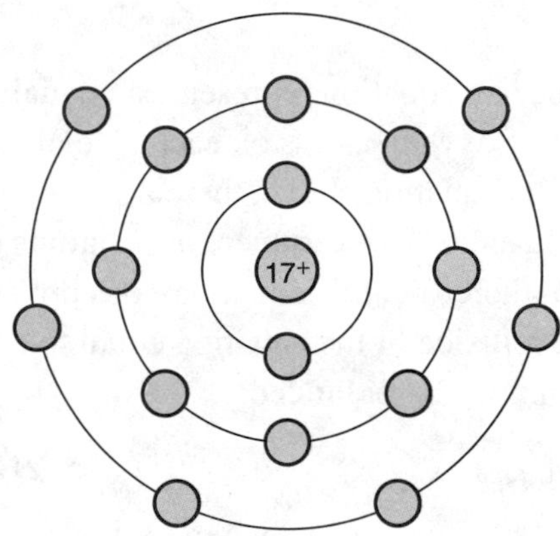

FIGURE 4.7 A chlorine atom. The number of electrons in the outer shell is 7. How many electrons are needed to complete the outer shell?

When an atom of sodium (Na) is made to react with an atom of chlorine (Cl), the gain and loss of electrons as described above takes place. The single electron in the outer shell of Na transfers onto the outer shell of Cl, forming a chemical bond resulting in the compound sodium chloride (NaCl). The bond formed between Na and Cl is an **ionic bond** (Figure 4.8). Ionic compounds do not form molecules. Each sodium ion is attracted to several chlorine ions, and each chlorine ion is attracted to several sodium ions, forming collections of oppositely charged ions.

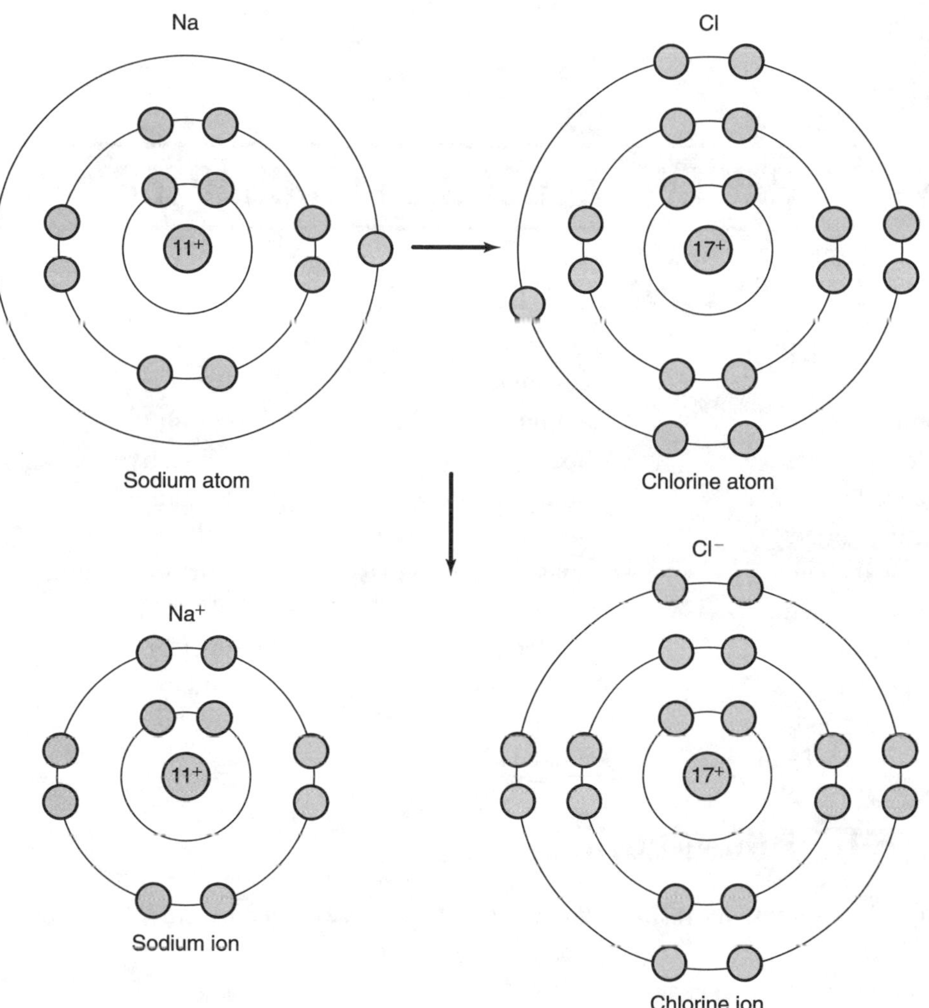

FIGURE 4.8 The ionic bonding of sodium and chlorine. Notice that the atoms of sodium and chlorine are electrically neutral. The sodium ion and the chlorine ion are charged particles.

REVIEW EXERCISES FOR CHAPTER 4

WORD-STUDY CONNECTION

activation energy
atom
atomic mass
atomic number
balanced equation
chemical bond
chemical change
chemical reaction
coefficient
compound
covalent bond
diatomic molecule
electron
element
empirical formula
ion
ionic bond
isotope
mass
matter
molecular formula
molecule
neutron
nucleus
product
properties
proton
radioactive
reactant
shell
structural formula
subscript
superscript
symbol

SELF-TEST CONNECTION

PART A. Completion. *Write in the word that correctly completes each statement.*

1. Anything that takes up space is called _____.

2. The pull of gravity on the mass of a substance is known as _____.

3. The smallest part of an element that retains the properties of the element is the _____.

4. Each element has specific _____ that make it different from any other element.

5. A substance made of two or more kinds of atoms is a (an) _____.

6. An element is made of the same kind of _____.

7. Each element has been given a symbol that begins with a _____ letter.

8. The symbol for sodium is _____.

9. All chemical activity depends on the structure of _____.

10. Atomic particles that have mass but no charge are called _____.

11. Negatively charged atomic particles are called _____.

12. A charged atom is known as a (an) _____.

13. The number of protons in the nucleus of an atom is termed the _____.

14. Electrons occupy spaces called energy levels or _____.

Study the diagram and then answer questions 15–17:

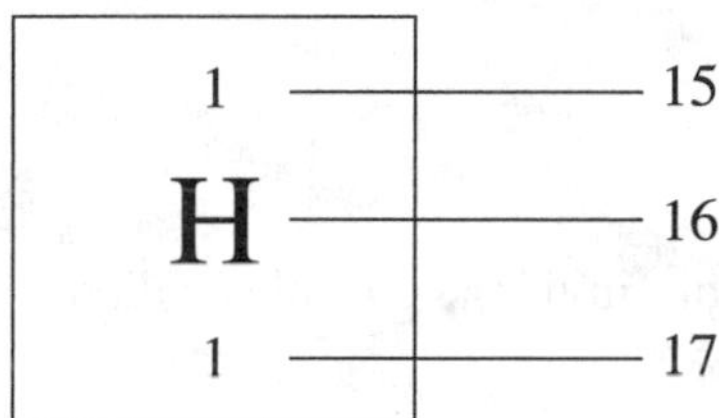

15. The item at 15 is the _____.

16. The item at 16 is the symbol for _____.

17. The item shown at 17 is the _____.

18. The second energy level of an atom is completed when it holds _____ electrons.

19. A calcium atom has 20 protons in its nucleus. The total number of its electrons is _____.

20. The electric charge on a neutron is _____.

21. The first energy level of an atom is completed when it holds _____ electrons.

22. When two or more different kinds of atoms combine chemically, the result is called a _____.

23. Atoms are held together by a force of attraction known as a _____.

24. Carbon has an atomic number of 6. The number of electrons in its outer shell is _____.

25. The empirical formula for this compound: $\mathrm{H{-}\overset{\displaystyle H}{\overset{|}{\underset{\displaystyle H}{\underset{|}{C}}}}{-}H}$ is _____.

PART B. Multiple Choice. *Circle the letter of the item that correctly completes each statement.*

1. The amount of matter in a substance is known as
   (a) gravity
   (b) energy
   (c) mass
   (d) space

2. The ability to do work is known as
   (a) force
   (b) energy
   (c) strength
   (d) pressure

3. The building blocks of matter are invisible particles called
   (a) air
   (b) elements
   (c) mass
   (d) atoms

4. The smallest part of a compound that retains the properties of the compound is a (an)
   (a) molecule
   (b) atom
   (c) substance
   (d) element

5. Water is a (an)
   (a) element
   (b) atom
   (c) formula
   (d) compound

6. The symbol for iron is
   (a) I
   (b) Ir
   (c) F
   (d) Fe

7. The center of an atom is its
   (a) core
   (b) nucleus
   (c) hub
   (d) centrum

8. Mass is properly defined as the pull on matter of
   (a) gravity
   (b) force
   (c) pressure
   (d) energy

9. Charged particles that rotate around the nucleus of an atom are called
   (a) electricity
   (b) protons
   (c) electrons
   (d) energy

10. The atomic mass number of an element is obtained by adding together the number of
   (a) protons and electrons
   (b) electrons and neutrons
   (c) electrons and nuclei
   (d) protons and neutrons

11. Isotopes of an element differ in the number of
   (a) protons
   (b) neutrons
   (c) electrons
   (d) clcctrical particles

12. Of the following, a true statement about an atom is
   (a) an atom has excess electric charges
   (b) an atom has an unstable nucleus
   (c) the energy shells are visible inside the nucleus
   (d) an atom is electrically neutral

13. Most hydrogen atoms lack a (an)
   (a) proton
   (b) neutron
   (c) nucleus
   (d) electron

14. A pair of shared electrons forms a (an)
    (a) covalent bond
    (b) ionic bond
    (c) unstable bond
    (d) radioactive bond

15. The force that holds atoms together is known as a (an)
    (a) atomic bond
    (b) chemical bond
    (c) isotopic bond
    (d) covalent bond

16. The number of known elements is approximately
    (a) 120
    (b) 115
    (c) 110
    (d) 112

17. The sodium ion is represented by the symbol
    (a) So^+
    (b) So^-
    (c) Na^+
    (d) Na^-

18. In a chemical formula, each element is represented by its
    (a) atomic number
    (b) chemical symbol
    (c) atomic mass
    (d) electric charge

19. Covalent bonding occurs between atoms in
    (a) calcium
    (b) sodium
    (c) salt
    (d) water

20. The proportions in which atoms combine are shown by numbers called
    (a) superscripts
    (b) subscripts
    (c) exponents
    (d) primes

21. Chlorine has an atomic number of 17. The number of electrons in its outer shell is
    (a) 1
    (b) 3
    (c) 5
    (d) 7

22. In a structural formula, the line between two atoms, as with C—C, represents a (an)
    (a) double bond
    (b) diatomic bond
    (c) covalent bond
    (d) ionic bond

23. A diatomic molecule consists of
    (a) 2 atoms of the same element
    (b) 4 covalent bonds
    (c) a double atomic number
    (d) 2 atoms of different elements

24. New substances formed as the result of a chemical reaction are known as
    (a) reactants
    (b) energy
    (c) diatoms
    (d) products

25. The smallest numbers of atoms in a compound are shown by a formula known as
    (a) ionic
    (b) structural
    (c) empirical
    (d) molecular

PART C. Modified True-False. *If a statement is true, write "true" for your answer. If a statement is incorrect, change the underlined expression to one that will make the statement true.*

1. Solids, liquids, and gases are collectively known as <u>mass</u>.

2. Hydrogen is best classified as <u>a compound</u>.

3. Iron is made of the same kind of <u>molecules</u>.

4. The symbol for potassium is Po.

5. Positively charged particles in the nucleus of an atom are proteins.

6. The mass of a neutron is approximately equal to the mass of a proton.

7. The word *planetary* is used to describe the behavior of protons.

8. An atom is electrically charged.

9. If an atom has 8 protons, it will have 10 electrons.

10. The symbol ^{18}O indicates the atomic number of a form of oxygen.

11. A number placed in front of a formula in order to balance an equation is called a multiplier.

12. Oxygen and hydrogen are examples of diatomecious molecules.

13. As the results of chemical changes, substances called reactants are formed.

14. An equation indicates a statement of activity.

15. Structural formulas show the kinds, numbers, and stability of atoms in a molecule.

16. An element is made of the same kind of molecules.

17. The formula $2H_2O_2$ shows that this substance has two atoms of oxygen.

18. An empirical formula shows atoms in their simplest proportion.

19. In formulas, the superscript shows the proportions in which atoms combine.

20. The symbol for the oxygen atom is O^-.

21. A calcium atom becomes a charged particle if its outer shell loses protons.

22. Ionic bonds are formed by loss and gain of electrons.

23. A double bond results from the sharing of one pair of electrons.

24. The atomic mass of each isotope of an element is the same.

25. All elements have isotopes.

CONNECTING TO CONCEPTS

1. Can air be considered a form of matter? Defend your "yes" or "no" answer with scientific reasoning.
2. Energy is defined as the ability to do work. Explain the meaning of this definition.
3. Why should a student of biology learn the basic principles of chemistry?

ANSWERS TO SELF-TEST CONNECTION

PART A

1. matter
2. weight
3. atom
4. characteristics or properties
5. compound
6. atoms
7. capital
8. Na
9. atoms or atomic number
10. neutrons
11. electrons
12. ion
13. atomic number
14. shells
15. atomic number
16. hydrogen
17. atomic mass
18. 8
19. 20
20. zero or neutral
21. 2
22. compound
23. chemical bond
24. 4
25. CH_4

PART B

1. **(c)**
2. **(b)**
3. **(d)**
4. **(a)**
5. **(d)**
6. **(d)**
7. **(b)**
8. **(a)**
9. **(c)**
10. **(d)**
11. **(b)**
12. **(d)**
13. **(b)**
14. **(a)**
15 **(b)**
16 **(d)**
17 **(c)**
18 **(b)**
19 **(d)**
20. **(b)**
21. **(d)**
22. **(c)**
23. **(a)**
24. **(d)**
25. **(c)**

PART C

1. matter
2. element
3. atoms
4. K
5. protons
6. true
7. electrons
8. neutral
9. 8
10. atomic mass or mass number
11. coefficient
12. diatomic
13. products
14. equality
15. arrangement
16. atoms
17. 4
18. true
19. subscript
20. ion
21. electrons
22. true
23. two
24. different
25. true

CONNECTING TO LIFE/JOB SKILLS

We live in a world of chemistry. You may wish to find out how the various branches of chemistry contribute to the abundance of health information and manufactured products. You may be interested in researching chemistry as a field of study with a possibility toward a career as a **research chemist** or a **chemical engineer**. Your library media center can offer the research tools of books, magazines, and computers.

Chronology of Famous Names in Biology

1805 **John Dalton** (England)—known as the "father of modern atomic theory"; presented ideas about the atom based on experimentation. His concepts are closely related to what we know today about atoms.

1863 **John Newlands** (England)—presented the concept of repeating properties of atoms in every group of eight elements.

1869 **Dimitry Mendeleyev** (Russia)—devised the first periodic table of the elements

1897 **J. J. Thompson** (England)—discovered electrically charged particles in the atom by using cathode rays and naming these particles electrons.

1909 **Robert Millikan** (United States)—measured the charge on an electron.

1911 **Ernest Rutherford** (England)—determined that the atomic nucleus is positively charged.

1912 **Henry Moseley** (England)—determined atomic numbers by using X rays.

1913 **Neils Bohr** (Denmark)—proposed a model of the atom as having a positively charged nucleus and electrons in orbits around the nucleus.

1916 **Gilbert Newton Lewis** (United States)—demonstrated electron pairing in covalent bonds.

1924 **Louis de Broglie** (France)—suggested that atomic particles can have wave-like characteristics as does light.

1932 **James Chadwick Newton** (England)—discovered the neutron.

1942 **Enrico Fermi** (United States)—directed the first controlled chain reaction caused by the fission of atomic nuclei.

5

The Basic Chemistry of Cells

WHAT YOU WILL LEARN

In this chapter you will explore how energy is released from fuel molecules and how it is used to build other compounds.

SECTIONS IN THIS CHAPTER

- The Cell as a Chemical Factory
- The Role of Enzymes in Living Cells
- Cellular Respiration
- Review Exercises for Chapter 5
- Connecting to Life/Job Skills
- Chronology of Famous Names in Biology

The Cell as a Chemical Factory

Most of the activities of the cell involve chemical changes. Small molecules may be joined together to form larger ones. Complex substances may be broken down into their smaller units. Materials are changed from one form to another, used up, or synthesized (put together) during biochemical activities that take place in cells. The cell is likened to a "chemical factory" that uses some of the elements present in the nonliving environment. Of the elements in the living material of the cell, carbon, hydrogen, oxygen, and nitrogen are present in the greatest amounts. Sulfur, phosphorus, magnesium, iodine, iron, calcium, sodium, chlorine, and potassium are found in smaller quantities. These elements are present in inorganic and organic compounds that are utilized by the cell.

INORGANIC COMPOUNDS

Inorganic compounds are compounds that do not have the elements carbon and hydrogen in chemical combination. The inorganic compounds in greatest percentages in cells are water, mineral salts, inorganic acids, and bases. For example, hydrochloric acid (HCl), an inorganic acid, is produced by the gastric glands in the stomach, acidifying stomach contents.

ORGANIC COMPOUNDS

Organic compounds are compounds that contain the elements carbon and hydrogen in chemical combination. Organic compounds are produced by living plants and animals and can be synthesized in the laboratory, as well. The special bonding property of carbon permits it to form compounds that are structured as long chains of atoms or as rings of atoms. The types of organic compounds that are contained in the living material of the cell and used in its chemical activities are carbohydrates, lipids, proteins, and nucleic acids.

CARBOHYDRATES

Carbohydrates are organic compounds that include all starches and sugars. Carbohydrates contain the elements carbon, hydrogen, and oxygen. Usually, for every two atoms of hydrogen there is one atom of oxygen. In other words, hydrogen and oxygen are present in the ratio of 2:1.

All starches and sugars are built from the basic unit $C_6H_{12}O_6$, *glucose*, a **monosaccharide** or *single sugar*. Figure 5.1 shows the structural formula for glucose. Examples of other monosaccharides are *fructose* (fruit sugar) and *ribose* (found in nucleic acids).

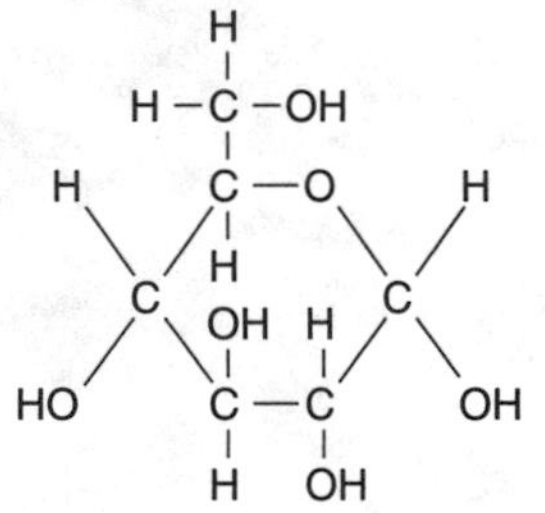

FIGURE 5.1 Structural formula for the monosaccharide glucose

Within living cells, molecules of monosaccharides (single sugars) may combine chemically. When two molecules of single sugars are combined chemically, a *double sugar* or **disaccharide** is formed. As the two molecules of single sugars are joined, a molecule of water is produced in addition to the disaccharide. When four molecules of single sugars are joined chemically, two molecules of water are formed. The chemical joining of several small molecules to produce a larger molecule is called *synthesis*. The formation of water during synthesis is known as *dehydration synthesis*. Dehydration synthesis makes possible the close "packaging" of complex molecules. Figure 5.2 shows in structural formula format the formation of the disaccharide maltose.

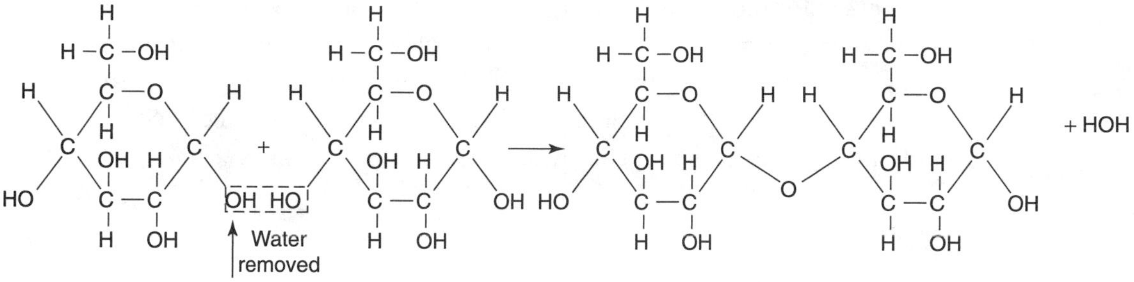

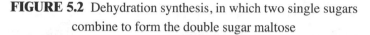

FIGURE 5.2 Dehydration synthesis, in which two single sugars combine to form the double sugar maltose

Carbohydrates are used by cells primarily as sources of energy. Remember that energy fuels the chemical work of the cell. Within living cells, many monosaccharides chemically combine by dehydration synthesis to form a complex carbohydrate known as a **polysaccharide**. In plant cells, polysaccharides are used in cell structures such as walls. The polysaccharide in cell walls is *cellulose*. In some animal cells, glycoproteins (carbohydrate and protein compounds) form recognition sites on the membranes of certain cells.

LIPIDS

The lipids are a group of organic compounds that include the fats and fatlike substances. A lipid molecule, like a carbohydrate molecule, contains the elements carbon, hydrogen, and oxygen. Unlike the carbohydrates, however, in lipid molecules the ratio of hydrogen to oxygen is much greater than 2:1. A lipid molecule is made up of two basic units: an *alcohol* (usually glycerol) and a class of compounds called *fatty acids*.

Some lipids are formed during the process of dehydration synthesis. One molecule of glycerol is chemically joined to three molecules of fatty acid with the accompanying loss of water. Figure 5.3 shows in structural formula the formation of a lipid.

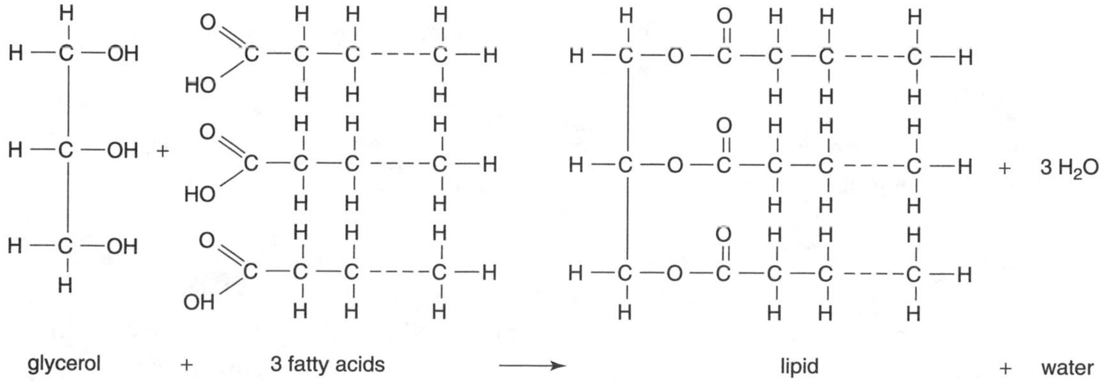

FIGURE 5.3 Formation of a lipid

Lipids are sources for biologically usable energy. Most lipid molecules provide twice as much energy per gram as do carbohydrate molecules. However, the energy in lipid molecules is not as easy to extract as that of glucose. Besides being useful for energy, lipids are essential to the structure of cells. The plasma membrane is composed of lipid molecules as is the myelin sheath of some nerve fibers.

PROTEINS

All proteins contain the elements carbon, hydrogen, oxygen, and nitrogen. In addition to these elements, some proteins contain sulfur. A protein molecule is constructed from building block units known as **amino acids**. Figure 5.4 shows a generalized structural formula of an amino acid.

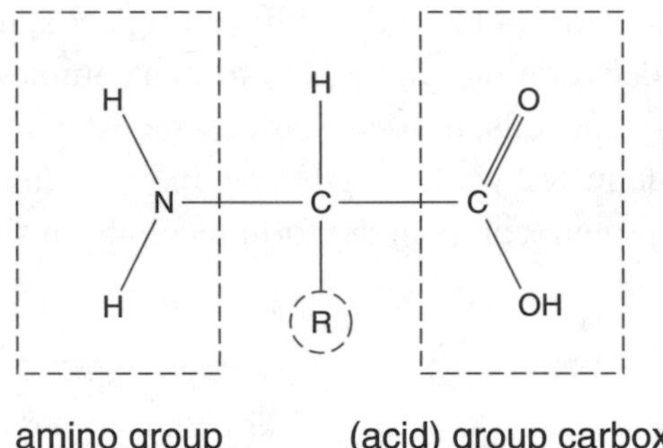

FIGURE 5.4 Generalized structure of an amino acid. R represents a variable group and is the basis for the variety of amino acids.

Notice the parts of an amino acid. The **amino group** at the left end of the molecule contains the elements hydrogen and nitrogen in the ratio of 2:1. At the right end of the amino acid molecule is a carboxyl or acid group, which contains the elements carbon and oxygen and a hydroxyl (OH) group. The center of the molecule contains carbon, hydrogen, and an "R" group, that is, a variable group, thereby allowing for variety among the amino acids.

Twenty amino acids are found in living cells; these are called essential amino acids. However, some proteins are composed of special amino acids, usually formed by a change in a common amino acid, that supplement the basic set of twenty essential amino acids.

Like complex carbohydrates, proteins are formed by dehydration synthesis when two or more amino acids are joined chemically. *Dipeptides* are formed when two amino acids are joined together (Figure 5.5). **Polypeptides** are formed by the dehydration synthesis of a number of amino acid groups. A protein is usually made up of one or more polypeptides.

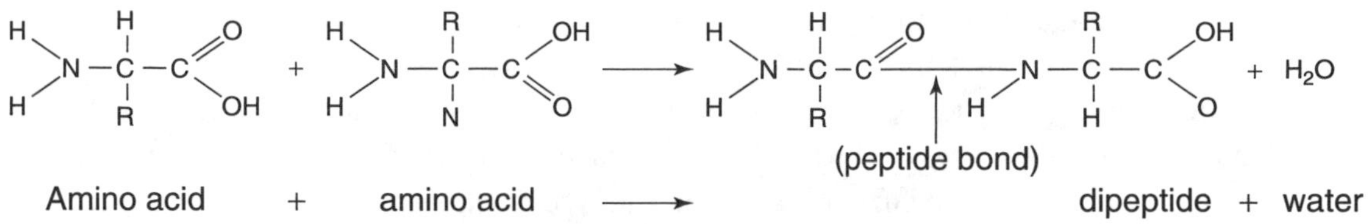

(peptide bond)

Amino acid + amino acid ⟶ dipeptide + water

FIGURE 5.5 Formation of a dipeptide

The number of different proteins is enormous. Variety in proteins is made possible by variability in structure. Although proteins have a basic structural similarity, each kind is different because of the number, types, and order of amino acids that compose it.

Proteins are the most abundant of the organic compounds in body cells. They compose all of the fibrous structures in the body, including hair, nails, ligaments, the microfilaments in cells, and the myofibrils of muscles. They also form part of hemoglobin, certain hormones such as insulin, and thousands of enzymes that control biochemical processes of cells. Proteins are assimilated into protoplasm and are vital to the formation of DNA molecules. Proteins are also necessary in forming antibodies, molecules that constitute an important part of the immune system which functions to ward off disease, and in regulating the water balance and acid-base balance in the body.

NUCLEIC ACIDS

Nucleic acids are a group of organic compounds that are essential to life. These are the compounds that pass hereditary information from one generation to another, making possible a remarkable continuity of life within the various species of living things. **Deoxyribonucleic acid (DNA)** molecules are the particular type of nucleic acid out of which genes are made. Genes are the bearers of hereditary traits from parent to offspring.

Another important characteristic of nucleic acids is their ability to carry information from genes in the cell nucleus to certain structures in the cytoplasm that direct major biochemical processes. For example, the building of proteins is controlled by the group of nucleic acids known in general as **ribonucleic acid (RNA)**. The structures and functions of the three types of RNA molecules and the structure and function of DNA are presented in detailed discussion in Chapter 16.

The Role of Enzymes in Living Cells

Cells are always engaged in chemical activity. The major difference between living things and nonliving matter is that living systems carry out vital chemical activities on a continuous and controlled basis. The control of chemical processes in cells requires the work of **enzymes**.

Enzymes are **organic catalysts**. A catalyst is a molecule that controls the rate of a chemical reaction but is itself *not* used up in the process. Enzymes control the rate of chemical reactions that take place in cells, tissues, and organs. Each chemical reaction that occurs in a living system requires the assistance of a specific enzyme. It is important to note that the same enzyme can often catalyze reactions in both directions (reactant → product and product → reactant); this flexibility results in the reversibility of many biochemical reactions.

THE STRUCTURE AND NATURE OF ENZYMES

Enzymes are large complex proteins made up of one or more polypeptide chains. Enzymes may be entirely protein or they may have nonprotein parts known as coenzymes. Frequently, vitamins function as coenzymes. For example, riboflavin, popularly called vitamin B_2, is important in a coenzyme function during the cellular process of respiration.

Scientists affix the ending *-ase* to the names of many enzymes. Two examples are *maltase* and *invertase*.

Enzymes have several characteristics that are important to the chemistry of living cells. These characteristics are due to their protein nature. First, enzymes are made inactive by heat. In the human body enzymes work best at normal body temperature of 98.6°F, or 37°C. An increase in temperature of only a few degrees can render enzymes inactive. Second, the action of enzymes can be blocked by certain compounds. We call these enzyme blocks *poisons*. Hydrogen cyanide blocks one of the enzymes that is involved in cellular respiration. Third, enzymes are specific in their activity. Usually they catalyze one particular reaction. For example, the enzyme maltase will catalyze only the digestion of the sugar maltose into two glucose molecules (Figure 5.6).

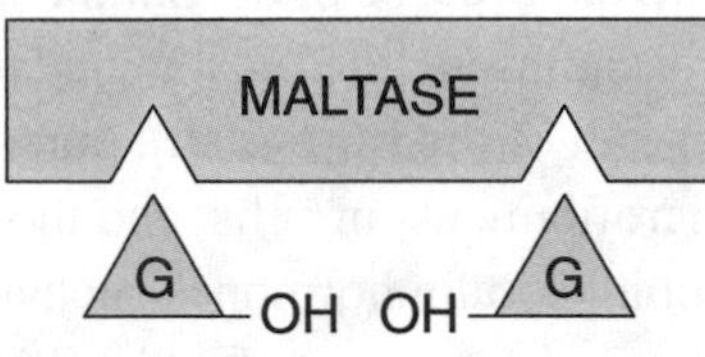

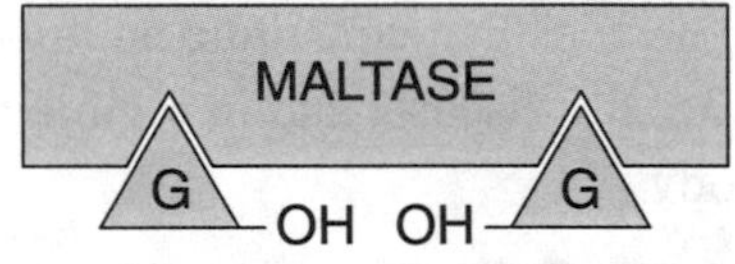

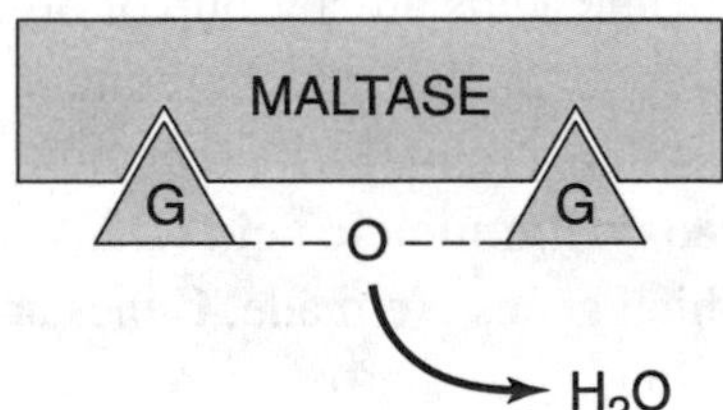

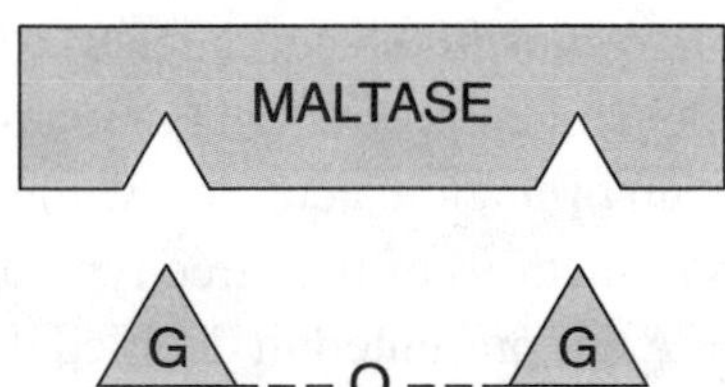

FIGURE 5.6 The enzyme maltase catalyzes the joining of two molecules of the single sugar glucose (G) to form one molecule of the double sugar maltose. Note that maltase is neither changed nor used up.

ENZYMES AND pH

The strength of an acid or base (alkali) is measured by its *pH* (see the accompanying table). An **acid** is a substance that releases hydrogen (H^+) ions in solution. The greater the number of hydrogen ions released, the stronger the acid. Thus pH is the measure of H^+ concentration. On the pH scale, the measure of acidity ranges from 0 to 6; the stronger the acid, the lower the pH. Therefore gastric juice, measuring 2 on the pH scale, is far more acidic than saliva, which has a pH of 6. **Neutral** substances, such as pure water, measure 7 on the pH scale.

A **base**, or alkali, is a substance that releases hydroxyl (OH^-) ions in solution. The strength of the base depends on the number of hydroxyl ions released in solution. The numbers between 8 and 14 on the pH scale represent bases of increasing strength; the higher the pH reading, the stronger the base. Therefore, 1 molar sodium hydroxide, which has a pH of 14, is a much stronger base than seawater, which measures 8 on the pH scale.

SCALE OF PH VALUES

pH	Example
0	1 Molar nitric acid
2	Gastric juice
2	Lemon juice
3	Vinegar
4	Tomato juice
5	Sour milk
6	Saliva
7	Pure water
8	Seawater
9	Baking soda
10	Great Salt Lake
11	Liquid soap
12	Washing soda
13	Oven cleaner
14	1 Molar sodium hydroxide

The pH influences enzyme activity. The stomach enzyme, pepsin, works most effectively in a strongly acidic environment. (The pH of gastric juice is 2.) Most enzymes, however, cannot function to catalyze reactions if the pH of their environmental solutions is not neutral. It seems as though the presence of too many hydrogen or hydroxyl ions interferes with the conforming shape of the enzymes.

HOW ENZYMES WORK

ENZYME-SUBSTRATE COMPLEX

Biochemical evidence provides insight into how enzymes work. The surface configuration of an enzyme is known as its **active site**. The **substrate** is the substance upon which the enzyme works. An enzyme affects the rate of reaction of the substrate molecule that fits the enzyme's activity site. In order for this to happen, a close physical association must take place between enzyme and substrate. This association is called the **enzyme-substrate complex** (Figure 5.7). While the enzyme-substrate complex is formed, the internal energy state of the substrate molecule is changed, bringing about a chemical reaction. As soon as the reaction is completed, the enzyme and the products separate.

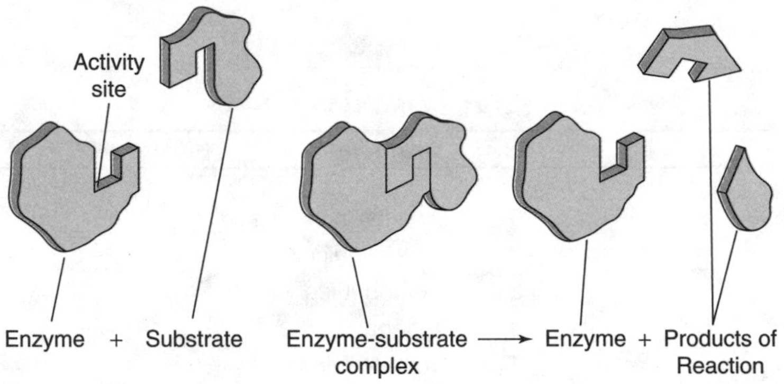

FIGURE 5.7 Enzyme-substrate complex

Each enzyme works at its own maximum rate. The rate at which an enzyme can catalyze all of the substrate molecules that it can in a given time is called the **turnover number**. Different enzymes have different turnover numbers. However, the general rate of enzyme action is influenced by certain environmental conditions such as temperature and relative amounts of enzyme and substrate.

"LOCK-AND-KEY" MODEL

Traditionally, a simple analogy has been used to explain the *specificity* of enzymes. Specificity refers to the characteristic of enzymes that permits a particular enzyme to form a complex with a specific substrate molecule only. The "lock-and-key" analogy explains enzyme specificity: the substrate is viewed as a padlock, and the enzyme as the key able to unlock it. When unlocked (in the analogy) or acted upon by the key, the padlock comes completely apart. The key remains unchanged and ready to work again on another padlock of the same type (Figure 5.8).

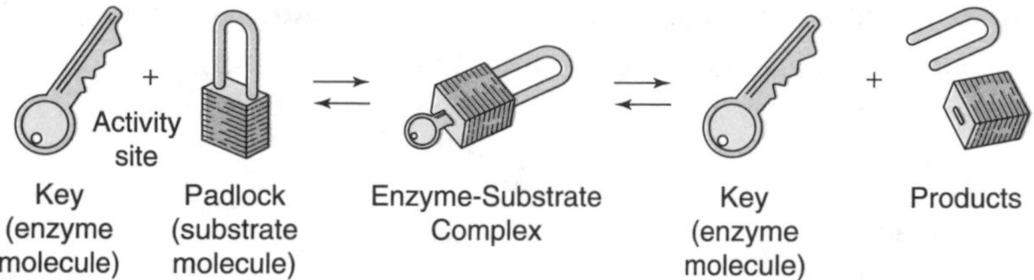

FIGURE 5.8 "Lock-and-key" analogy of an enzyme and its substrate

INDUCED FIT THEORY

Newer biochemical evidence, however, supports the *induced fit* hypothesis as an explanation of the specificity of enzymes. Biochemists now tell us that the active site on the enzyme is not rigid, as suggested in the "lock-and-key" analogy. As the substrate attaches to the enzyme's active site, the site changes shape to fit the substrate. The substrate, now joined to the enzyme's active site, becomes stressed, and the stress weakens certain chemical bonds in the substrate. The weakened bonds give away, and the substrate becomes broken into parts. At the end of this reaction, the substrate and the enzyme separate. The enzyme is unaltered, but the substrate has been chemically changed.

CONDITIONS AFFECTING THE ACTION OF ENZYMES

Enzymes catalyze some chemical reactions. Enzymes can help small molecules join together to form larger ones, and enzymes can also assist in the breaking down of compounds into smaller units. However, enzymes do not cause reactions to take place. They speed up the rates of reaction between molecules, thus decreasing reaction time. The rate at which an enzyme works is not fixed, but varies according to the environmental conditions of the cell and the reacting substances. Three factors affect enzyme activity.

TEMPERATURE

In order for enzymes to work, they must come into contact with their substrate molecules. Random motion of molecules brings enzymes and substrates into close physical association. As the temperature increases, the random motion of molecules, and thus the rate at which enzyme and substrate molecules meet, also increase. Normal body temperature of human beings is 98.6°F (37°C), and enzymes work well at this temperature. If the temperature is increased to 40°C, enzymes in human cells become misshapen and lose their function. Distorted enzymes are said to be "denatured." The temperature at which an enzyme works best is the **optimum** temperature. Study Figure 5.9. Can you determine the optimum temperature for this enzyme?

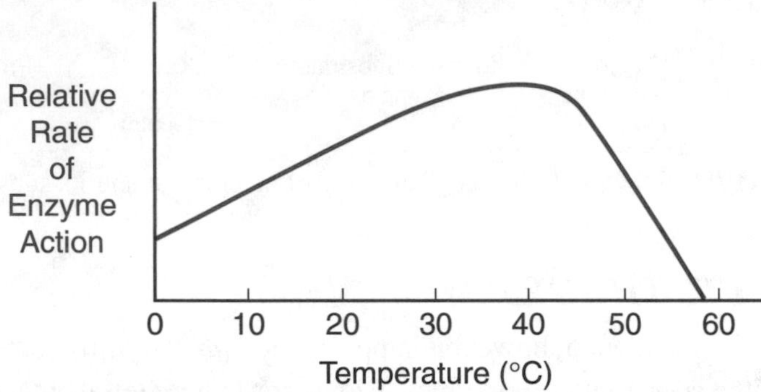

FIGURE 5.9 Rates of enzyme action at various temperatures.

Clearly, the optimum temperature for this enzyme is 37°C.

AMOUNTS OF ENZYME AND SUBSTRATE

The rate of enzyme action also varies according to the number of free substrate molecules. Let us suppose that in a cell there are a certain number of enzyme molecules. To this system a large number of substrate molecules are added. The rate of enzyme action will tend to increase when the concentration of substrate is increased (Figure 5.10). The increase in rate of enzyme activity will continue for a while and then level off as long as the enzyme concentration remains constant. The amount of substrate (or amount of enzyme) is referred to as the *concentration*.

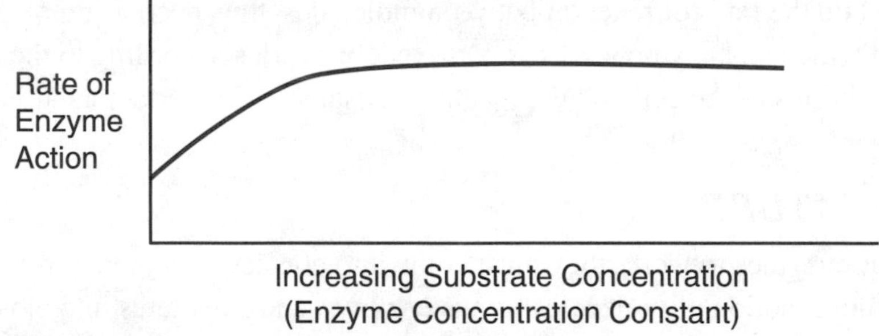

FIGURE 5.10 The graph shows the pattern of enzyme action rates when the concentration of substrate is greater than the concentration of enzyme.

pH AND ENZYME ACTIVITY

You have learned that the acidity or alkalinity of a solution is measured on the pH scale. On the scale, a pH of 7 indicates a neutral solution. Readings below 7 indicate an acid environment in which there is an excess of H+ ions. Readings above 7 indicate alkaline conditions in which OH– ions are present in excess. Within cells, many enzyme-controlled reactions work best at pH 7. Other enzymes, however, work best within different pH ranges. Figure 5.11 shows the ranges of pH in which two enzymes work best.

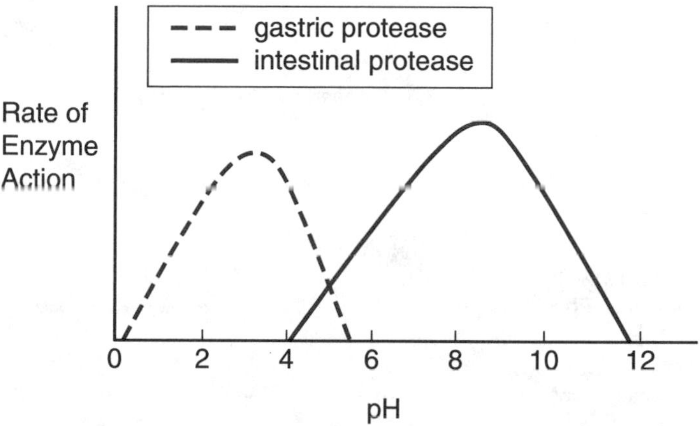

FIGURE 5.11 Optimum pH environments for two enzymes

Cellular Respiration

Respiration includes all of the processes used continuously by cells to produce usable energy. Energy enables cells to do the work of building up some molecules (synthesis) and taking others apart. Synthesis is a process that consumes energy, while decomposition (taking molecules apart) releases energy. Cellular respiration is a pathway of decomposition; it is a series of reactions that break down sugars, releasing energy along the way. Each step in the cellular respiration pathway is catalyzed by a specific enzyme.

Living cells require a constant supply of energy to fuel the chemical activities that sustain life. Glucose is the major supplier of the cell's energy. The cell is able to extract energy from glucose in small packets. The released energy is then stored in the third phosphate bond of **adenosine triphosphate** (ATP). When this unstable and excitable bond is broken, energy and inorganic phosphate are released and **adenosine diphosphate** (ADP) remains.

TYPES OF CELLULAR RESPIRATION

Energy is produced by the cell during two major phases: **anaerobic respiration**, also known as **glycolysis**, and **aerobic respiration**.

ANAEROBIC RESPIRATION

As is implied by the name, oxygen is not used during anaerobic respiration. Figure 5.12 provides a summary of the events that take place during this phase of cellular respiration.

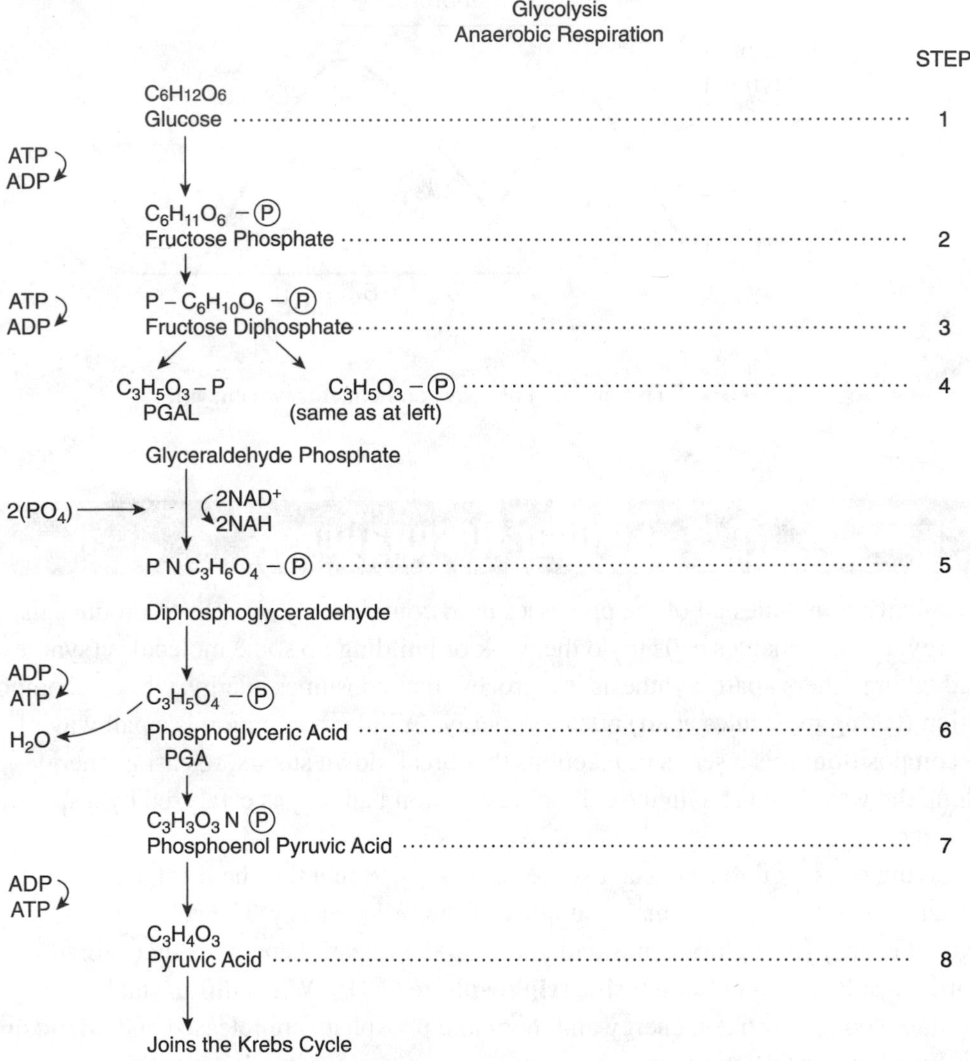

FIGURE 5.12 Anaerobic respiration, or glycolysis

Anaerobic respiration takes place in the cytoplasm of cells. It begins with a molecule of glucose. It becomes activated by energy supplied by ATP. With the assistance of enzymes, glucose is converted through a series of steps to **pyruvic acid**. At each step either a hydrogen atom is given up or a molecule of water is formed. Every time a hydrogen bond is broken, energy within the molecule becomes a bit more concentrated. This energy is ultimately released to ADP molecules and stored between phosphate bonds. Whenever ADP accepts energy, inorganic phosphate from the cellular fluid is attached to the ADP molecule upgrading it to ATP. There is a net gain of two ATP molecules.

AEROBIC RESPIRATION

The **aerobic phase** of cellular respiration takes place inside the mitochondria. As the name implies, oxygen is used, serving the important function of the final hydrogen carrier. Figure 5.13 provides a summary of the events that take place during aerobic respiration, also known as the Krebs citric acid cycle.

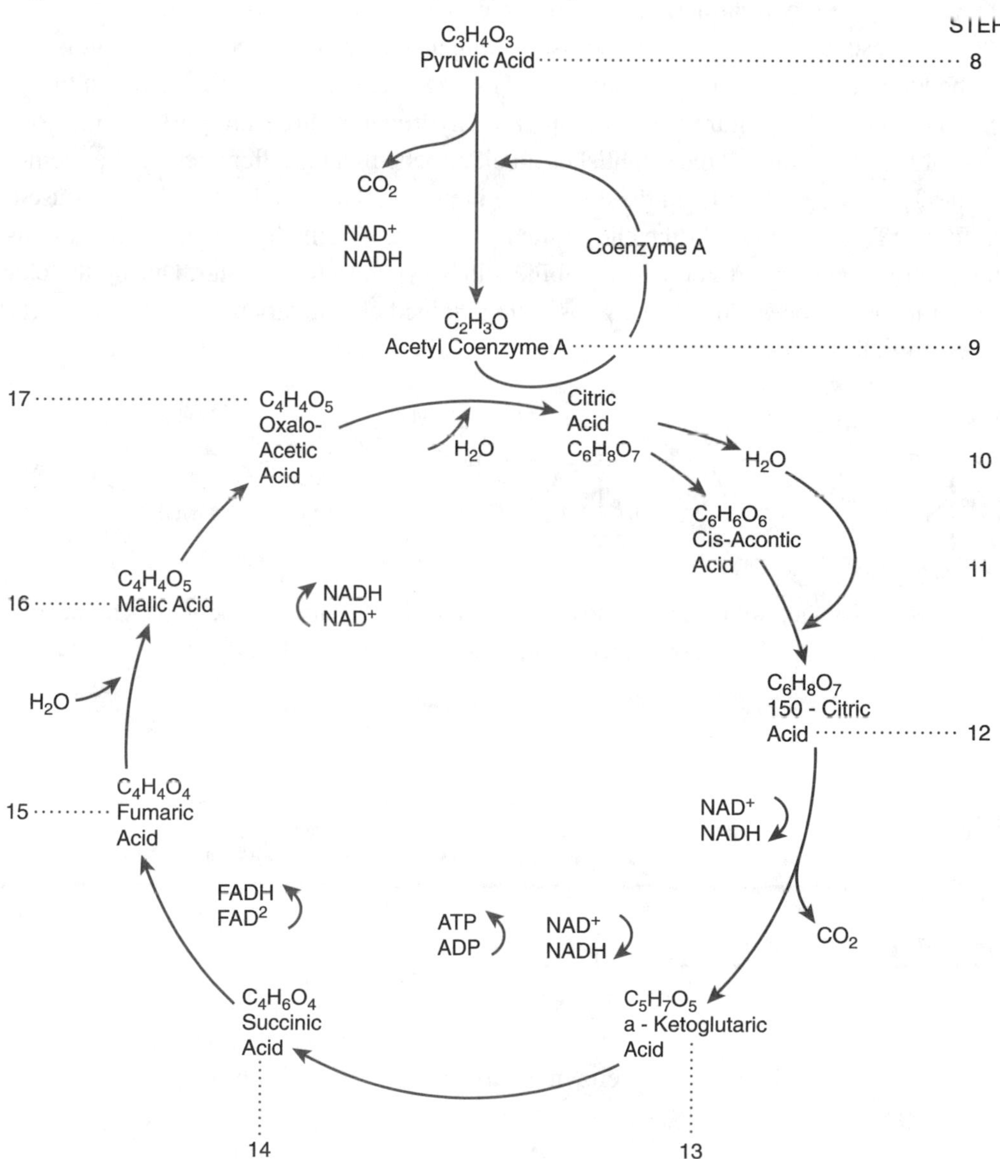

FIGURE 5.13 Krebs citric acid cycle

Aerobic respiration begins with pyruvic acid, which is quickly converted to acetyl coenzyme A. Through a cycle of chemical changes, fuel molecules are broken apart bit-by-bit, releasing a great deal of energy that is stored in ATP molecules. Hydrogen atoms from compounds formed during glycolysis and aerobic respiration then enter the next phase—oxidative phosphorylation.

OXIDATIVE PHOSPHORYLATION

The coenzyme nicotinamide adenine dinucleotide (NAD) receives the hydrogen atoms. These hydrogen atoms are passed along an electron transport chain that is embedded in a mitochondrial membrane. Hydrogen electrons are passed down the chain of acceptors (Figure 5.14). Meanwhile, hydrogen protons are pushed to the outside of the membrane. This establishes an electrochemical gradient across the membrane. Energy resulting from the difference in potential across the membrane is used to form ATP molecules. At the end of the transport chain, the hydrogen electron joins the proton. These hydrogen ions combine with oxygen to form water. During the electron transport process, inorganic phosphate is joined chemically to ADP forming ATP. Therefore, this phase of aerobic respiration is called *oxidative phosphorylation*.

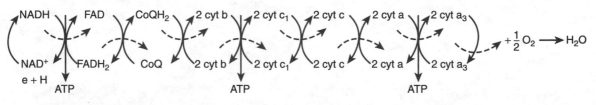

FIGURE 5.14 Steps of oxidative phosphorylation including an electron, or cytochrome (cyt), transport system, leading to the formation of ATP

REVIEW EXERCISES FOR CHAPTER 5

WORD-STUDY CONNECTION

acid
active site
adenosine diphosphate (ADP)
adenosine triphosphate (ATP)
aerobic respiration
disaccharide
enzyme
enzyme-substrate complex
glucose
glycolysis
inorganic
pH
phospholipid
polypeptide
polysaccharide
protein
pyruvic acid
respiration

alkaline
amino acid
anaerobic respiration
base
carbohydrate
carboxyl group
catalyst
deoxyribonucleic acid (DNA)
dipeptide
Krebs citric acid cycle
lipid
monosaccharide
nucleic acid
organic
organic catalyst
oxidative phosphorylation
peptide bond
ribonucleic acid (RNA)
substrate
turnover number

SELF-TEST CONNECTION

Part A. Completion. *Write in the word that correctly completes each statement.*

1. In living cells, the elements present in the greatest amounts are oxygen, hydrogen, nitrogen, and _____.

2. Water is an example of a (an) _____ compound.

3. Stomach contents are acidified by the compound _____ acid.

4. Living plant and animal cells make _____ compounds.

5. In living cells the elements carbon and _____ are found in chemical combination.

6. The special _____ property of carbon permits it to form long chains or rings.

7. Carbohydrates include all _____ and sugars.

8. Carbohydrates contain the elements carbon, hydrogen, and _____.

9. Single sugars are classified as _____.

10. Double sugars are classified as _____.

11. Glucose is an example of a _____ sugar.

12. Complex carbohydrates are known as _____.

13. The parts of plants in which cellulose is found are the _____.

14. Dehydration synthesis indicates that _____ is formed when molecules are chemically joined.

15. Maltose is an example of a _____ sugar.

16. Maltase is an example of a (an) _____.

17. Glycoproteins are compounds made from _____ and proteins.

18. The word *synthesize* means _____ together chemically.

19. Glycerol and fatty acids form organic compounds known as _____.

20. In carbohydrate molecules the ratio of hydrogen to carbon is _____.

21. The compounds specialized for passing hereditary information from one generation to the next are the ._____ acids.

22. The initials DNA stand for _____.

23. Compounds known as ._____ control the chemical work of the cell.

24. A catalyst controls the _____ of a chemical reaction.

25. Ribonucleic acid functions in the building of _____.

Part B. Multiple Choice. *Circle the letter of the item that correctly completes each statement.*

1. The ending *-ase* is used for the names of
   (a) enzymes
   (b) disaccharides
   (c) proteins
   (d) catalysts

2. Glucose and fructose are examples of
   (a) double sugars
   (b) disaccharides
   (c) single sugars
   (d) polyssacharides

3. Amino acids are correctly associated with the compounds known as
   (a) polysaccharides
   (b) proteins
   (c) lipids
   (d) carbohydrates

4. Polypeptides are correctly associated with the compounds known as
   (a) polysaccharides
   (b) proteins
   (c) lipids
   (d) disaccharides

5. When two amino acids are joined together, the compound formed is a
   (a) polypeptide
   (b) disaccharide
   (c) polysaccharide
   (d) dipeptide

6. Substances that block the action of enzymes are known as
   (a) protein blocks
   (b) coenzymes
   (c) poisons
   (d) polypeptides

7. In solutions, pH is a measure of the
   (a) acidity
   (b) longevity
   (c) strength
   (d) activity

8. A base or alkali releases ions of
   (a) Na
   (b) H
   (c) Cl
   (d) OH

9. On the pH scale the neutral reading is
   (a) 1
   (b) 3
   (c) 5
   (d) 7

10. Maltase is best classified as a (an)
    (a) single sugar
    (b) acid
    (c) enzyme
    (d) dipeptide

11. The most abundant organic compounds in the body are
    (a) carbohydrates
    (b) proteins
    (c) organic acids
    (d) lipids

12. Carbohydrates are used by cells as sources of
    (a) energy
    (b) water
    (c) protons
    (d) elements

13. Glycerol and fatty acids join to form
    (a) solutions
    (b) proteins
    (c) carbohydrates
    (d) lipids

14. All of the processes by which cells produce usable energy are known collectively as
    (a) dehydration synthesis
    (b) active transport
    (c) respiration
    (d) oxidative phosphorylation

15. ATP molecules are best associated with the
    (a) "lock and key" theory
    (b) storage of energy
    (c) enzyme-substrate complex
    (d) pH scale

16. Anaerobic respiration takes place in the
    (a) nasal passages
    (b) cytoplasm of cells
    (c) cell membranes
    (d) mitochondria

17. Aerobic respiration takes place in the
    (a) cell sap
    (b) cytoplasm of cells
    (c) cell membranes
    (d) mitochondria

18. As the result of anaerobic respiration, glucose is converted into
    (a) pyruvic acid
    (b) fatty acid
    (c) amino acid
    (d) hydrochloric acid

19. During oxidative phosphorylation, hydrogen ions join with oxygen to form
    (a) hydrogen peroxide
    (b) water
    (c) ATP
    (d) amino acid

20. The function of ATP molecules is the storage of
    (a) water
    (b) phosphate
    (c) energy
    (d) coenzymes

21. Another term for glycolysis is
    (a) glucose synthesis
    (b) protein synthesis
    (c) aerobic respiration
    (d) anaerobic respiration

22. The Krebs citric acid cycle results in the formation of molecules of
    (a) DNA
    (b) RNA
    (c) ATP
    (d) ADP

23. The carboxyl group is best associated with molecules of
    (a) amino acid
    (b) hydrochloric acid
    (c) fatty acid
    (d) nucleic acid

24. A formula that shows the bonding and location of elements in a molecule is known as a (an)
    (a) empirical formula
    (b) structural formula
    (c) molecular formula
    (d) basic formula

25. Organic catalysts are
    (a) hormones
    (b) substrates
    (c) nucleic acids
    (d) enzymes

Part C. Modified True-False. *If a statement is true, write "true" for your answer. If a statement is incorrect, change the* <u>underlined</u> *expression to one that will make the statement true.*

1. Most of the activities of the cell involve <u>physical</u> changes.
2. Sulfur and sodium are found in <u>large</u> quantities in the cell.
3. Living cells synthesize <u>inorganic</u> compounds.
4. Inorganic compounds <u>do not</u> have the elements carbon and hydrogen in chemical combination.
5. Nucleic acids are <u>organic</u> compounds.
6. In cells, the elements hydrogen and oxygen are present in the ratio of <u>4:1</u>.
7. Fructose, the sugar in fruit, is best classified as a <u>double</u> sugar.
8. Maltose is best classified as a <u>double</u> sugar.
9. Cellulose is used in the structure of plant cell <u>membranes</u>.
10. Polypeptides are formed by the chemical joining of a number of <u>fatty acid</u> groups.
11. Enzymes control the <u>rate</u> of chemical reaction.
12. Pure water has a reading of <u>14</u> on the pH scale.
13. Substances with readings between 8 and 14 are <u>acidic</u> on the pH scale.
14. The term *turnover number* refers to the speed with which <u>phosphates</u> work.
15. Denaturing <u>increases</u> the function of protein molecules.
16. On the pH scale, sour milk would have a reading between <u>4 and 6</u>.
17. pH <u>has no</u> influence over enzyme activity.
18. The substance upon which an enzyme works is the <u>complex</u>.
19. The *induced fit* hypothesis explains the <u>generality</u> of enzymes.

20. When molecules of ADP accept energy, they are upgraded to ACD molecules.

21. The aerobic phase of cellular respiration takes place inside the ribosomes.

22. The Krebs citric acid cycle is the anaerobic phase of respiration.

23. Emil Fisher used the "lock and key" analogy to explain the relationship between substance and hormones.

24. Fuel molecules provide cells with structure.

25. Carbon, hydrogen, oxygen, and nitrogen are best described as inorganic compounds.

CONNECTING TO CONCEPTS

1. We breathe in air. Oxygen from the air is filtered out by the lungs and sent to all the cells in the body. From what you have learned in the section on respiration, explain how oxygen is used in your cells.

2. Normal human body temperature is 98.6°F (37°C). How may a 3° increase in body temperature affect the action of enzymes in cells?

3. Nutritionists suggest that we should drink 6–8 glasses of water a day. What are some uses of water in the body?

ANSWERS TO SELF-TEST CONNECTION

PART A

1. carbon
2. inorganic
3. hydrochloric
4. organic
5. hydrogen
6. bonding
7. starches
8. oxygen
9. monosaccharides
10. disaccharides
11. single
12. polysaccharides
13. cell walls
14. water
15. double
16. enzyme
17. carbohydrates
18. join
19. lipids
20. 2:1
21. nucleic
22. deoxyribonucleic acid
23. enzymes
24. rate
25. protein

PART B

1. (a)
2. (c)
3. (b)
4. (b)
5. (d)
6. (c)
7. (a)
8. (d)
9. (d)
10. (c)
11. (b)
12. (a)
13. (d)
14. (c)
15. (b)
16. (b)
17. (d)
18. (a)
19. (b)
20. (c)
21. (d)
22. (c)
23. (a)
24. (b)
25. (d)

PART C

1. chemical
2. small
3. organic
4. true
5. true
6. 2:1
7. single
8. true
9. walls
10. amino
11. true
12. 7
13. basic or alkaline
14. enzymes
15. decreases
16. true
17. has
18. substrate
19. specificity
20. ATP
21. mitochondria
22. aerobic
23. substrate and enzyme
24. energy
25. elements

CONNECTING TO LIFE/JOB SKILLS

Understanding the significance and interactions of biological compounds in living organisms is part of the job of the pharmacological researcher. These scientists can work in discovery, or initial compound development. They may be involved in developing computer models of living systems to virtually test compounds. They may work to understand how basic chemical compounds will alter genetic material or cells in culture. Some pharmacological researchers test how drugs affect different types of bacteria. Working with live animals and with human subjects, some drug research identifies safety issues, mechanisms of delivery, and dosages associated with specific medications.

Chronology of Famous Names in Biology

1902 **Emil Fischer** (Germany)—won the Nobel Prize in Chemistry for laying the foundation of enzyme chemistry. He synthesized fructose and glucose. All knowledge of purine groups is attributable to Fischer.

1907 **Eduard Buchner** (Germany)—won the Nobel Prize for his work in alcoholic fermentation of yeast cells. He discovered the enzyme zymase in yeast.

1922 **Otto Meyeroff** and **A. V. Hill** (United States)—won the Nobel Prize for the lactic acid theory of muscle contraction.

1923 **David Kellin** (England)—was the first to link cytochrome to cellular respiration.

1937 **Sir Hans Krebs** (Germany)—discovered the citric acid cycle in cellular respiration.

1948 **Albert Lehninger** and **Eugene Kennedy** (United States)—discovered that the synthesis of ATP occurs in the mitochondria.

1951 **Maurice Wilkins** and **Rosalind Franklin** (England)—used X-ray diffraction to study DNA.

1952 **Hugh Huxley** (England)—elucidated the biochemical processes involved in muscle contraction.

1954 **Linus Pauling** (United States)—won the Nobel Prize in Chemistry for the major breakthrough in understanding the molecular structure of matter.

1962 **M. F. Perutz** and **J. C. Kendrew** (England)—won the Nobel Prize for discovering the structure of the muscle protein myoglobin.

6

Bacteria and Viruses

WHAT YOU WILL LEARN

In this chapter you will explore a group of organisms composed of a single cell that differs in structure and chemistry from most other cells.

SECTIONS IN THIS CHAPTER

- Prokaryotes
- Classification of Bacteria
- Importance of Bacteria
- Viruses
- Prions
- Review Exercises for Chapter 6
- Connecting to Life/Job Skills
- Chronology of Famous Names in Biology

Prokaryotes

BACTERIA

OVERVIEW

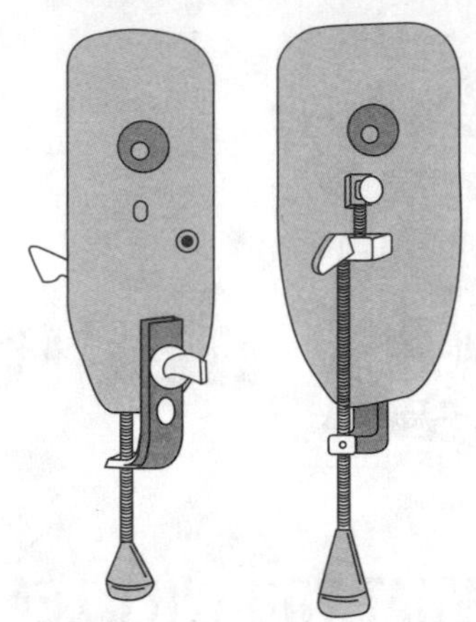

FIGURE 6.1 Van Leeuwenhoek's microscope. How does it compare with today's light microscopes?

In 1684, the Dutchman Antonie van Leeuwenhoek made a number of very simple magnifying instruments (Figure 6.1). These were forerunners of our modern light microscopes. Using these carefully ground lenses, he examined numerous types of biological specimens, including pond water, blood, and the scrapings from his teeth. In the latter, he saw organisms that we now call *bacteria*. He did not give these organisms a name, but made careful drawings of them in his notebook. Van Leeuwenhoek is credited with being the first person to see bacteria.

A **bacterium** (singular of **bacteria**) is an organism composed of a single cell. The bacterial cell is different from the cells of higher organisms in basic structure and in chemical composition. A bacterium does not have a nucleus bound by a membrane and is therefore classified as a prokaryote (Figure 6.2). (Derived from the Greek word *karyon*, prokaryote means "before nucleus.")

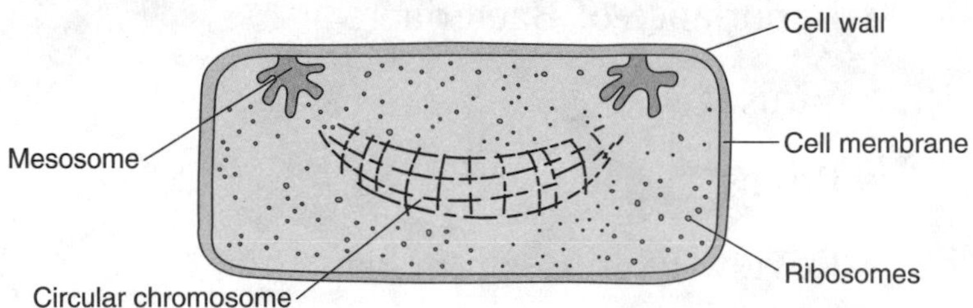

FIGURE 6.2 A prokaryotic cell

Until recently, bacteria and cyanobacteria (blue-green algae) were classified in the kingdom Monera. Newer evidence obtained from biochemical research has prompted investigators to divide the kingdom Monera into two new kingdoms, the Archaeobacteria and the Eubacteria. **Archaeobacteria** is the kingdom of primitive bacteria. The species of Archaeobacteria show marked chemical differences from the true bacteria, which are classified in the kingdom **Eubacteria**. Since all bacteria are prokaryotes, it is important to understand the characteristics of these cells.

A **prokaryote**, as indicated above, is a cell that does not have a nucleus or other membrane-bound organelles; it lacks mitochondria, an endoplasmic reticulum, Golgi bodies, and lysosomes. A **eukaryote** is a cell that contains an organized nucleus and other membrane-bound organelles.

GENERAL CHARACTERISTICS OF PROKARYOTES

CELL WALL

Present in most prokaryotic cells, the cell wall is an important structure providing shape to the cell and protecting it against water-pressure damage. (However, *mycoplasmas*, the smallest living cells, do not have a cell wall.) Ranging from 5 to 80 nanometers in diameter, the walls of prokaryotic cells get their tensile strength from **murein**. Murein is composed of an enormous number of polysaccharide molecules held together by short chains of amino acids. Muramic acid, one of the major molecules in murein, never occurs in the cell walls of eukaryotic cells. Penicillin is an effective drug against bacteria because it inhibits the synthesis of murein and thus prevents the reproduction of bacteria. Penicillin does not act, however, against eukaryotic cells.

CELL MEMBRANE

The cell membrane of the prokaryotic cell is very much like that of the eukaryotic cell except that the prokaryotic cell membrane lacks cholesterol and other steroids. In the Archaeobacteria, the cell membrane is composed of modified branched fatty acids. Straight-chain fatty acids comprise the structure of the plasma membrane in the Eubacteria. The surface area of the cell membrane in some prokaryotes is greatly increased by *convolutions* (folds and loops). The convoluted cell membranes have incorporated in their structure electron transport systems and enzymes necessary for the chemical events taking place during respiration. The circular DNA molecule is attached to a site on the cell membrane called the *mesosome*.

CYTOPLASM

The cytoplasm of the prokaryotes does not have the complex fine structure characteristic of the eukaryotes. Bacterial cytoplasm appears to be highly granulated due to the presence of a large number of ribosomes which are smaller in size than those found in nucleated cells. The ribosomes—the only cellular organelles in most prokaryotes—function just the same as their larger counterparts in the eukaryotic cells by being the sites of protein synthesis. The cell membrane pushes inward, causing a pocket known as a *mesosome*. An outer membrane system is present in cells of the cyanobacteria. These outer membranes contain chlorophyll and accessory pigments.

GENETIC MATERIAL

DNA (deoxyribonucleic acid) is the substance of heredity. In contains genetic codes (genes) for characteristics that are passed on from one generation to the next. In eukaryotic cells, DNA molecules are linked in chromosomes in the nucleus of cells. You know that prokaryotic cells do not contain a membrane-bound nucleus.

Prokaryotic cells contain a circular molecule of naked DNA (not enclosed by a membrane). The DNA chromosome is replicated (reproduced) before cell division. Then the cell wall and the plasma membrane grow inward, dividing the cell in two. This type of cell division is known as **binary fission**. As you have learned, the mesosome is formed by a loop in the plasma membrane and seems to function in the separation of the circular chromosomes during cell division.

OTHER CHARACTERISTICS OF PROKARYOTES

Other differences separate nonnucleated cells from cells with a nucleus. Most prokaryotic cells are quite small, ranging in size from 0.5 to 10 micrometers. The average eukaryotic cell is much larger, although some nucleated cells measure only 7 micrometers in diameter. Some bacteria produce a slimy capsule made of carbohydrate that surrounds the cell wall and protects the bacterial cell from being engulfed by phagocytic white blood cells. Some prokaryotic cells move by means of a stiff, inflexible, rodlike structure called a **flagellum**. Also, prokaryotic cells are not able to join together to form tissues.

Classification of Bacteria

As stated previously, because of distinct differences in the chemical composition of their organelles, bacteria are now divided into two separate kingdoms or two separate domains: the Archaeobacteria and the Eubacteria. The Archaeobacteria are the "ancient or first bacteria"; the Eubacteria, the "true bacteria." All the species contained in both of these domains or both of these kingdoms are prokaryotes.

GROUPS OF BACTERIA

Many biologists refer to "divisions" or "groups" of bacteria, because the concept of species is difficult to apply to these organisms. By accepted definition, a species is a group of organisms so closely related that they can mate and produce viable offspring. Since bacteria usually reproduce by binary fission, and since the evolutionary history is difficult to determine, many "species" of bacteria do not easily meet the requirements of the definition. However, since *species* is a convenient word to show close structure and functional relationship, we shall continue to use it when necessary.

GRAM STAINING

In 1884 the Danish microbiologist Hans Christian Gram developed a technique for staining bacteria that is useful in distinguishing groups of organisms, particularly among the Eubacteria. The procedure of **gram staining** identifies cells based on their ability to absorb certain dyes. Bacterial cell walls that are made of peptidoglycans (amino sugars) absorb the purple dye gentian violet with such tenacity that washing the cells with alcohol does not remove the dye. Such purple-staining cells are designated as *gram-positive*. Cell walls composed of lipopolysaccharides, however, do not hold the purple dye and are easily bleached with alcohol. These cells stain red with dyes such as safranin or carbol fuchsin. Cells that take up the red counterstain are known as *gram-negative*.

Gram staining is one means by which bacteria are identified and classified. As a matter of fact, the composition of the cell wall affects the behavior of bacteria. Gram-positive bacteria are more susceptible to the effects of antibiotics than are gram-negative organisms. Gram-positive organisms are also more susceptible to the lysing effects of the enzyme lysozyme, which is found in human secretions such as tears, saliva, and nasal discharges.

Gram-positive bacteria include the genera (plural of genus) *Staphylococcus* and *Streptococcus*, which can affect the skin and other body systems; *Clostridium*, which includes a species that can cause tetanus; *Bacillus*, which can cause tuberculosis; and *Listeria* and *Enterococcus*, which can both cause meningitis and gastrointestinal disorders. Examples of gram-negative bacteria include the genera *Pseudomonas*, *Legionella*, and *Hemophilus*, all of which cause respiratory disorders; *Escherichia*, *Salmonella*, and *Heliobacter*, which are associated with gastrointestinal illnesses; and *Neisseria*, which has species that cause meningitis and gonorrhea.

KINGDOM ARCHAEOBACTERIA

Bacteria classified in the kingdom Archaeobacteria exhibit unusual characteristics. All are **anaerobes**, and all are unable to survive in an environment of oxygen. Rather, they live in harsh environments—stagnant marsh water, boiling hot springs, the salt water of lakes and ponds, and the anaerobic sludge at the bottom of lakes.

BIOCHEMICAL UNIQUENESS OF ARCHAEOBACTERIA

The archaeobacteria differ biochemically from all other organisms living on Earth today. First, their cell membranes are composed of branched lipid molecules unlike the straight-chain fatty acids of other cells. Second, in their metabolic processes, enzymes different from those in other cells are used in the synthesis of lipids and in RNA synthesis activities. Biologists use these and other differences as reasons to place the archaeobacteria in a separate kingdom.

MAJOR GROUPS OF ARCHAEOBACTERIA

The largest group of archaeobacteria are the **methanogens**. As the name implies, these bacteria produce methane gas. They live where anaerobic conditions exist: in airless marshes, anaerobic lake bottoms, sewage treatment ponds, and even in the lower bowels of cattle and sheep, where anaerobic conditions prevail. The methanogens use hydrogen gas to reduce carbon dioxide in their energy-making process, producing methane gas as a byproduct. The reaction can be summarized as follows:

$$4H_2 + CO_2 \rightarrow CH_4 + 2H_2O$$

The methanogens must live where the inorganic gases carbon dioxide and hydrogen are available for their use and where oxygen is not present. In addition, they require three organic compounds for nutrition: methanol (H_3COH), formate (HCOO—), and acetate (H_3C—COO—). Evidence indicates that these five simple molecules, two inorganic and three organic, were present on early Earth at the time when oxygen was absent from the atmosphere. Thus it is believed that the methanogens are descendants from an ancient line of bacteria unrelated to the Eubacteria.

The **salt-loving bacteria** belong to the genus *Halobacterium*. The members of this genus are able to produce their own energy by carrying out a simple type of chemosynthesis. Embedded in the plasma membranes of the halophiles (salt-loving bacteria) are patches of a purple pigment, bacteriorhodopsin. In an atmosphere free of oxygen, this pigment is able to capture light energy and use it to form ATP molecules, the energy-storing molecules of cells.

Other species included in the kingdom Archaeobacteria include the **thermophiles** and the **thermoacidophiles**. As the name implies, thermophiles thrive in extremes of heat. They live in hot springs and can survive in temperatures near the boiling point of water, 90°C (194°F). Thermoacidophiles live in harsh environments of extreme heat and extreme acid. Of interest to scientists is the fact that the internal pH (acid-base measure) of these bacteria remains neutral.

KINGDOM EUBACTERIA

The eubacteria, referred to as the "true bacteria," comprise a large number of subgroups. Members of the kingdom Eubacteria live under less harsh conditions than do the archaeobacteria. A large number of prokaryotes are classified in this kingdom because of strong biochemical similarities in their cell walls, which are composed of amino sugars known as peptidoglycans. Their cell (plasma) membranes are similar in that they are built from straight-chain fatty acids rather than the short, branched chains found in the archaeobacteria.

CHARACTERISTICS OF THE EUBACTERIA

All of these bacteria have thick, rigid cell walls. Some species are *nonmotile* (nonmoving), while others are *motile*, using flagella (Figure 6.3) or a sling motion to move from place to place.

Shape is one means of identifying eubacteria. The rod-shaped bacteria are known as *bacilli* (bacillus, sing.), the round bacteria as *cocci* (coccus, sing.) and the spiral-shaped as *spirilla* (spirillum, sing.) (Figure 6.4). Some species typically remain attached. The *diplococci* occur in pairs, the *streptococci* in chains, and the *staphylococci* in clusters.

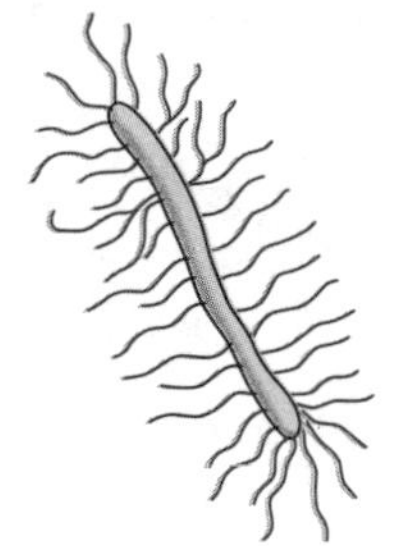

FIGURE 6.3 A flagellated bacterium

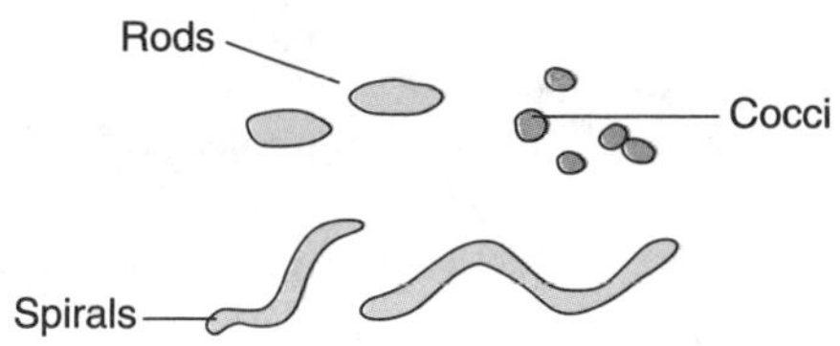

FIGURE 6.4 Three shapes of bacteria

Reproduction in the eubacteria takes place by *binary fission*. There is no mitosis as in the eukaryotic cells. Replication of the circular chromosome is followed by equal splitting of the cytoplasm. The reproductive potential of bacteria is enormous because cell division is possible every 20 minutes. In this short span of time, a parent cell divides into two identical daughter cells. It has been estimated that, if the rate of reproduction were not slowed by environmental changes, one bacterium could give rise to 500,000 descendants by the end of 6 hours.

Eubacteria are able to survive unfavorable environmental conditions in the rather unique way of forming *endospores*. The spore is a vegetative cell containing DNA and little else. It becomes surrounded by a thick and almost indestructible cell wall secreted by the cytoplasm. In this condition, the endospore is able to withstand boiling, freezing, drying, treatment with disinfectants and other extremes of environment.

Capsule formation occurs in certain disease-producing species of eubacteria. The cytoplasm of the bacterium secretes a mucoid coat of polysaccharide materials which surrounds the cell wall. The capsule is resistant to phagocytosis (engulfing by white blood cells) and may even, itself, release toxins into the tissues of the host.

Most eubacteria are **heterotrophs** that have to depend on sources outside of their own bodies for organic food materials. Many species obtain food by absorbing nutrients from dead organic matter; organisms that feed in this way are known as **saprophytes**, or, in new terminology, *saprobes*. Bacteria that live on or inside of the bodies of animals or plants and cause disease are **parasites**.

A number of groups in the eubacteria are **autotrophs** and are able to change inorganic materials into organic compounds. Among these are the **photosynthetic bacteria**, which, like green plants, use light energy to produce food, and the **chemosynthetic bacteria**, which are capable of oxidizing the inorganic compounds of ammonia, nitrites, sulfur, or hydrogen gas into high-energy organic compounds without the need of light energy.

Nitrifying bacteria are excellent examples of the chemosynthetic bacteria. These bacteria live in little *nodules* (bumps) on the roots of leguminous plants such as peas, beans, peanuts, alfalfa and clover (Figure 6.5). In two separate steps, the nitrifying bacteria convert ammonia, released into the soil by the breakdown of proteins from the bodies of dead plants and animals, to nitrites and then they change the nitrites to nitrates:

$$(1)\ 2NH_4^+ + 3O_2 \rightarrow 2NO_2^- + 4H^+ + \text{energy}$$

$$(2)\ 2NO_2^- + O_2 \rightarrow 2NO_3^- + \text{energy}$$

Rod-shaped bacteria belonging to the genus *Nitrobacter* are examples of nitrifying bacteria. Bacteria of decay, responsible for the breakdown of proteins in dead organic matter, belong to the genera *Pseudomonas* and *Thiobacillus*. Species belonging to the nitrifying and denitrifying genera aid in the recycling of nitrogen (Figure 6.6).

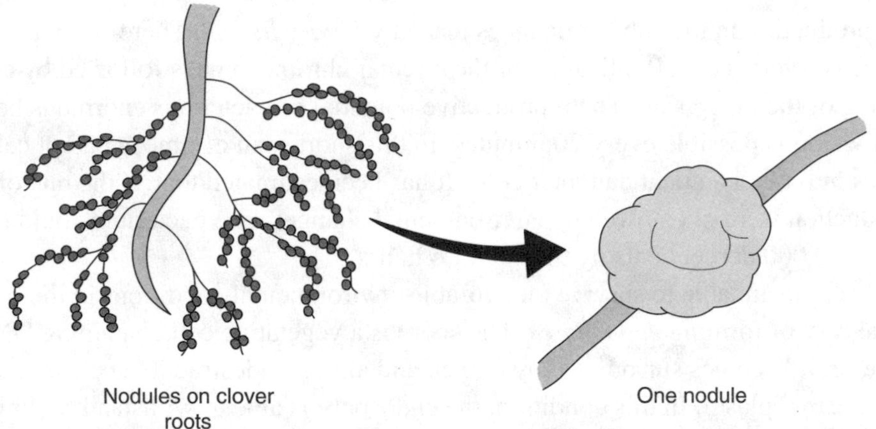

FIGURE 6.5 Nodules on the roots of leguminous plants

Sulfur bacteria are also chemosynthetic bacteria. Sulfur, an element necessary for life, is present in soil. It can be free or combined, and part of both inorganic and organic compounds. The decomposition of rock, the breakdown of organic products, and rainwater are the major sources of soil sulfur. Living in the soil, the "sulfur" bacteria oxidize free sulfur in sulfates, producing energy for themselves and providing compounds of sulfur that can be used by plants.

$$2S + 3O_2 + 2H_2O \rightarrow 2SO_2 + 4H^+ + \text{energy}$$

Photosynthetic bacteria (Endothiobacteria) use pigment systems which are contained in the cell membrane to capture light energy and use it in the synthesis of organic molecules. Unlike green plants, however, photosynthetic bacteria do not have the pigment chlorophyll *a* and they do not produce molecular oxygen. In fact, most photosynthetic bacteria do not need or use oxygen.

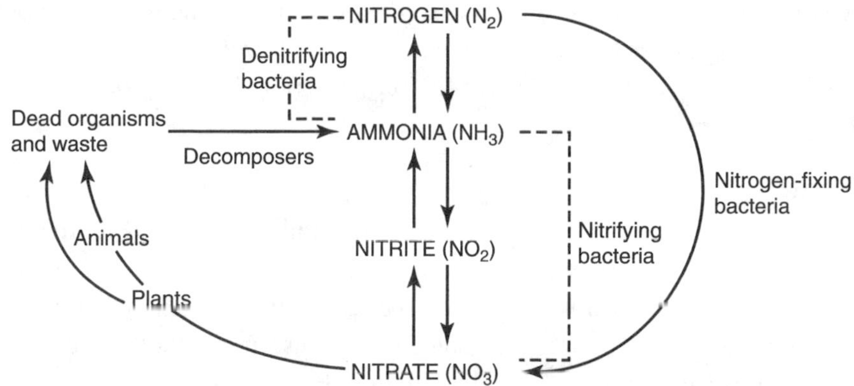

FIGURE 6.6 Nitrogen cycle

Most groups of the eubacteria are **aerobic**, using molecular oxygen in the process of breaking down carbohydrates to carbon dioxide and water. **Obligate aerobes** are organisms that can live only in an environment that provides free or atmospheric oxygen. An example of an obligate aerobe is *Bacillus subtilis*.

Some bacteria are **obligate anaerobes** and derive their energy by fermentation. These organisms cannot live in an environment of free oxygen. Usually, the obligate anaerobes are disease producers; included in this group are *Clostridium tetani*, the causative organism of tetanus, and *Clostridium botulinum*, the bacterium that induces food poisoning. Still other bacteria are **facultative anaerobes**. These are basically aerobic bacteria, but they can live and grow in an environment that lacks free oxygen.

FOUR IMPORTANT GROUPS OF EUBACTERIA

Myxobacteria—Slime Bacteria

The Myxobacteria are also known as the slime bacteria. The cells develop as a colony and move together in a flowing mass of slime resembling the movements of amoebae. The individual cells are long, thin rods. Myxobacteria live in the soil and obtain nutrients by absorbing dead organic matter (saprobes). At times the slime mass, or *pseudoplasmodium*, forms reproductive structures called **fruiting bodies** that look like small mushrooms. The fruiting bodies produce *cysts* which are much like the endospores of other bacteria in regard to their ability to withstand adverse environmental conditions. When environmental conditions improve, the encysted cells become metabolically active and reproduce by binary fission.

Spirochetes

Most bacteria belonging to this group are anaerobic; many are disease producers. These bacteria are long, thin and curved, moving with a wriggling, corkscrew-like motion, made possible by an axial filament (Figure 6.7). In some ways, the spirochetes resemble protozoa, but they are nonnucleated. They do not form spores or branches. They reproduce by transverse fission. The spirochete *Treponema pallidum* causes syphilis.

Rickettsiae

The rickettsiae are very small gram-negative intracellular parasites (Figure 6.8). They were first described in 1909 by Harold Taylor Ricketts, who found them in the blood of patients suffering from Rocky Mountain spotted fever. The rickettsiae are nonmotile, non-spore-forming, nonencapsulated organisms. In length, these organisms range from 0.3 to 1 micrometer. They live in the cells of ticks and mites and are transmitted to humans through insect bites. The rickettsiae are responsible for several febrile (fever-producing) diseases in humans, such as typhus fever, trench fever, and Q fever.

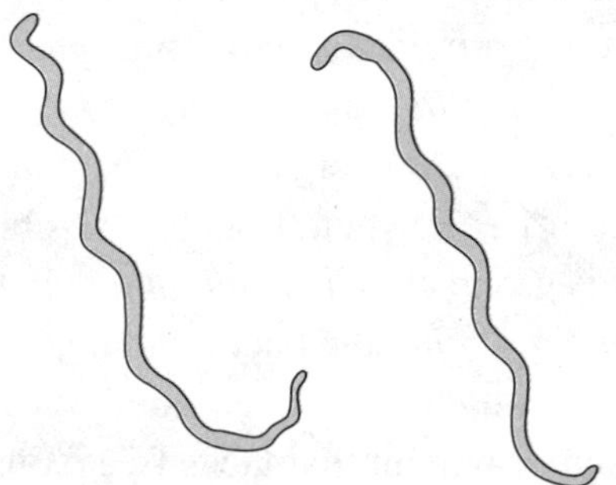

FIGURE 6.7 Spirochetes are highly motile but they lack polarity.

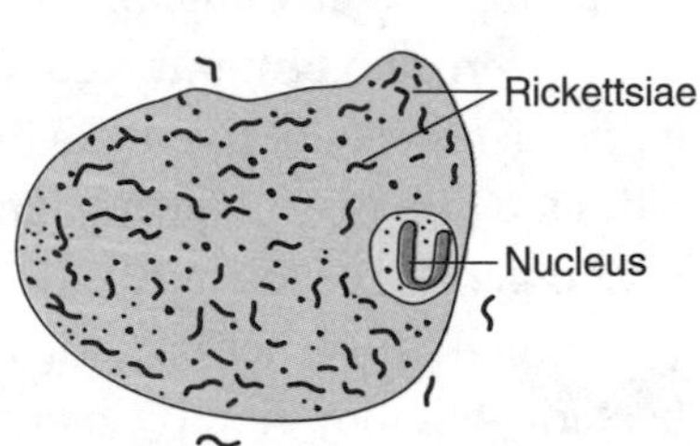

FIGURE 6.8 Rickettsiae in tissue cells

Actinomycetes

The Actinomycetes (also known as the Actinomycota) have characteristics of both the eubacteria and the fungi. Like the latter, their body structure takes the form of branching multicellular filaments. They live in soil where they obtain nutrition saprophytically. Some species are anaerobes. These organisms are mostly nonmotile and break down waxes and lipids in dead plants and animals. Some species cause diseases in animals and humans. *Actinomyces bovis* causes “lumpy” jaw in cattle. *Mycobacterium tuberculosis* causes tuberculosis in humans and *Mycobacterium leprae* causes leprosy. Some other actinomycetes have important medical and commercial value—primarily as sources of *antibiotics*. Antibiotics in common use that are produced by the actinomycetes are tetracycline, erythromycin, neomycin, nystatin, and chloramphenicol.

CYANOBACTERIA

A controversial phylum in kingdom Eubacteria is Cyanobacteria. Some biologists classified these one-celled photosynthetic organisms as blue-green algae—thus the former name Cyanophyta. Now some biologists regard these organisms as bacteria and call them *cyanobacteria*. Others avoid the issue and refer to this group simply as the blue-greens.

CHARACTERISTICS OF THE BLUE-GREENS

In structure, the cyanobacteria are prokaryotes. As in the bacteria, the cytoplasm of the blue-greens is crowded with ribosomes. In addition, the cytoplasm of most of these organisms stores particles of protein materials and special carbohydrate molecules known as *polyglucans*. The cells of the blue-greens are larger than those of bacteria but smaller than eukaryotic cells. The cell wall is composed of murein, as in the bacteria, but is surrounded by a gelatinous material composed of pectin and glycoproteins. The cyanobacteria have no visible means of locomotion (no flagella), but some species move with a gliding motion, as do the myxobacteria.

The blue-greens derive their name from the pigment molecules that are contained in flattened membranous vesicles called *thylakoids*. Several types of pigments have been isolated from the cyanobacteria. As in the higher plants, chlorophyll *a* is the pigment used for photosynthesis. Present also in these organisms is the blue pigment *phycocyanin*, the red pigment *phycoerythrin*, and several types of carotenoids (yellow pigments). Not all of the cyanobacteria are blue-green. Some are black, brown, yellow, red, and bright green. At times the Red Sea takes on a reddish hue due to a species of cyanobacteria that inhabits these waters in great numbers and contains large amounts of phycoerythrin.

The cyanobacteria may live singly or occur in filaments or colonies. Although the cells of some species live in close association with one another, they usually carry on independent life functions. Interestingly enough, cytoplasmic bridges called *plasmadesmata* join the cytoplasm of some of the cells of the filamentous groups. At certain times the plasmadesmata may serve as the passageways for special materials.

Some of the filamentous blue-greens have the ability to fix atmospheric nitrogen. Under anaerobic conditions, special cells called *heterocysts* produce the enzyme *nitrogenase*, which has the ability to convert atmospheric nitrogen into usable nitrates.

Fresh water is the habitat of most of the cyanobacteria, but a few species are marine, and some species live in tropical soils where the nitrogen content is poor. Other habitats include microfissures in desert rocks, where light penetrates and a small amount of water is trapped. Cyanobacteria can also be found on tree bark, on damp rocks, and on flower pots. They reproduce by binary fission or by fragmentation of filaments.

IMPORTANCE OF THE BLUE-GREENS

The cyanobacteria are an important part of every environment. In the process of photosynthesis they produce oxygen as a by-product which helps to replenish the atmosphere. They also produce food for invertebrate and vertebrate species that feed on *phytoplankton*, the floating plants that live on the surfaces of oceans and lakes.

Cyanobacteria can also be harmful to humans and fish in their role as water pollutants. The rapid overpopulation of blue-greens such as *Oscillatoria* coats lakes with a slimy, smelly mass that is toxic to fish. The "blooms," aided by excessive nitrate compounds in lakes, have contributed to a serious decline in the freshwater fish populations in some areas.

PROCHLORON

The Prochloron were discovered in 1976 by R. A. Levin. These cells are prokaryotic and seem to be an evolutionary link between the cyanobacteria and the eukaryotic algae. Their pigment system seems to resemble that of the green algae and higher green plants. The only known representatives are those that live in association with a group of marine invertebrates known as the tunicates.

Importance of Bacteria

BACTERIA AND DISEASE

As we have already mentioned, some species of bacteria cause disease in humans, animals, and plants; these bacteria are termed **pathogenic**. A disease is any condition that interrupts the normal functioning of body cells preventing completion of a particular biochemical task. When pathogenic bacteria invade body tissues, they introduce a particular set of circumstances that change the environment of the cells.

There are three significant ways in which the cellular environment can be altered resulting in a disease condition. One way is by sheer numbers of organisms: enormous numbers of bacteria will affect adversely the functioning cells. For example, *Escherichia coli*, gram-negative rods that normally live in the human intestine, will cause disease if present in increased numbers. Another way is by the destruction of cells and tissues. A third change brought about by pathogenic bacteria is directly related to their production of **toxins**. Toxins are poisonous substances that inhibit the metabolic activities of the host cells.

HELPFUL BACTERIA

Contrary to popular belief, most species of bacteria are helpful to humans. The nitrogen-fixing bacteria enable plants to obtain the nitrates necessary for protein synthesis. The bacteria of decay release ammonia and nitrates from dead organic matter into the soil. Bacteria that live in the human intestines synthesize several vitamins and contribute to the synthesis of a digestive enzyme. Manufacturers of vinegar, acetone, butanol, lactic acid, and certain vitamins depend upon the action of bacteria in the production processes of these products. The retting of flax and hemp is a process in which bacteria are used to digest the pectin compounds that hold together the cellulose fibers. Once these fibers are free they can be used to make linen, textiles, and rope.

Bacteria are useful in the preparation of skins for leather and in the curing of tobacco. Manufacturers of dairy products use bacteria to ripen cheese and to improve the flavor of special items such as Swiss cheese. Farmers depend on bacteria in their fermenting of silage that is used for cattle feed. The pharmaceutical industry produces antibiotics such as aueromycin, teramycin, and streptomycin from bacteria.

Viruses

There is debate among scientists about whether viruses qualify as living organisms. Virus particles are not cells and do not exhibit the characteristics of life, as do cells. Lacking the enzymes and other cell machinery for carrying out metabolic processes, they remain inert and cannot function as living systems. Reproduction is the only function of viruses. To reproduce they must invade (enter) living cells. Viruses are parasites.

STRUCTURE OF VIRUSES

The virus particle is known as a *viron*. It consists merely of a protein coat called a *capsid*, a nucleic acid core, and tail fibers (Figure 6.9).

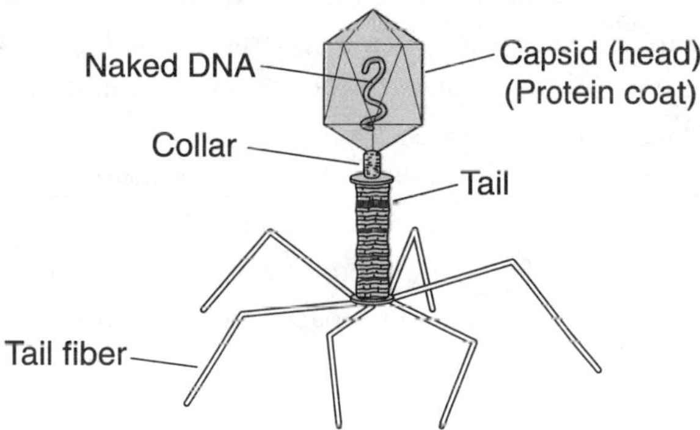

FIGURE 6.9 Structure of a virus

In shape, most virus particles are either helical (spiral) or polyhedral (manysided). The capsid of the helical viruses is composed of subunits called *capsomeres*. The tobacco mosaic virus (Figure 6.10), the first virus to be described, and the virus that causes influenza are helical viruses. A few viruses are shaped like a cube, and a few are brick-shaped. The T-even bacteriophage (a virus that enters a bacterium) is polyhedral and has a tail with extending fibers. The smallpox virus is brick-shaped.

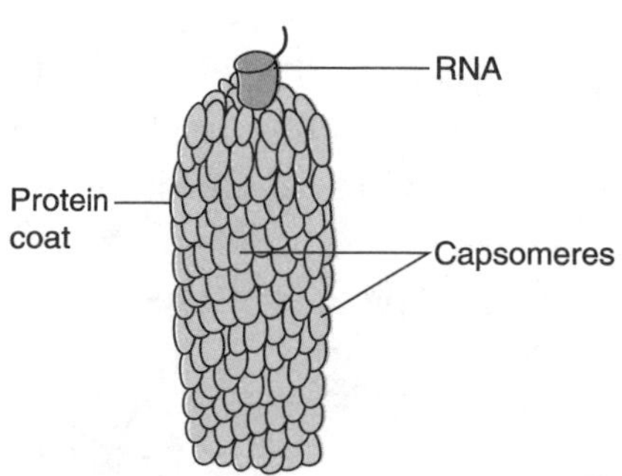

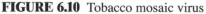

FIGURE 6.10 Tobacco mosaic virus

In some viruses, including those that cause influenza and herpes, a cytoplasmic membrane surrounds the protein coat. This surrounding *envelope* may come from the plasma membrane of the host cell or may be synthesized by the host's cytoplasm. Virologists have found that the envelope contains proteins that are virus-specific.

VIRAL NUCLEIC ACID

Viral nucleic acid may be a single molecule consisting of as few as five genes or may be multimolecular and have as many as several hundred genes. Viral nucleic acid may be single-stranded or double stranded; it may be circular or linear. Some viral nucleic acid is made of DNA (deoxyribonucleic acid); some has only an RNA (ribonucleic acid) core. Viruses never contain both DNA and RNA. For example, the polio virus contains only RNA, while the herpes viruses contain only DNA.

REPRODUCTION OF VIRUSES

Viruses generally reproduce in one of two ways, the lytic cycle or the lysogenic cycle. In the lytic cycle, a virus infects (enters) a host cell. The nucleic acid of the virus captures the nucleic acid of the host cell and uses it to make more virus nucleic acid and virus proteins. New virus particles, complete with the nucleic acid core and the protein coat, are assembled inside of the host cell until no more can fit. The host then bursts, releasing the newly assembled virions. This generalized pattern of virus reproduction varies with different viruses that typically infect bacteria, plant, or animal cells. A virus that infects a bacterial cell is known as a *bacteriophage*, or simply as a *phage*.

Alternatively, a virus may enter the lysogenic cycle following the incorporation of its DNA into host DNA. This viral DNA does not disable or destroy its host at this point; rather it becomes part of a living reproductive host cell, often changing the traits of the cell. Over time, this viral DNA can simply remain part of the host DNA, or it may be activated by some trigger like stress or compromised host immunity. Activation of this viral DNA can result in the production of new viruses and destruction of the host cell (the lytic cycle).

REPRODUCTIVE PATTERN

A simplified version of the replication of a virus is shown in Figure 6.11. The virus (*A*) enters a cell (*B*). The viral capsid (protein coat) is shed as soon as the virus gains entry to the cell. Thus the viral DNA is naked. At (1) *replication* of the viral DNA has taken place. The virus uses the nucleotides and other machinery of the cell to make its own DNA. At (2) the process of *transcription* replicates viral RNA. Then through *translation* capsid proteins are formed. At (3) the assembly of viral particles and their exit from the cell occur.

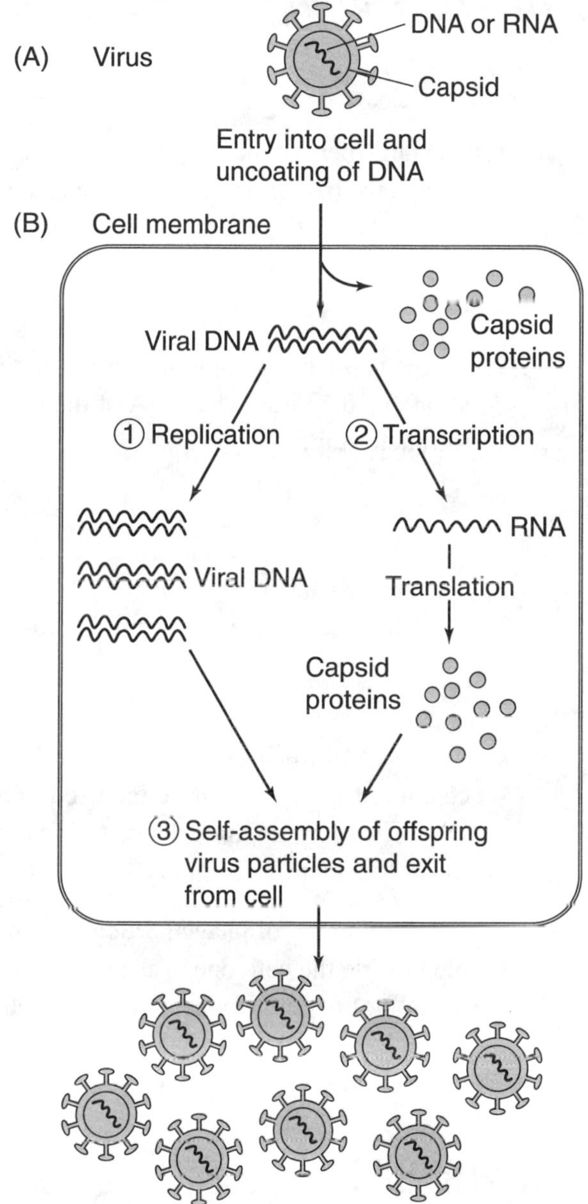

FIGURE 6.11 A simplified version of the replication of virus particles

LYTIC CYCLE OF VIRAL REPRODUCTION

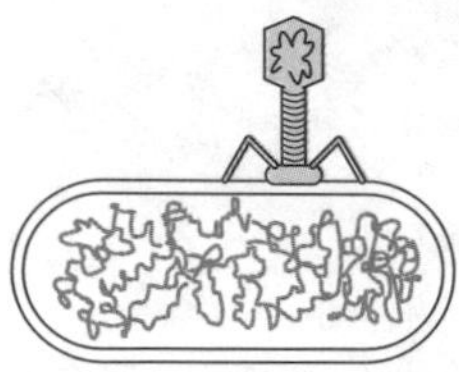

1. The tail fibers of the T4 phage become attached to a specific place (receptor site) on the surface of the bacterial cell *Escherichia coli*.

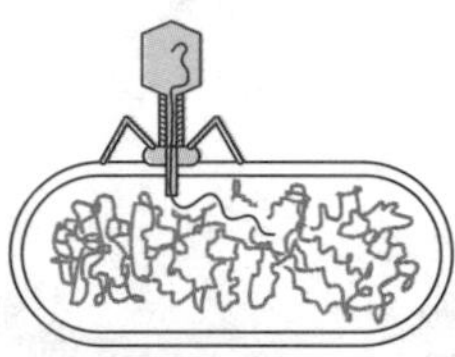

2. A tube is poked through the wall of the *E. coli* cell by the sheath of the virus. The DNA of the phage is injected into the bacterium.

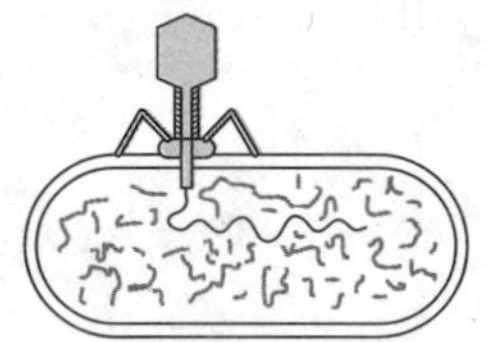

3. The empty capsid of the phage is left outside the bacterium. The DNA of the bacterium is broken apart.

4. The DNA of the bacterial cell, as well as its nucleotides, is captured by the viral DNA and used to replicate viral DNA. In addition, three distinct kinds of protein are used to put together the T4 phage heads, tails, and tail fibers.

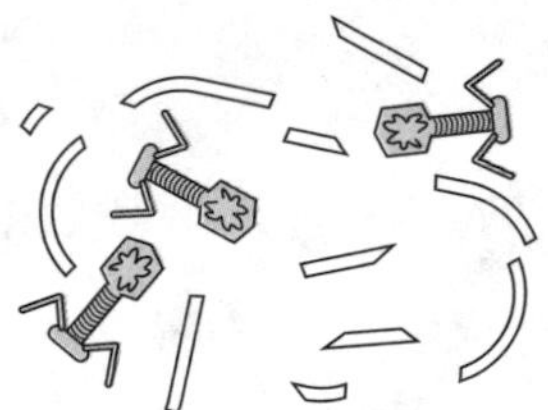

5. A great number of phage particles are made. Once assembled inside the cell, one of the phage genes directs the production of the enzyme lysozyme, which digests a hole through the wall of the bacterium, releasing the newly formed phage particles into the bloodstream.

IMPORTANCE OF VIRUSES

Viruses are notorious for their disease-producing potential in animals and plants. In human beings, viruses cause such diseases as measles, the common cold, smallpox, rabies, chicken pox, yellow fever, influenza, viral pneumonia, encephalitis, infectious mononucleosis, AIDS, and several types of hepatitis, as well as fever blisters and venereal herpes. It is believed that some cancers are caused by viruses. Virus diseases of plants include tobacco mosaic disease, tobacco necrosis, and rice dwarf.

VIROIDS

In the 1960s, it was discovered that several plant diseases are caused by *viroids*, particles smaller than viruses. The viroid contains a short RNA chain and lacks the protective protein coat. New viroids are produced in the nucleus of the host cell. Diseases such as the stunting of chrysanthemums, spindle tuber of potatoes, and disease of citrus trees are caused by viroids.

Prions

Unlike viruses and viroids, prions are made of only protein. Prions do not have DNA or RNA. They are infectious particles that can alter the way host proteins function. Prions have been found to cause mad cow disease, or bovine spongiform encephalopathy (BSE), and its human counterpart, Creutzfeldt-Jakob disease.

REVIEW EXERCISES FOR CHAPTER 6

WORD-STUDY CONNECTION

acrobic
anaerobe
Archaeobacteria
autotroph
bacillus
bacteriophage
bacterium
binary fission
blue-greens
capsid
capsomere
chemosynthetic bacteria
coccus
convolution
cyanobacteria
endospore
Eubacteria
eukaryote
facultative anaerobe
flagellum
fruiting body
Gram staining
heterocyst
heterotroph
mesosome
methanogens
motile
murein
mycoplasma
nodule
obligate aerobe
obligate anaerobe
parasite
pathogenic
peptidoglycans
phage
photosynthetic bacteria
phytoplankton
plasmadesmata
prion
Prochloron
prokaryote
spirillum
spricochete
saprobe
thermoacidophile
thermophile
toxin
viron
viroid

SELF-TEST CONNECTION

PART A. Completion. *Write in the word that correctly completes each statement.*

1. The prokaryotes are cells that lack _____ organelles. (2 words)
2. The cell membrane of the prokaryotes lacks the steroid _____.
3. The DNA molecule is attached to a place on the cell membrane of bacteria called a _____.
4. Mycoplasmas, unlike other prokaryotic cells, do not have a _____ (2 words)
5. Peptidoglycans are necessary for _____ formation in bacteria.
6. The only cellular organelle in bacteria is the _____.
7. A chromosome makes an identical copy of itself in a process called _____.
8. The methanogens belong to the kingdom _____.
9. A group of interbreeding organisms is called a _____.
10. The _____ are referred to as the “true” bacteria.
11. Bacteria reproduce by the method known as _____.
12. In newer terminology, a saprophyte is referred to as a _____.
13. Rod-shaped bacteria are known as _____.
14. *Nitrobacter* are examples of _____ bacteria.
15. Bacteria that require molecular oxygen for respiration are known as _____.
16. Bacteria that are indifferent to oxygen are called _____.
17. Gram-positive bacteria absorb a _____ colored dye.
18. Toxins inhibit the _____ activities of cells.
19. True bacteria can survive unfavorable environmental conditions by forming _____.
20. The _____ are evolutionary links between prokaryotic and eukaryotic cells.
21. In some cyanobacteria cells known as _____ can fix atmospheric nitrogen.

22. The habitat of most blue-greens is _____ water.

23. A virus particle is known as a _____.

24. A virus that infects bacteria is called a _____.

25. Floating green plants that inhabit surfaces of lakes and oceans are known collectively as _____.

PART B. Multiple Choice. *Circle the letter of the item that correctly completes each statement.*

1. Bacteria are best described as
   (a) eukaryotic cells
   (b) photosynthetic cells
   (c) prokaryotic cells
   (d) filamentous cells

2. A chain of cells is known as
   (a) filament
   (b) mass
   (c) chord
   (d) convolution

3. The DNA of prokaryotes is structured into
   (a) scattered chromatin granules
   (b) histone chains
   (c) several nucleoids
   (d) a single chromosome

4. Muramic acid is isolated from the bacterial structure known as
   (a) mesosome
   (b) cell membrane
   (c) cell wall
   (d) chromosome

5. The function of the ribosomes is to
   (a) synthesize cell walls
   (b) carry hereditary traits
   (c) direct replication of the mesosome
   (d) translate mRNA into protein

6. Prokaryotic cells contain
   (a) DNA only
   (b) RNA only
   (c) both DNA and RNA
   (d) no nucleic acids

7. An incorrect statement concerning bacteria is that they
   (a) build ATP molecules
   (b) lack multienzyme systems
   (c) lack Golgi bodies
   (d) assemble proteins

8. Survival of the eubacteria is increased by the ability to
   (a) replicate DNA
   (b) synthesize proteins
   (c) generate cells walls
   (d) form endospores

9. Encapsulated bacteria resist
   (a) reproduction
   (b) endospore formation
   (c) phagocytosis
   (d) DNA replication

10. Nitrifying bacteria
    (a) convert ammonia into nitrates
    (b) convert ammonia into nitrogen gas
    (c) release ammonia from decaying bodies
    (d) synthesize legumes

11. The bacteria that live as parasites in the cells of ticks and mites are
    (a) spirochetes
    (b) actinomycetes
    (c) mycoplasmas
    (d) rickettsiae

12. It is true that viruses
    (a) reproduce
    (b) ingest
    (c) synthesize ATP
    (d) reduce H_2

13. Tunicates are
    (a) blue-green algae
    (b) fruiting bodies
    (c) a type of virus
    (d) marine invertebrates

14. Bacteriorhodopsin is correctly associated with the
    (a) chlorophyta
    (b) prochloron[illegible]
    (c) halophiles
    (d) viroids

15. Capsomeres are subunits of
    (a) bacteria cell walls
    (b) viral coats
    (c) clover nodules
    (d) slime mold

16. Thermophiles live best
    (a) on the bottom of bogs
    (b) in hot springs
    (c) in salt pools
    (d) in stagnant marshes

17. Lysozyme is a (an)
    (a) human secretion
    (b) small vacuole
    (c) enzyme
    (d) hydrogen carrier molecule

18. Fruiting bodies are best associated with
    (a) spirochetes
    (b) Eubacteria
    (c) Prochlorophyta
    (d) Myxobacteria

19. The organism that causes syphilis is a
    (a) spirochete
    (b) eubacterium
    (c) prochlorophyte
    (d) myxobacterium

20. Rickettsiae are best classified as
    (a) molds
    (b) bacteria
    (c) fungi
    (d) protists

21. Plasmadesmata are
    (a) shrinking cell walls
    (b) oversized vacuoles
    (c) cytoplasmic bridges between cells
    (d) several attached cell membranes

22. Heterocysts are most closely associated with
    (a) nitrogen fixation
    (b) cell protection
    (c) water storage
    (d) binary fission

23. Actinomycetes are a type of
    (a) virus
    (b) rock
    (c) fungus
    (d) bacterium

24. The protein coat of a virus is known as a
    (a) desmid
    (b) plastid
    (c) capsid
    (d) capsule

25. A virus
    (a) can reproduce itself independently
    (b) can reproduce only within living cells
    (c) does not contain either DNA or RNA
    (d) can be considered a type of cell

PART C. Modified True-False. *If a statement is true, write "true" for your answer. If a statement is incorrect, change the underlined expression to one that will make the statement true.*

1. At one time bacteria were classified as one-celled plants.

2. Archaeobacteria are visible to the naked eye.

3. The smallest living cells are bacteria.

4. In addition to the circular chromosome the only other organelle found in bacteria is the Golgi body.

5. A micrometer is 1/100 of a millimeter.

6. Mycoplasmas have twice as much DNA as the true bacteria.

7. Photosynthetic bacteria do not need light energy to synthesize food molecules.

8. During favorable conditions, bacteria can reproduce every 2 hours.

9. Disease-producing bacteria may be surrounded by a polysaccharide cell wall.

10. Small bumps on roots of clover plants that house bacteria are known as tumors.

11. Round bacteria are called spirilla.

12. *Escherichia coli* live normally in the human heart.

13. When oxygen is not available, facultative anaerobes obtain energy by way of the Krebs cycle.

14. Obligate anaerobes must live in an environment free of carbon dioxide.

15. Bacterial cells can build their own ATP molecules.

16. Gram-positive bacteria are more susceptible to the effects of antibiotics.

17. Most species of bacteria are harmful.

18. Retting is a process in which viruses are used to digest the pectin in flax plants.

19. Hemp fiber is used to make linen.

20. Polyglucans are molecules of protein.

21. The enzyme nitrogenase is produced under aerobic conditions.

22. The envelope surrounding the coat of a virus comes from the host.

23. Hepatitis is a viroid disease.

24. Viroids are smaller than viruses.

25. Viroids are known to cause several animal diseases.

CONNECTING TO CONCEPTS

1. What are the distinguishing characteristics of prokaryotic cells?
2. Why is the classification "species" not appropriate for bacteria?
3. Why are virus particles not considered to be cells?
4. In what ways are bacteria harmful to cells?

ANSWERS TO SELF-TEST CONNECTION

PART A

1. membrane-bound
2. cholesterol
3. mesosome
4. cell wall
5. cell wall
6. ribosome
7. replication
8. Archaeobacteria
9. species
10. Eubacteria
11. binary fission
12. saprobe
13. bacilli
14. nitrifying
15. aerobes
16. facultative anaerobes
17. purple
18. metabolic
19. endospores
20. prochlorons
21. heterocysts
22. fresh
23. virion
24. bacteriophage
25. phytoplankton

PART B

1. **(c)**	6. **(c)**	11. **(d)**	16. **(b)**	21. **(c)**
2. **(a)**	7. **(b)**	12. **(a)**	17. **(c)**	22. **(a)**
3. **(d)**	8. **(d)**	13. **(d)**	18. **(d)**	23. **(d)**
4. **(c)**	9. **(c)**	14. **(c)**	19. **(a)**	24. **(c)**
5. **(d)**	10. **(a)**	15. **(b)**	20. **(b)**	25. **(b)**

PART C

1. true
2. invisible
3. mycoplasmas
4. ribosome
5. 1/1000
6. one half
7. Chemosynthetic
8. 20 minutes
9. capsule
10. nodules
11. cocci
12. intestine
13. fermentation
14. oxygen
15. true
16. true
17. helpful
18. bacteria
19. rope
20. carbohydrate
21. anaerobic
22. true
23. virus
24. true
25. plant

CONNECTING TO LIFE/JOB SKILLS

Veterinary microbiology is a field worth researching. Infectious diseases caused by bacteria and viruses take a heavy toll on animal life. Epidemics may result in death or the spread of infection from sick wild animals to domesticated ones. In an increasing number of cases, the spread of animal infection is transmitted to humans. Ticks, fleas, and mosquitoes carry infections from sick animals to humans. Use your library media center to research educational opportunities in the field. On the Internet, www.sciam.com is a good place to find out about job opportunities in veterinary microbiology.

Chronology of Famous Names in Biology

1684 **Antonie van Leeuwenhoek** (Netherlands)—was the first person to see bacteria.

1796 **Edward Jenner** (England)—demonstrated the use of cowpox as an immunizing vaccine against smallpox.

1850 **Ignaz Semmelweis** (Germany)—discovered that attending physicians carried the organisms that cause childbed fever from patient to patient on their unwashed hands.

1861 **Louis Pasteur** (France)—made major contributions to the understanding of bacteria; helped to prove the germ theory of disease; discovered a method for the prevention of rabies.

1867 **Joseph Lister** (England)—developed antiseptic principles in the practice of sugery.

1884 **Robert Koch** (Germany)—discovered the bacteria that cause tuberculosis and cholera. He formulated Koch's postulates, the steps to determine if a bacterium causes a disease.

1884 **Hans Christian Gram** (Netherlands)—developed differential staining techniques for bacteria which have become known as the Gram stain.

1892 **Dimitri Iwanowski** (Russia)—discovered the tobacco mosaic virus.

1909 **Harold Taylor Ricketts** (United States)—was the first person to describe rickettsiae, found in the blood of patients suffering from Rocky Mountain spotted fever.

1929 **Alexander Fleming** (England)—discovered lysozyme in human secretions.

1935 **Wendell Stanley** (United States)—isolated the tobacco mosaic virus.

1946 **Joshua Lederberg** and **Edward Tatum** (United States)—were the first to demonstrate genetic recombination in bacteria.

1975 **Carl R. Woese** (United States)—developed the idea that the prokaryotes include two distinct, unrelated kingdoms.

1976 **R. A. Levin** (United States)—discovered Prochloran, a group of prokaryotes that have a greater resemblance to chloroplasts than to cyanobacteria.

1997 **Stanley B. Pruisner** (United States)—discovered prions, a new biological principle of infection.

The Protist Kingdom: Protozoa, Fungus-Like Protists, and Plant-Like Protists

7

WHAT YOU WILL LEARN

In this chapter you will read about the astounding number of one-celled organisms classified as Protists and will learn how their life functions are carried out within a single cell.

SECTIONS IN THIS CHAPTER

- Major Groups of Protists
- Importance of Protists to Humans
- Review Exercises for Chapter 7
- Connecting to Life/Job Skills
- Chronology of Famous Names in Biology

Major Groups of Protists

All of the species assigned to the kingdom Protista are eukaryotic. Many protists carry out their lives within a single cell as free-living organisms. However, some protist species are organized into **colonies** where each cell carries out its own life functions and where, also, there may be some simple division of labor among the cells in the grouping. An impressive variety of species are classified as protists, and they probably descended from diverse evolutionary lines. The protists themselves represent evolutionary modification and are probably the ancestors of the modern fungi, plants and animals.

Most modern classification schemes divide the Protista into three major groups: the protozoa, or animal-like protists; the fungus-like protists; and the plant-like protists.

KINGDOM: PROTISTA

All protists are eukaryotic (they have nuclei and membrane-bound organelles); they may contain single or multiple nuclei; they may be unicellular or multicellular; they may be autotrophs or heterotrophs; they may be stationary or mobile.

Animal-like Protists		
Mastigophorans	Protists with flagella	*Trypanosoma*
Sarcodines	Protists with pseudopods	*Amoeba*
Sporozoa	Parasitic protists	*Plasmodium*
Ciliates	Protists with cilia	*Paramecium*
Fungus-like Protists		
Myxomycota	Heterotrophic, amoeboid mass called *plasmodium*	Plasmodial slime molds
Acrasiomycota	Heterotrophic, separate cells	Cellular slime molds
Plant-like Protists		
Euglenophytes	Unicellular, photosynthetic, single flagellum	*Euglena*
Chrysophytes	Unicellular, photosynthetic, chlorophyll *a* and *c*	Diatoms, golden brown algae
Dinoflagellates	Unicellular, two spinning flagella, chlorophyll *a* and *c*	*Gessnerium*

PROTOZOA, THE ANIMAL-LIKE PROTISTS

Protozoa, meaning "first animals," are one-celled heterotrophs. Species of protozoa number in the thousands. They live in fresh water, salt water, dry sand, and moist soil. Some species live as parasites on or inside of the bodies of other organisms. Reproduction in the protozoans is usually described as being asexual by means of mitosis, but recent research has revealed that many protozoa augment asexual reproduction with a sexual cycle. Usually, the sexual cycle occurs during periods of adverse environmental conditions, and the cell arising from the fusion of gametes (*zygote*) can resist unfavorable conditions. The thick wall and the decreased metabolic rate of the *cyst* permits survival during periods of cold, drought, or famine.

The protozoa are divided into four phyla, based primarily on the methods of locomotion.

MASTIGOPHORA

The Mastigophora are protozoa that have one or more flagella. This phylum, also known as the Zoomastigina or zooflagellates, includes a rather heterogeneous group of organisms. Some species are free-living and inhabit fresh or salt water; of these, some are free-swimming; others glide over the surface of rocks; and some are sessile, attached to available submerged surfaces. Other Mastigophora species live in a **symbiotic** relationship with organisms of other species, each helping the other with a particular life function. For example, several species live in the intestines of termites, cockroaches and woodroaches, where they digest cellulose for these insects. The genus *Trypanosoma* includes parasites that cause debilitating diseases in human beings. *Trypanosoma gambiense* (Figure 7.1) is the zooflagellate that causes African sleeping sickness. Humans are infected with the trypanosome by the bite of an infected tsetse fly.

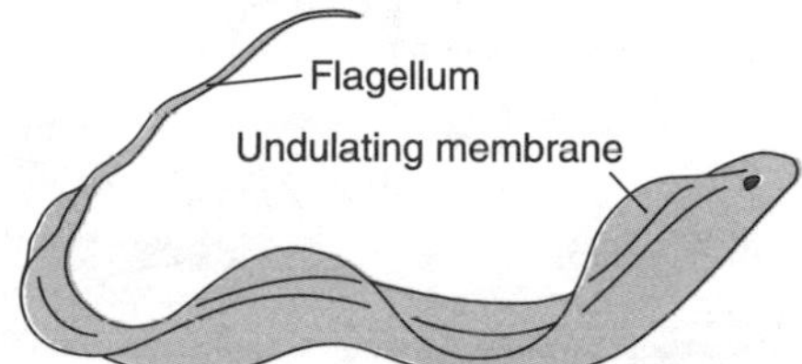

FIGURE 7.1 The causative organism of African sleeping sickness, *Trypanosoma gambiense*

SARCODINA

The members of the phylum Sarcodina are described as being amoeboid. *Amoeba proteus* (Figure 7.2) is the type species. Species included in the Sarcodina move by means of **pseudopods**, flowing extensions of the flexible and amorphous body. The pseudopods also serve in food-catching. Most of the Sarcodines live in fresh water. A **contractile vacuole**, an organelle designed to expel excess water from the protist cell body, plays an important role in maintaining water balance. Food is temporarily stored in a food vacuole where it is digested by the action of enzymes.

The Foraminifera and the Radiolaria are groups of marine Sarcodines that secrete hard shells of mineral compounds around themselves. When they die, their shells become an important constituent of the bottom mud of the ocean floor. The Foraminifera have built the limestone and chalk deposits that date back to the Cambrian period; the Radiolarians, the siliceous rocks dating back to the Precambrian period.

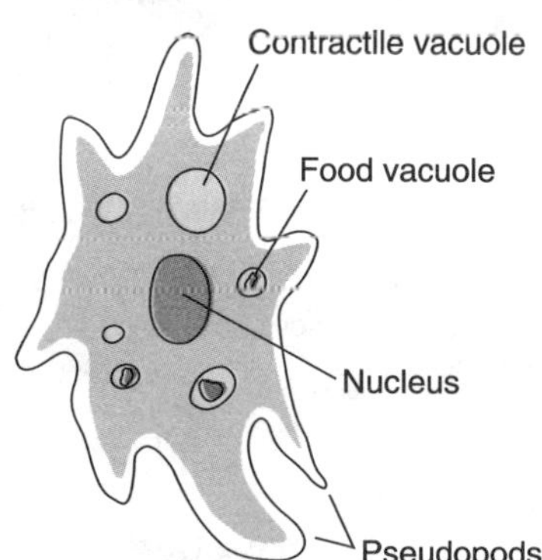

FIGURE 7.2 *Amoeba proteus*

SPOROZOA

The Sporozoa are parasitic spore-formers. The adult forms are incapable of locomotion, although immature organisms may move by means of pseudopodia. Some species of sporozoa go through a complicated life cycle requiring different hosts during different life stages. For example, the species *Plasmodium vivax*—the agent that causes **malaria**—requires two hosts: the *Anopheles* mosquito and a human (Figure 7.3).

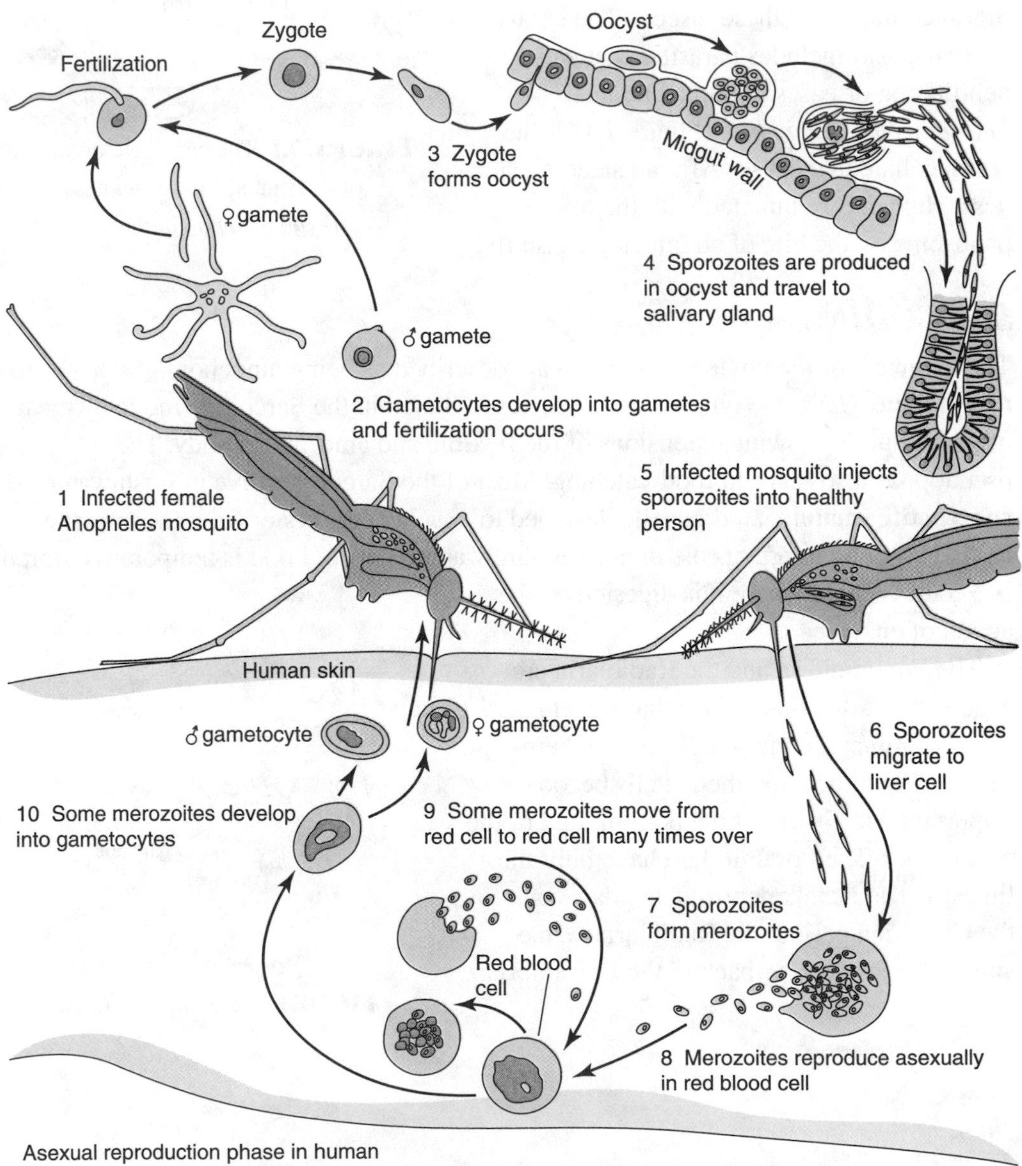

FIGURE 7.3 Life cycle of *Plasmodium vivax*. Human infection results in malaria.

CILIATA

Of the four protozoan divisions, the phylum Ciliata has the greatest number of species. Species belonging to this phylum have *cilia*, short cytoplasmic strands that are used for locomotion and in some cases to sweep food particles into an opening called the *oral groove*. Protozoans included in this phylum inhabit fresh water and salt water. Some are free-swimming; some creep; others are sessile; and some are parasitic in other animals.

The cytoplasm in ciliates is differentiated into rigid outer ectoplasm and a more fluid inner endoplasm. A *pellicle* lies just inside of the cell membrane. Some species respond to adverse environmental stimuli by discharging elongated threads called *trichocysts* which serve as defense mechanisms or a means of anchoring the protist to floating pond material while feeding. Characteristic of the ciliata is the presence of two kinds of nuclei. The *macronucleus* controls metabolic activities, while the smaller *micronucleus* directs cell division (Figure 7.4). shows the structure of *Paramecium*, a typical representative of the ciliata. Figure 7.5 shows some other Ciliata.

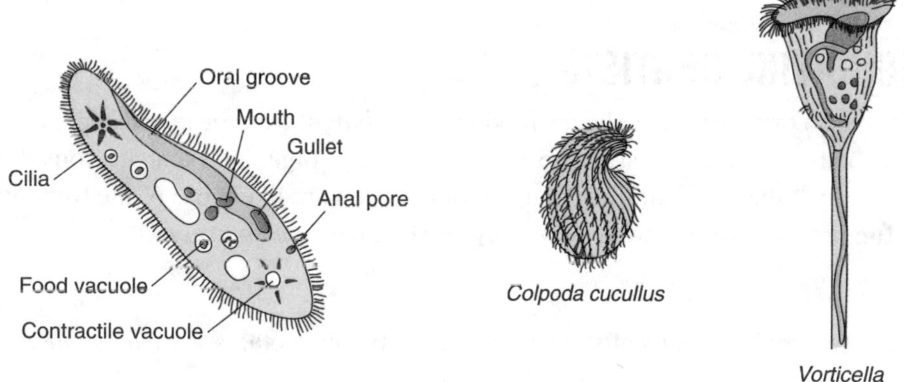

FIGURE 7.4 *Paramecium*

FIGURE 7.5 Some other examples of ciliates

Conjugation is a form of sexual reproduction that is occasionally demonstrated by the ciliates. Two organisms will join together at the oral groove. The micronucleus of each will undergo meiosis, producing several cells. All but two of these in each organism disintegrate. One of these haploid micronuclei remains in each cell, while the other migrates into the other cell, fusing with the stationary gamete. The new nucleus—which is now diploid and contains a new genetic combination—goes through cell division producing a new macronucleus and a new micronucleus (Figure 7.6).

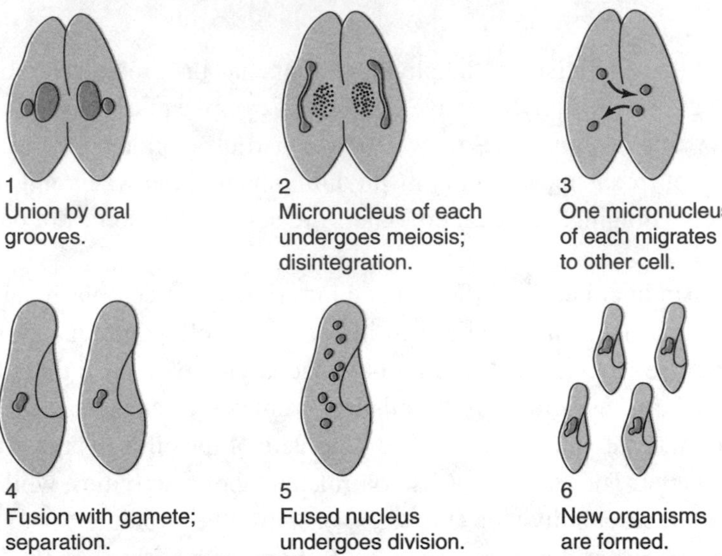
1
Union by oral
grooves.
2
Micronucleus of each
undergoes meiosis;
disintegration.
3
One micronucleus
of each migrates
to other cell.
4
Fusion with gamete;
separation.
5
Fused nucleus
undergoes division.
6
New organisms
are formed.

FIGURE 7.6 Conjugation in *Paramecium*

FUNGUS-LIKE PROTISTS

A *fungus* is an organism that obtains food by *absorbing* it from dead organic matter or from the body of a living host. There are two groups of organisms that are considered protists and yet their way of life and mode of nutrition are like those of the true fungi. These fungus-like protists are the *Protomycota* and the *Gymnomycota*.

SLIME MOLDS

There are two major groups of slime molds: the Myxomycota, or the *true slime molds*, and the Acrasiomycota, the *cellular slime molds*.

Myxomycota—*True Slime Molds*

The true slime molds live on the forest floor, where they grow in damp soil, on or around rotting logs, and on decaying vegetation. They appear to be shapeless globs of slime of varying colors: white, yellow, or red.

The life cycle of the true slime molds begins with a multinucleate mass known as a *plasmodium*. This plasmodium glides about in amoeboid fashion, engulfing bacteria and small bits of organic material. The nuclei that make up the plasmodium are diploid. At some time in its life cycle the plasmodium stops moving and undergoes change, developing stalk-like structures with rounded knobs on top. These are *fruiting bodies*, and they support the structures known as *sporangia* (*sporangium*, sing.).

A sporangium is a structure that contains *spores*. The spores go through meiosis, producing flagellated gametes. The gametes fuse and form a *zygote* which is not flagellated, but instead, resembles an amoeba. This amoeba-like organism glides along the soil, engulfing food materials in phagocytic fashion. Its diploid nucleus goes through a series of mitotic divisions not accompanied by division of the cytoplasm. In this way the multinucleate plasmodium develops. The life cycle then repeats (Figure 7.7).

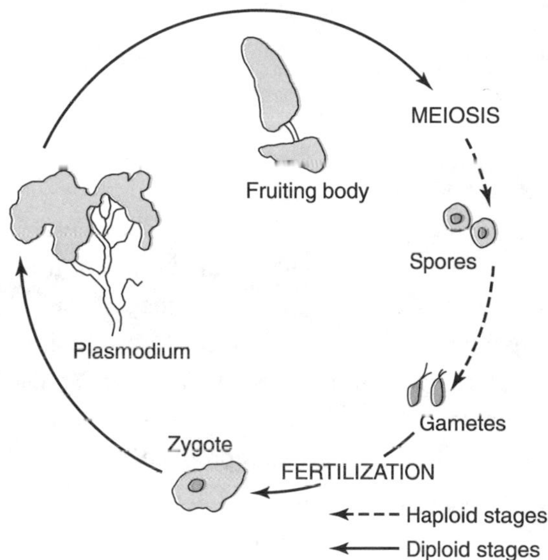

FIGURE 7.7 Life cycle of a true slime mold

Acrasiomycota—Cellular Slime Molds

The life cycle of the cellular slime molds, often called the *social amoebae*, is much different from that of the true slime molds. The fruiting bodies of the social amoeba are called *sorocarps*. The spores are carried by wind, water, insects or other means to new environments. Those spores placed in favorable environments will germinate. As the spore wall disintegrates, an amoeba, using pseudopods, pushes its way out onto the soil. These amoebae resemble other amoebae in structure. Each has a cell membrane, a nucleus, nucleolus, contractile vacuoles, food vacuoles, mitochondria and endoplasmic reticulum.

The free-living haploid amoebae feed on organic matter from the soil and grow to an optimum size. They then divide by mitosis and cytokinesis, producing daughter cells which, like the parent, have one haploid nucleus. Then their behavior changes. They stop feeding and begin to move in a directed manner toward definite centers or collecting points where they form closely packed groups and take on the formation of a compact mass of cells.

Next they become grouped together forming a *pseudo-plasmodium* or slug. A very thin sheath of polysaccharide material surrounds the mass, but the cells maintain their individuality. The finger-shaped pseudoplasmodium moves across the soil substrate slowly but in a directed and coordinated way. Eventually, the front end of the pseudoplasmodium stops moving and the rear segment moves underneath the front end to form a mound of cells. At this time, differentiation of cells begins to take place. Stalk cells which give rise to a fruiting body are formed, and the life cycle repeats (Figure 7.8).

PLANT-LIKE PROTISTS

There are four major groups of plant-like protists.

THE EUGLENOPHYTA

The euglenoids are represented by the organism *Euglena* (Figure 7.9). This unicellular organism has both plant-like and animal-like characteristics. It has chlorophyll *a* and *b* and some carotenoids, and it is able to carry on photosynthesis. However, *Euglena* lacks a cell wall, swims by using a flagellum, has a light-sensitive red-orange eyespot known as the *stigma* and a large contractile vacuole. *Euglena* also has a *pyrenoid body* that functions in the synthesis of *paramylum*, a carbohydrate storage product that is peculiar to the euglenoids. These organisms reproduce asexually by longitudinal mitotic cell division. However, during mitosis, the cell membrane remains intact and the nucleolus persists.

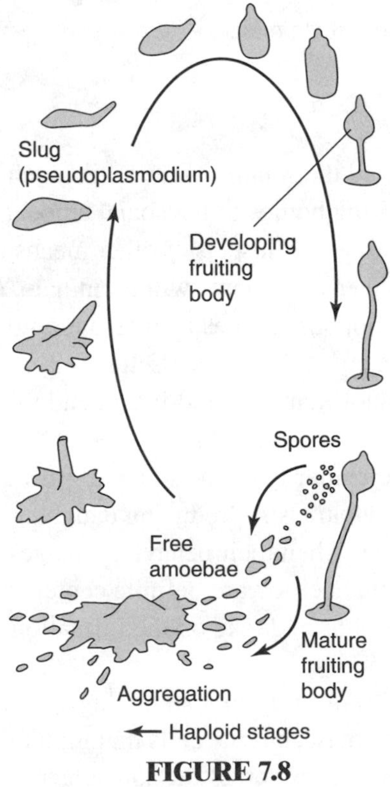

FIGURE 7.8
Life cycle of a cellular slime mold

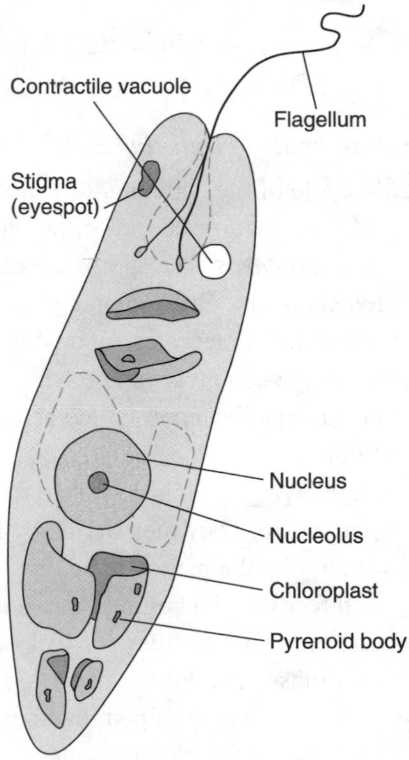

FIGURE 7.9 *Euglena*

THE CHRYSOPHYTA

FIGURE 7.10
A chrysophyte

The yellow-green algae, the golden-brown algae, and the diatoms are included in this group. Most species in the Chrysophyta group are single-celled and reproduce asexually. There are a few simple multicellular forms. The members of this group share certain characteristics. They are pigmented with chlorophyll *a* and *c* (lacking *b*) and contain the carotenoid *fucoxanthin*, which gives them the golden color. These protists live in salt water, fresh water, and in damp places between rocks. They use a polysaccharide called chrysolaminaran instead of starch. Many of the Chrysophyta have two flagella of unequal length at the anterior end of the body (Figure 7.10); but some have only one flagellum and others, none.

The *diatoms* are diploid cells that lack flagella. Their cell walls, composed of two pieces, one fitting over the other, are impregnated with silica and pectin and often have intricate patterns in the forms of pits and ridges. When the organisms die, their shells fall to the bottom of the water where they disintegrate and form *diatomaceous earth*, a substance used as an abrasive in silver polish and detergents, as the packing in air and water filters, and in paint removers, deodorizing oils, and fertilizers.

DINOFLAGELLATES

The *dinoflagellates* are small protists and usually unicellular. Most of these organisms have two unequal flagella, one extending longitudinally from the posterior end of the cell, the other encircling the central part of the cell. Some dinoflagellates extend trichocysts like the *Paramecium*; others have *nematocysts, stinging cells* common in the coelenterates.

The dinoflagellates have a most unusual nucleus. The chromosomes do not have centromeres and even during interphase remain in evidence as short, thickened rods. During mitosis, the nuclear membrane and the nucleolus remain, and no spindle is formed. Most species reproduce asexually by cell division and make up one of the two main groups of phytoplankton. Some species—*Noctiluca*, for example—are bioluminescent, giving off light like a firefly.

SOME ATYPICAL PROTISTS

Caryoblastea is a phylum of protists containing only one species, the giant multinucleate amoeba *Pelomyxa palustris*. This organism lacks mitochondria and other organelles characteristic of eukaryotic cells. The nuclei divide without mitosis. *Pelomyxa palustris* lives as a parasite inside the cell of a host organism.

Microspora is a phylum of heterotrophic protists. They are small parasites that lack mitochondria and live inside the cells of their hosts, which include arthropods, chordates, and all classes of vertebrates. After a microsporidian infects a cell, a multicellular plasmodium forms. The plasmodium then undergoes multiple fission, producing offspring. The mass that results resembles a single-cell tumor called a *xenoma*.

Importance of Protists to Humans

The quality of human life is in part determined by the ability to be free of disease. The disease-producing protists affect the health of large populations of people throughout the world. One such protist-initiated disease is amoebic dysentery, a painful condition of bleeding ulcers caused by a type of amoeba that parasitizes human intestines. This protist-pathogen is passed to people through contaminated food and water. We have already mentioned two pathogenic protists carried by insects—*Trypanosoma gambiense* and *Plasmodium vivax*.

The red-pigmented dinoflagellate *Gonyaulax* live in the waters of the Gulf of Mexico off the Florida coast. At times this organism reproduces explosively and uncontrollably coloring the waters red. This so-called "red tide" poisons millions of fish and does harm to people who eat these fish.

Many other protist species are helpful to humans. Slime molds help to keep an ecological balance by feeding on decayed plant and animal matter. These organisms also have a great deal of value as research specimens for investigators who are trying to ferret out the secrets of cell specialization and differentiation.

Plankton is the mass of green that floats on rivers, lakes, and oceans. In reality, protists—microscopic "plant" life (phytoplankton) and microscopic "animal" life (zooplankton)—intermingle with mutual benefit in this floating mass, which serves vital roles in the food chains of aquatic species.

REVIEW EXERCISES FOR CHAPTER 7

WORD-STUDY CONNECTION

chytrid	nematocyst	social amoeba
cilia	oral groove	sorocarp
colony	paramylum	porangium
conjugation	pellicle	stigma
contractile vacuole	plankton	symbiosis
cyst	plasmodium	trichocysts
diatom	protest	tsetse
diatomaceous earth	protozoa	zooflagellate
dinoflagellate	pseudoplasmodium	zooplankton
fungus	pseudopod	zoospore
hypochytrid	pyrenoid body	zygote
macronucleus	rhizoid	
micronucleus	slime mold	

SELF-TEST CONNECTION

PART A. Completion. *Write in the word that correctly completes the statement.*

1. One-celled protists that resemble animal cells are the _____.
2. Another name for a "self-feeder" is a (an) _____.
3. Amoebae move by means of false feet known as _____.
4. The usual mode of reproduction in protozoa is _____.
5. The relationship in which two organisms of different species live together and neither is harmed by the association is known as _____.
6. Digestion in *Amoeba proteus* takes place in the _____.
7. The Sporozoa are harmful to organisms of other species and are therefore classified as _____.
8. *Anopheles* is the genus name of a _____.
9. The organelle that expels excess water from the protist is the _____.
10. In *Paramecium*, the nucleus that controls metabolic activity is the _____.
11. The body of the paramecium is prevented from being totally flexible by the _____.
12. The form of reproduction in which like gametes fuse is called _____.
13. The plasmodium is a stage in the life cycle of _____.
14. A spore-producing structure is known as _____.
15. Sorocarps are correctly associated with the _____ slime molds.
16. In *Euglena*, the stigma is sensitive to _____.
17. The cell walls of the diatoms are impregnated with pectin and _____.
18. Diatomaceous earth forms from the _____ of diatoms.
19. The number of flagella usually found in dinoflagellates is _____.
20. The number of flagella usually present in diatoms is _____.

PART B. Multiple Choice. *Circle the letter of the item that correctly completes each statement.*

1. Classification systems are best described as being
   (a) unchanging
   (b) fixed
   (c) unreliable
   (d) artificial

2. The evolution of protists involved the development of
   (a) membrane-bound organelles
   (b) a discrete nucleus
   (c) functioning mitochondria
   (d) an active Golgi

3. Protists live
   (a) on land
   (b) in the sea
   (c) in fresh water
   (d) in all of these

4. The primary purpose of the protozoan cyst is to
   (a) produce new organisms
   (b) survive unfavorable conditions
   (c) increase the metabolic rate
   (d) aid in better nutrition

5. *Trypanosoma gambiense* is the organism that causes
   (a) botulism
   (b) cholera
   (c) African sleeping sickness
   (d) malaria

6. Limestone and chalk deposits are found in the shells of dead
   (a) diatoms
   (b) Sarcodina
   (c) Foraminifera
   (d) zooflagellates

7. *Plasmodium vivax* is the organism that causes
   (a) botulism
   (b) dysentery
   (c) African sleeping sickness
   (d) malaria

8. Trichocysts in ciliates serve mainly
   (a) as defense mechanisms
   (b) to gather food
   (c) to expel water
   (d) for movement

9. Structures used for locomotion in the protists included all of the following except
   (a) pseudopodia
   (b) legs
   (c) cilia
   (d) flagella

10. The best way to prevent malaria is by destroying the breeding places of
    (a) *Plasmodium vivax*
    (b) *Trypanosoma vivax*
    (c) tsetse flies
    (d) *Anopheles* mosquitos

11. Paramecia move by means of
    (a) flagella
    (b) cilia
    (c) trichocysts
    (d) pseudopodia

12. Flagellated gametes produced by chytrids are the
    (a) zygotes
    (b) zooflagellates
    (c) zoospores
    (d) zymogens

13. Rhizoids are structures specialized for
    (a) reproduction
    (b) absorbing nutrients
    (c) locomotion
    (d) providing rigidity

14. The plasmodium of true slime molds is best described as
    (a) multinucleate
    (b) uninucleate
    (c) binucleate
    (d) prokaryotic

15. In part of the life cycle of the true slime molds, sporangia are supported by structures known as
    (a) hyphae
    (b) mycelia
    (c) rhizoids
    (d) fruiting bodies

16. During mitosis in *Euglena*
    (a) the nucleolus disappears
    (b) chromosomes do not form
    (c) the nuclear membrane persists
    (d) transverse fission occurs

17. The yellow-green algae lack
    (a) chlorophyll *a*
    (b) chlorophyll *b*
    (c) chlorophyll *c*
    (d) caratenoids

18. An organism that lives inside of another and does harm to its host is known as a (an)
    (a) autotroph
    (b) symbiont
    (c) saprophyte
    (d) parasite

19. *Noctiluca* is noted for its
    (a) reproduction mode
    (b) bioluminescense
    (c) swimming ability
    (d) apparent immortality

20. Nematocysts are used for
    (a) swimming
    (b) slinging
    (c) gliding
    (d) stinging

PART C. Modified True-False. *If a statement is true, write "true" for your answer. If a statement is incorrect, change the underlined expression to one that will make the statement true.*

1. Protists are <u>prokaryotes</u>.

2. Chromosomes in the eukaryotes are best described as <u>circular</u>.

3. Protists descended from the same evolutionary lines.
4. Protozoa are unicellular autotrophs.
5. Protozoa are classified according to methods of nutrition.
6. *Trypanosoma gambiense* is carried by the insect known as the house fly.
7. *Trypanosoma gambiense* is best classified as a dinoflagellate.
8. *Plasmodium vivax* requires two hosts: fly and human.
9. Adult sporozoans are motile.
10. The pseudoplasmodium of social amoebae moves in a directed way.
11. In *Paramecium*, reproduction is controlled by the macronucleus.
12. Chytrids are best classified as fungi.
13. The nuclei in the plasmodium of the true slime molds is haploid.
14. The pyrenoid body in *Euglena* stores glycogen.
15. The chloroplasts enable *Euglena* to carry out heterotrophic nutrition.
16. The pigment fucoxanthin causes the green color in the chrysophyta.
17. *Euglena* swims by means of a flagellum.
18. The diatoms have no commercial value.
19. Water balance in the amoeba is controlled by the food vacuole.
20. The chromosomes in the dinoflagellate nucleus lack structures called chromatids.

CONNECTING TO CONCEPTS

1. Why are protozoa classified as animal-like protists?
2. Study Figure 7.8. Discuss each stage in the life cycle of a true slime mold.
3. What is the function of chlorophyll in *Euglena*?
4. Of what commercial value are the diatoms?

ANSWERS TO SELF-TEST CONNECTION

PART A

1. protozoa
2. autotroph
3. pseudopodia
4. asexual
5. symbiosis
6. food vacuole
7. parasites
8. mosquito
9. contractile vacuole
10. macronucleus
11. pellicle
12. conjugation
13. slime molds
14. sporangium
15. cellular
16. light
17. silica
18. cell walls
19. two
20. zero

PART B

1. **(d)**
2. **(a)**
3. **(d)**
4. **(b)**
5. **(c)**
6. **(c)**
7. **(d)**
8. **(a)**
9. **(b)**
10. **(d)**
11. **(b)**
12. **(c)**
13. **(b)**
14. **(a)**
15. **(d)**
16. **(c)**
17. **(b)**
18. **(d)**
19. **(b)**
20. **(d)**

PART C

1. eukaryotes
2. linear
3. diverse
4. heterotrophs
5. locomotion
6. tsetse
7. zooflagellate
8. mosquito
9. nonmotile
10. true
11. micronucleus
12. protists
13. diploid
14. paramylum
15. autotrophic
16. yellow
17. true
18. great
19. contractile
20. centromeres

CONNECTING TO LIFE/JOB SKILLS

Microbiology is a field of study and employment that has grown very rapidly over the past decade. Microbiology is an extensive science that branches off into a number of specialized fields, including *medical microbiology*, *clinical microbiology*, *molecular microbiology*, *agricultural microbiology*, *plant microbiology*, and *industrial microbiology*. All of these fields provide employment to research scientists, physicians, university teachers, hospital clinical workers, various levels of technologists and technicians, and science writers. For information, contact the American Society of Microbiology, 1325 Massachusetts Ave. N.W., Washington, D.C. 20005; (202) 737-3600.

Chronology of Famous Names in Biology

1667 **Antonie van Leeuwenhoek** (Netherlands)—was the first person to see and describe protozoa.

1786 **Otto Frederick Muller** (Denmark)—wrote the first treatise on protozoa, titled *Animalcula Infusoria*.

1836 **Christian G. Ehrenberg** (Germany)—wrote a beautifully illustrated book describing 69 protozoa accurately.

1841 **Felix Dujardin** (France)—discovered protoplasm in protozoa.

1845 **Friedrich Stein** (Austria)—named the orders and suborders of protozoa.

1880 **Richard Hertwig** (Germany)—discovered chromatin in the protozoan nucleus.

1959 **John T. Bonner** (United States)—made a number of investigative studies that elucidated basic facts about the behavior and structure of cellular slime mold.

1962 **John T. Bonner** (United States)—discovered that cAMP directs the movement of the social amoeba toward a collection center.

1969 **Robert Whittaker** (United States)—reorganized the classification system into five kingdoms. He assigned the one-celled eukaryotes to the kingdom Protista.

1995 **B. Franz Lang** and **Gertrand Burger** (Canada)—discovered that the protist mitochondrial genome of *Reclinamonas americana* is identical in structure and function to that of bacteria. This research establishes an evolutionary link between bacteria and the one-celled eukaryotes.

8

The Fungi

WHAT YOU WILL LEARN

In this chapter you will review the kingdom Fungi. Although plant-like, fungi cannot make their own food. Fungi are important to protists, plants and animals.

SECTIONS IN THIS CHAPTER

- General Features of Fungi
- Major Divisions of Fungi
- Special Nutritional Relationships
- Importance of Fungi to Humans
- Review Exercises for Chapter 8
- Connecting to Life/Job Skills
- Chronology of Famous Names in Biology

General Features of Fungi

Most **fungi** are eukaryotic, multicellular, and multinucleate organisms. Yeasts are unicellular forms. The cells of fungi are different from those of other species because the boundaries separating the cells are either entirely missing or only partially formed. Thus fungi are primarily **coenocytic** organisms; this means that the cells have more than one nucleus in a single mass of cytoplasm. However, the characteristic that most distinguishes the fungi from other organisms is their mode of nutrition.

KINGDOM: FUNGI

An absorptive mode of nutrition permits fungi to live as decomposers. The ecosystem depends on the decomposing processes of fungi. Fungi reproduce by ejecting spores that are produced either sexually or asexually. Some fungi are parasitic.

Chytridiomycota	Small, unicellular organisms, flagellate spores, aquatic	Chytrids
Oomycota	Water molds; aquatic; cell wall made of cellulose	*Albugo candida*, white rust
Ascomycota	Sac fungi; marine, fresh-water; some live in association with algae	Morels, yeast, powdery mildews
Zygomycota	Zygote fungi; live on land, saprobes	Rhizopus (black bread mold)
Basidiomycota	Club fungi: club forms at end of each reproductive hypha	Mushrooms
Deuteromycota	Imperfect fungi: atypical fungal lifestyles	Ringworm, athlete's foot

NUTRITION

All organisms require food, from which they extract energy. Unlike green plants, fungi (fun-guy) cannot synthesize their own food from inorganic molecules. Nutritionally, the fungi are classified as **heterotrophs** because they must obtain their food from ready-made organic compounds. Animals, including humans, are heterotrophs.

Fungi cannot engulf (surround) their food as do amoebae; nor can they eat food as do animals that are equipped with mouths. Most fungi are **saprophytes**, obtaining nutrition by absorbing dead organic matter. Some fungi, however, are **parasitic**, absorbing their nutrition from living hosts. In either case, fungi must live very near their food supply in order to stay alive. Specialized fungal structures secrete digestive (hydrolytic) enzymes into the food substrate. The organic molecules of the substrate are made smaller by these enzymes and thus can be absorbed by the fungus.

BODY ORGANIZATION

The fungus begins its life as a *spore* (Figure 8.1). A **spore** is a microscopic cell that, under favorable conditions, will develop into a new individual. The cell wall of the spore is thick and tough, resistant to adverse environmental conditions. However, the spore is light in weight and is transported to new habitats by currents of air. When a spore settles on its substrate in favorable environmental conditions, its cell wall breaks open and the spore commences to sprout or *germinate* (Figure 8.2). The germinating spore absorbs food and grows an elongated thread called a **hypha**. As growth continues, many hyphae develop until they appear as a tangled mass of threads. When many hyphae appear, the body of the fungus is now called a **mycelium** (Figure 8.3). A hypha gives off a chemical that makes other hyphae grow away from it. In this way, competition for food is reduced among hyphae. In general, the function of the mycelium is to absorb food and to produce new fungus plants.

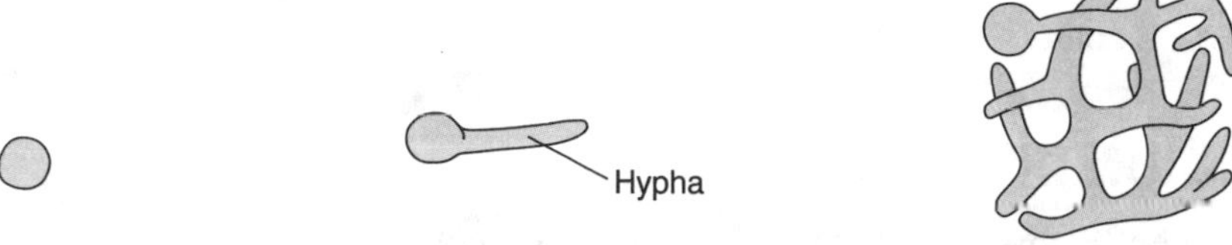

FIGURE 8.1 Spore **FIGURE 8.2** Germinating spore **FIGURE 8.3** Mycelium

A microscopic study of a hypha provides interesting information about its internal structure. Some hyphae are multinucleate with many nuclei sharing the same mass of cytoplasm without dividing membranes (Figure 8.4). Other hyphae are divided into compartments by **septa**. These compartments contain one or two nuclei.

Parasitic fungi have structures called **haustoria** that are specialized to penetrate the cell walls of plants. The haustoria grow into plant cell cytoplasm and absorb nutrients directly from the cells which they parasitize (Figure 8.5).

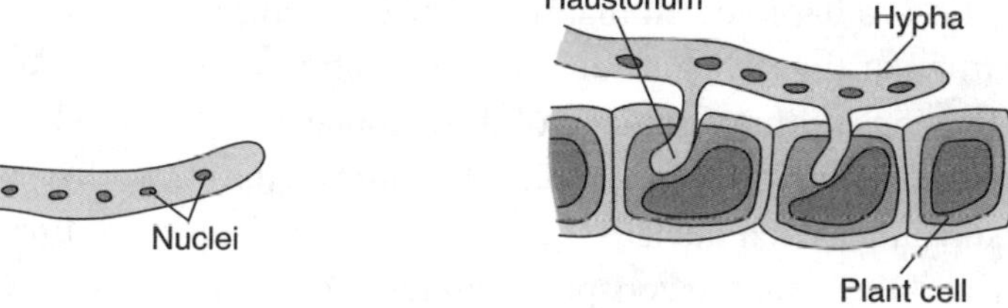

FIGURE 8.4 Internal view of a hypha **FIGURE 8.5** Haustoria

METHODS OF REPRODUCTION

Reproduction in fungi usually occurs by the production of spores. Special reproductive hyphae called *sporangiophores* have growing at their tips *sporangia* (sing., *sporangium*), or spore cases. It is within the spore case that reproductive spores form.

Some fungi also reproduce vegetatively. *Vegetative reproduction* is the process by which a new individual is produced from a part of the parent's body without involving sex cells. New individuals produced vegetatively are identical to the parent. Broken fragments of mycelia produce new fungus individuals identical to themselves.

In most fungi asexual reproduction (vegetative and sporulation) is augmented by a sexual cycle. The sexual phase begins with the fusion of *gametangia* (sing., *gametangium*), special gamete-producing structures. This is followed by a fusion of special nuclei or gametes resulting in the formation of a zygote.

Major Divisions of Fungi

Mycota is the only phylum in the kingdom Fungi. It is divided into six divisions: Chytridiomycota, Oomycota, Ascomycota, Zygomycota, Basidiomycota, and Deuteromycota.

DIVISION CHYTRIDIOMYCOTA: CHYTRIDS

Until recently, this group of fungi was known as class Protomycota and classified with the fungus-like protists. Newer methods in biochemical research have revealed, however, that the chytrids show an evolutionary relationship between the protists and the fungi. The chytrids have some enzymes and metabolic pathways that are present only in fungi and are not found in fungus-like protists.

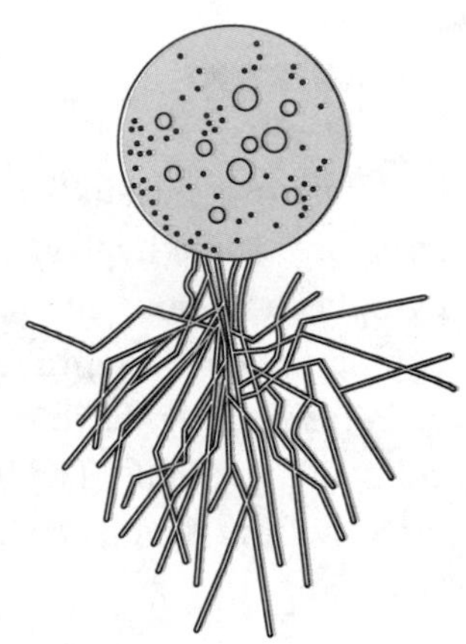

FIGURE 8.6 *Chytridium* (chytrid). The branched hyphae absorb nutrients from the surrounding aquatic environment.

The chytrids are small, water-dwelling, unicellular organisms not readily visible. Some chytrids live in association with a host organism or cell; others are saprobes, obtaining nutrition by absorbing dead organic matter. One form of chytrid is a haploid cell that lives within another cell. At a particular time, the nucleus goes through a series of mitotic divisions without division of the cytoplasm. When the nuclear divisions are completed, a little of the cytoplasm surrounds each nucleus. Each new cell develops a flagellum, and these motile zoospores are released into the surrounding water. (See Figure 8.6.)

Chytrids are the only fungi with a flagellated stage. Some of the zoospores may fuse and go through sexual reproduction. Others merely develop into vegetative cells, and then the process repeats. Significant fungal characteristics of chytrids are a saprophytic (absorptive) mode of nutrition, cell walls made of chitin, and hyphae.

DIVISION OOMYCOTA: WATER MOLDS

The Oomycota are primarily water molds, although some species live on land. This is the only fungus division in which the cell walls are made of cellulose, not of chitin, and in which the gametes are differentiated into male sperm and female egg cells. Another characteristic of the division is that the spores are flagellated and require free water for swimming.

Most of the Oomycota are *saprobes*, absorbing their nutrients from dead organic matter. Some species are parasitic and disease-producing. For example: *Albugo candida* causes white rust on cabbage and other leafy plants. *Saprolegnia*, a saprobe, grows as mold on the water-borne bodies of dead insects, fish, and frogs.

DIVISION ASCOMYCOTA: SAC FUNGI

Yeast is an example of a single-celled member of the Ascomycota. Yeast cells are small, oval structures that reproduce by budding. Most Ascomycota, however, are multicellular. This is the largest division of fungi, with about 30,000 species, including the powdery mildews, black and blue-green molds, and the truffles and morels.

The Ascomycota reproduce asexually by means of very fine spores known as *conidia*. The hyphae of the Ascomycota are divided into compartments by septa. Each compartment contains its own nucleus, but pores in the septa allow the migration of cell structures from one compartment to another.

The Ascomycota are so named because during part of the life cycle reproductive cells are held in a little sac or *ascus*. At a certain time, two hyphae grow together. Although their cytoplasm intermingles, the nuclei remain separate and do not fuse. The new hyphae that grow from this fused structure have nuclei of different genetic strains. The fused nuclei are known as a *dikaryon*. The dikaryon fuses with nonreproductive hyphae to form a fruiting body in which the asci form. At this time, within the cell that is to become the ascus, the two nuclei of the dikaryon fuse. They now go through a series of meiotic and mitotic divisions, resulting in eight haploid nuclei, which soon become surrounded by their own walls to form eight *ascopores* that disperse when the ascus ruptures.

DIVISION ZYGOMYCOTA

Most of the Zygomycota are land-dwelling organisms that inhabit the soil and carry out a saprobic way of life. Their hyphae are coenocytic and the cell walls are composed of *chitin*. Chitin is a tough, nitrogen-containing polysaccharide that is present in the exoskeletons of insects and in the cell walls of most species of fungi.

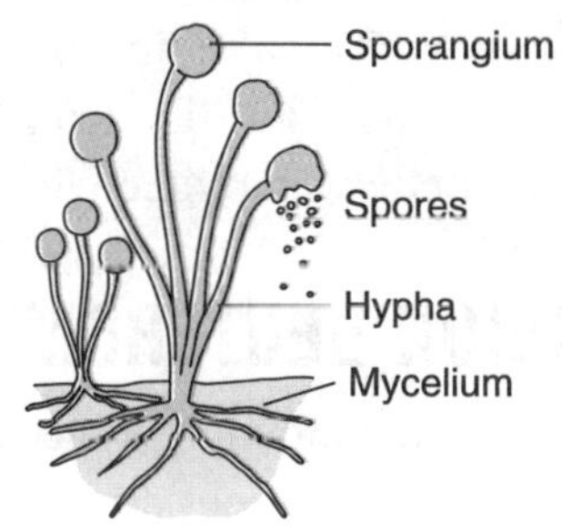

FIGURE 8.7 Bread mold and spores

The type species of this division is *Rhizopus stolonifer* (Figure 8.7), the black bread mold. The life cycle of *Rhizopus* begins with a germinating spore that is established on a favorable substrate such as a piece of bread in a moist, warm, and dark environment. Hyphae grow from the spore, eventually forming a mycelium. Some of the hyphae become specialized into rhizoids, downward growing threads that secure the mycelium to the substrate. The rhizoids have additional functions, also. They send out digestive juices that break down the large organic molecules of the substrate into smaller ones in the process of digestion. The rhizoids then perform their third function by absorbing the digested material. Other specialized hyphae grow upward. These are known as *sporangiophores* because they bear at their tips sporangia that produce haploid spores.

During part of its life cycle, *Rhizopus* goes through sexual reproduction. Hyphae of opposite mating strains from a **gametangium**, a structure that produces special nuclei that function as gametes. The gametangia of opposite mating strains fuse. Some of their nuclei pair off and fuse forming diploid nuclei. A thick wall covers the fused gametangia and their zygote nuclei to form a **zygospore**. The zygospore remains dormant for a while. When the zygospore germinates, meiosis takes place in the zygotes. An aerial hypha grows from the activated zygospore. This hypha is a sporangiophore and supports a sporangium which produces many haploid spores. The life cycle repeats.

DIVISION BASIDIOMYCOTA: CLUB FUNGI

The Basidiomycota include the mushrooms, bracket fungi, and smuts. The life cycle of the Basidiomycota begins when a haploid spore germinates, giving rise to hyphae. Hyphae of different mating types fuse forming mycelia that are dikaryotic. Most of the life history of the mushrooms and their relatives takes place in the dikaryotic stage.

The name of this division is derived from the **basidium** (club) that forms at the tip of each reproductive hypha. Inside each basidium, the two haploid nuclei fuse, forming a diploid nucleus. This nucleus undergoes meiosis, thereby producing four haploid nuclei. The haploid nuclei move to the outer edge of the basidium. Each nucleus becomes surrounded by an elongated cell wall in a structure known as the *basidiospore*. The basidiospore is maintained on a delicate stalk that separates it from the rest of the fungus. Dissemination of the spore takes place when the stalk breaks and is carried away by the wind.

DIVISION DEUTEROMYCOTA: IMPERFECT FUNGI

This division contains the species known as "Fungi Imperfecti." The species that puzzle taxonomists are put into this division. Most of these organisms have life cycles that are not typical of fungi. Since sexual reproduction has not been observed in these species, they are called imperfect fungi. The organisms that cause the human diseases of ringworm, athlete's foot and thrush are classified as Deuteromycota.

Special Nutritional Relationships

Fungi are often found associated with other species in special relationships. When two different species of organisms live together, the relationship is called **symbiosis**. If the relationship is of mutual benefit to both species, it is called **mutualism**. When one species benefits and the other does not but is not harmed by the association, the condition is known as **commensalism**. When one species lives at the expense of another, doing harm to its *host*, the relationship is **parasitism**. Disease-producing organisms are parasites.

LICHENS

Lichens are pioneer organisms that can inhabit bare rock and other uninviting substrates. They live on the barks of trees and even on stone walls. A lichen is a combination of two organisms—an alga and a fungus—that live together in a mutualistic relationship. The alga carries on photosynthesis, while the fungus absorbs water and mineral matter for its partner. The fungus also anchors the lichen to the substrate. Scientists have determined that the alga in this partnership can live alone. The fungus, which may belong to the Ascomycota, the Deuteromycota, or the Basidiomycota, cannot exist by itself and is therefore the dependent member of the team. New lichens are formed by the capture of an alga by a fungus. If the fungus kills the alga, the fungus also dies.

MYCORRHIZA

Several fungus species, including mushrooms, live in close association with plant roots. The mushroom absorbs minerals from the soil and passes them along to the plant on which it lives. This association is known as *mycorrhiza*, meaning "fungus root." Scientists are not sure how the plant helps the fungus, but it is known that most plant families grow better when in association with fungi.

Importance of Fungi to Humans

From the viewpoint of environment and ecology, fungi help to keep the natural environment in balance. Species dependent on dead organic matter as their source of nutrition assist with the breaking down of fallen leaves, the dead bodies of plants and animals, and animal wastes. Fungi have the ability to absorb moisture from the air, a characteristic that permits them to live in environments that do not have adequate soil water as required by other species.

Most fungi are not pathogenic to humans. However, some nonparasitic fungi do work against human interests. There are fungus species that live quite well digesting such unlikely substrates as the insulation on telephone wires, leather, polyvinyl plastics, cork, hair, and wax. Other species cause the mildew of clothing, wallpaper, and books. Wood-rotting fungi destroy the wood construction in houses and ships.

Parasitic fungi invade plants more readily than animal bodies. Diseases of food crops can have disastrous effects on human populations. The oomycete *Phytophthora infestans* caused the potato blight in Ireland, resulting in a devastating famine between the years 1845 and 1851, and *Plasmopara viticola*, the cause of downy mildew of grapes, nearly ruined the wine-making industry of France. Ergot, a disease of rye, is caused by the ascomycete *Claviceps purpurea*. Ergotism, the disease that affects humans who eat the infected rye, induces gangrene, nervous spasms, convulsions, and psychotic delusions. Ergot is the source of the hallucinogen LSD. Other plant diseases caused by Ascomycota are peach leaf curl, Dutch elm disease, chestnut blight, and apple scab. Chestnut tree blight was first noticed in 1904 by Herman W. Merkel of the New York Zoological Park in the Bronx, New York. Continuing into the present, generations of research mycologists (scientists who specialize in the study of fungi) have been trying to find ways to inhibit the growth of *Cryphonectria parasitica*. This pathogenic fungus has invaded American chestnut trees, killing off vast populations until just a few of these once-mighty trees remain. The eastern part of the United States is the natural range for these trees. In 2004, the DNA genome in the parasitic fungus was published. Research scientists are at work trying to find ways to decrease the virulence (disease-producing strength) of *C. parasitica*.

Some Ascomycota, however, are beneficial to humans. These include the truffles and the morels, edible fruiting bodies, that can cost anywhere from $250 to $1,400 a pound in New York. *Saccharomyces cerevisiae* is the species of yeast needed in the fermenting processes of malt, barley, and hops to make beer. The fermenting of grapes produces wine. Another species of yeast is used in making dough.

The Fungi Imperfecti include *Penicillium notatum*, the source of the antibiotic penicillin, as well as the species that help to ripen Roquefort and Camembert cheeses. Soy sauce is prepared by fermenting soybeans with the mold *Aspergillus oryzae*.

REVIEW EXERCISES FOR CHAPTER 8

WORD-STUDY CONNECTION

ascospore
ascus
basidiospore
basidium
blight
chitin
coenocytic
commensalism
conidia
dikaryon
gametangium
germinate
haustoria
host
hypha
lichen
mutualism
mycelium
mycorrhiza
parasitism
rhizoid
saphrophyte
saprobe
septa
sporangiophores
sporangium
spore
symbiosis
vegetative reproduction
zygospore

SELF-TEST CONNECTION

PART A. Completion. *Write in the word that correctly completes each statement.*

1. A cell that has several nuclei in a single mass of protoplasm is called a _____.
2. Fungi absorb dissolved food molecules from dead organic matter and thus are classified as _____.
3. Hydrolytic enzymes are the same as _____ enyzymes.
4. Vegetative reproduction does not involve _____ cells.
5. Conidia are cells known more commonly as _____.
6. An example of a unicellular ascomycete is _____.
7. Rhizoids secrete substances that serve the purpose of _____.
8. The combination of an alga and a fungus living in a mutualistic relationship is called a _____.
9. Ringworm is a _____ infection and not one caused by worms.
10. A cell that has two nuclei originating from different genetic strains is known as a _____.

PART B. Multiple Choice. *Circle the letter of the item that correctly completes each statement.*

1. A saprobe is an organism that
   (a) absorbs material from living cells
   (b) absorbs material from dead cells
   (c) ingests dead organic matter
   (d) ingests living organic matter
2. Septa are
   (a) compartments
   (b) double nuclei
   (c) dividing walls
   (d) masses of protoplasm

3. Gametangia are associated with
   (a) sporulation
   (b) vegetative reproduction
   (c) germination
   (d) sexual reproduction

4. The only fungus division in which the gametes are differentiated into sperm and egg is the
   (a) Oomycota
   (b) Zygomycota
   (c) Ascomycota
   (d) Basidiomycota

5. A dikaryon is a cell that has
   (a) a single nucleus and lots of cytoplasm
   (b) perforated cell membranes
   (c) two nuclei of different genetic strains
   (d) a micronucleus and a macronucleus

6. Zygospores are characteristic of the fungus division
   (a) Oomycota
   (b) Fungi Imperfecti
   (c) Basidiomycota
   (d) Zygomycota

7. Black bread mold is the type species for the fungus division
   (a) Oomycota
   (b) Basidiomycota
   (c) Deuteromycota
   (d) Zygomycota

8. Each class in the kingdom Fungi uses special methods to produce
   (a) hyphae
   (b) rhizoids
   (c) mycelia
   (d) spores

9. Fungi Imperfecti are so named because
   (a) sexual reproduction has not been observed in them
   (b) they cause human disease
   (c) many of their hyphae are missing
   (d) they do not reproduce by spores

10. The disease of rye which is devastating to humans is known as
    (a) morel
    (b) truffle
    (c) ergot
    (d) ringworm

PART C. Modified True-False. *If a statement is correct, write "true" for your answer. If a statement is incorrect, change the underlined expression to one that will make the statement true.*

1. Most of the fungi begin life as zygotes.
2. To germinate means to contaminate.
3. Flagellated swimming spores are characteristic of the fungus division Zygomycota.
4. A fungal disease of the American chestnut tree is commonly called rust.
5. A nitrogen-containing polysaccharide that composes the cell walls of most fungi is cellulose.
6. *Rhizopus stolonifer* is the scientific name for blue-green mold.
7. When two gametes fuse, a zygospore is formed.
8. An example of the division Basidiomycota is water mold.
9. The word *basidium* means mold.
10. Pioneer organisms that can inhabit bare rocks are lichens.

CONNECTING TO CONCEPTS

1. Why do fungi have to live near their food supply to stay alive?
2. What role is played by spores in the lives of fungi?
3. How do the water molds differ from other fungi?
4. Of what commercial value is yeast?

ANSWERS TO SELF-TEST CONNECTION

PART A

1. coenocyte
2. aprobes
3. digestive
4. sex
5. spores
6. yeast
7. digestion
8. lichen
9. fungus
10. dikaryon

PART B

1. **(b)**
2. **(c)**
3. **(d)**
4. **(a)**
5. **(c)**
6. **(d)**
7. **(d)**
8. **(d)**
9. **(a)**
10. **(c)**

PART C

1. spores
2. sprout
3. Oomycota
4. blight
5. chitin
6. black bread mold
7. zygote
8. Oomycota
9. club
10. true

CONNECTING TO LIFE/JOB SKILLS

The study of the fungus pathogen that causes chestnut tree blight was begun in 1904 and continues today, representing nearly 100 years of research. You can sharpen your research techniques by finding out what field scientists and laboratory investigations have revealed about the life history of *Cryphonectria parasitica*. A good place to start is the Internet at www.sciam.com, the *Scientific American* website.

Chronology of Famous Names in Biology

1904 **Herman W. Merkel** (United States)—was the first to notice the disease (blight) affecting American chestnut trees.

1906 **William A. Murrill** (United States)—gave the fungus infecting the American chestnut tree the name *Diaporthe parasitica*. This pathogenic fungus has been renamed *Cryphonectria parasitica*.

1929 **Alexander Fleming** (England)—discovered the antibiotic properties of the fungus *Penicillium notatus*.

1943 **Albert Hoffman** (Switzerland)—isolated LSD from ergot and discovered its powerful hallucinogenic properties.

1950 **Antonio Biraghi** (Italy)—discovered a fungal pathogen of the European chestnut tree that had lost its virulence. This discovery led to genetic research of chestnut tree blight.

1964 **Jean Grente** (France)—isolated the hypovirulent strain of the pathogenic fungus in the European chestnut tree.

1975 **Richard Lister** and **Eileen Moffitt** (United States)—discovered that the hypovirulent and virulent strains of *C. parasitica* show differences in their DNA.

1981 **Joseph R. Newhouse** (United States)—developed a technique for preparing the fungus *C. parasitica* for study via the electron microscope.

9

The Green Plants

WHAT YOU WILL LEARN

In this chapter you will be taken on a journey through the kingdom of the green plants. Some green plants are very small, consisting of a single cell. Other members of the Plantae, such as the giant sequoias in California, are enormous. You will learn why organisms of such diversity are classified in one kingdom.

SECTIONS IN THIS CHAPTER

- The Plant Kingdom
- Parts of Higher Plants
- Photosynthesis
- Plant Hormones
- Photoperiodicity
- Review Exercises for Chapter 9
- Connecting to Life/Job Skills
- Chronology of Famous Names in Biology

The Plant Kingdom

OVERVIEW

Taxonomists group all green plants in the **kingdom Plantae**. This kingdom consists of some single-celled species that include free-living cells, cells that live together in colonies, and some cells that adhere together in long filaments. However, most members of the Plantae are multicellular. Although *multicellular* means "having many cells," the concept of multicellularity involves more than numbers of cells. It embraces several ideas about cell structure and function. One such idea concerns *cell specialization* in which cells are "programmed" to carry out special tasks. Cell specialization in multicellular plants also brings with it a **division of labor** in which groups of cells in tissue formation work together to perform some special life function of benefit to the entire plant organism. Another condition necessary to all specialization is the evolution of cell structures that permit specialization and difference in cell function.

As implied in the term *green plants*, members of the kingdom Plantae contain the green pigment **chlorophyll**. Not only does chlorophyll color plant leaves and some stems green, it, more importantly, traps light energy that is used in the process of photosynthesis. As an outcome of photosynthesis nutrient molecules are made, serving as food for both plants and animals.

Most species of the kingdom Plantae are nonmotile, anchored to one place, and unable to move about, but a few of the lower plants are motile for at least part of the life cycle. However, the evolutionary trend exhibited in green plants is toward stationary organisms that carry out their life functions on land in locations where they remain for life. Lower plant species equipped to swim about live in salt and fresh water. Higher plants are *terrestrial* (land-dwelling). Another evolutionary trend exhibited by plants involves their reproductive processes and is called "ascendancy of the sporophyte generation." All plants reproduce by cycling through phases called gametophyte and sporophyte generations. During this alternation of generations, sporophyte cells go through meiosis to produce haploid spores, containing unpaired chromosomes, which are often dispersed by animals, wind, and weather. The spore undergoes mitosis to become a haploid gametophyte and may ultimately become a gamete (egg or sperm cell). These gametes unite to become a diploid zygote, with paired chromosomes, which will mature into a sporophyte plant, and then the process is repeated. Simpler, more primitive, plants have a dominant gametophyte generation, while higher plants maintain a dominant sporophyte generation.

LOWER PLANTS

The lower plants are not differentiated into roots, stems, and leaves. The simple body of lower plants is either a single cell or a flat sheet of simple cells. Lower plants do not have specialized tissues to carry water, to anchor the plants, or to grow new cells. As a rule, the sex cells of the lower plants are produced in rather simple sex organs that are not protected by a surrounding wall of cells. The zygotes do not develop into embryos that are contained in a female reproductive organ. Most of these plants live in fresh water, although there are a few saltwater forms. Some species live in damp soil or on the bark of trees.

The green algae, the brown algae, and the red algae comprise the three divisions of lower plants.

CHLOROPHYTA—GREEN ALGAE

Green algae belong to the **Chlorophyta**. The scientific name tells that these algae contain the green pigment chlorophyll. Of interest to botanists is the fact that members of the Chlorophyta are considered to be the ancestors of the higher land plants based on three points of evidence. First of all, green algae have chlorophyll *a* and *b* in the same amounts as cells in higher plants. Second, green algae store food in the same form as do the higher plants. Third, green algae, like the higher plants, have well-defined cell walls made of cellulose.

The Chlorophyta includes forms such as *Chlamydomonas* that typically live as single cells, simple and complex colonial forms, and forms like sea lettuce that typically undergo a life cycle known as alternation of generations (see page 181).

Chlamydomonas

Chlamydomonas is a unicellular autotroph that is a representative type of green algae. In some ways it resembles cells of the Protist Kingdom, because it lives in water, is motile, and has a pair of flagella (Figure 9.1). However, biochemical analysis of its chlorophylls, carotenoids, and stored food reveals that these compounds are identical to those in the cells of green land plants. Scientists believe that *Chlamydomonas* represents the general type of unicellular algae from which multicellular green plants evolved.

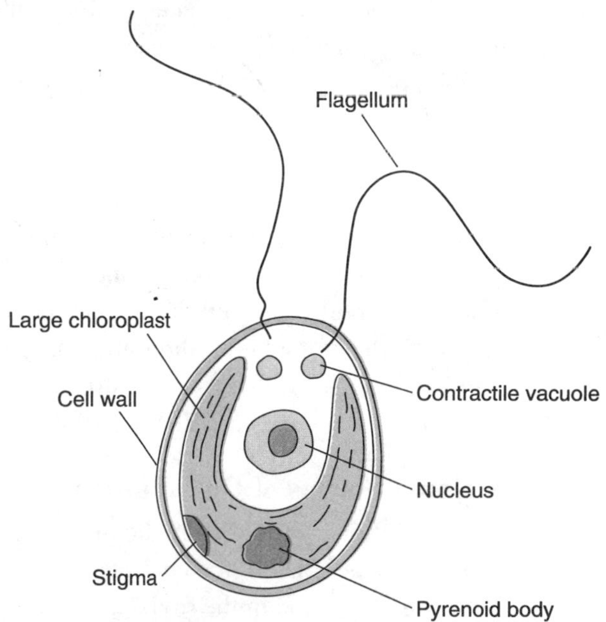

FIGURE 9.1 *Chlamydomonas*

Chlamydomonas is an oval-shaped cell with a haploid nucleus. Although its pigments and starch granules are identical with those of higher plants, its cell wall, unlike that of other green algae, is composed of glycoprotein, not cellulose. Sticking out from the anterior end of the organism are two flagella of equal length. The large cup-shaped chloroplast fills up nearly the entire volume of the cell. Inside the chloroplast there are membraneous sacs that contain the chlorophyll. At the bottom portion of the chloroplast is a rather large, circular **pyrenoid body** that functions in the synthesis and storage of starch. The **stigma**, a light sensitive eyespot, is located inside the chloroplast. Two contractile vacuoles that open and close alternately lie at the base of the flagella.

Before the onset of cell division, certain changes take place in the vegetative *Chlamydomonas* cell. The organism attaches itself to some object in the water and both of the flagella are resorbed into the cytoplasm. The once motile cell is now sessile. Mitosis and cytokinesis take place resulting in two identical daughter cells. These cells are retained within the membrane of the parent cell and quickly divide once more. Four identical flagellated *zoospores* are now released from the confines of the old cell wall. These cells grow and mature into vegetative *Chlamydomonas* organisms. This form of reproduction is asexual, and it is the usual mode of reproduction in *Chlamydomonas*.

Under unfavorable conditions (low nitrogen content in the water, for example), *Chlamydomonas* undergoes a primitive form of sexual reproduction. The vegetative *Chlamydomonas* cells go through several mitotic divisions, releasing several small flagellated cells into the water. These cells are known as *isogametes*, gametes that are indistinguishable from each other. The isogametes pair off, attached to each other end to end by the flagella. The cell wall slips away from each cell. Their cytoplasms fuse, forming a diploid zygote. A thick protective cell wall surrounds the zygote before it falls to the bottom of the pond. The zygote remains inactive and survives such an extreme condition as the drying of the pond. When environmental conditions improve, the zygote goes through a meiotic division forming zoospores that mature into haploid vegetative *Chlamydomonas* cells (Figure 9.2).

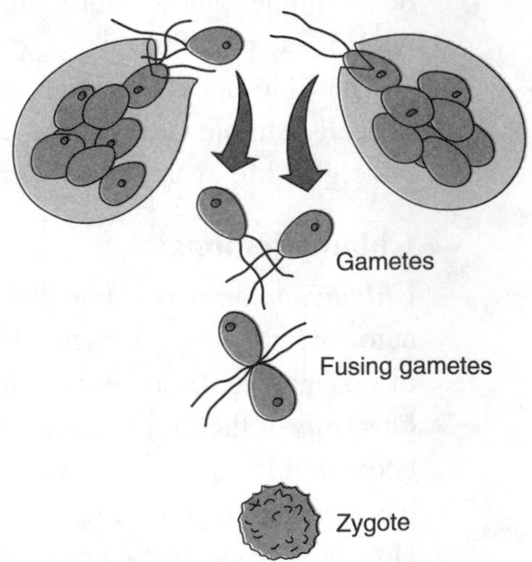

FIGURE 9.2 Gamete formation in Chlamydomonas

Colonial Forms of Chlorophyta

The evolutionary trend from the unicellular *Chlamydomonas* to cells that can survive only when associated in a colony can be traced quite efficiently in the Chlorophyta. ***Gonium*** is a genus in the division Chlorophyta. Each *Gonium* species lives as a simple colony of cells. The cells of *Gonium* look very much like the single cell of *Chlamydomonas*. However, instead of living singly, they are held together in a gelatinous disk. The colony swims as a unit and the cells divide at the same time. Sexual reproduction in *Gonium* follows the pattern set by *Chlamydomonas* in that isogametes fuse to form a zygote.

Pandorina, another colonial form of green algae, shows a bit more complexity than *Gonium*. In *Pandorina*, the colony has an anterior end and a posterior end as shown by the orientation in swimming. Cells in the *Pandorina* colony cannot live alone and the colony dies if broken apart. The male gametes of *Pandorina* are smaller than the female gametes. Since the gametes can be distinguished, this type of sexual reproduction is known as **heterogamy**. Since their only distinguishing characteristic is size, they are known as *anisogametes* (Figure 9.3).

Further colonial development is shown in the genus ***Eudorina***. The cells of the anterior portion are more dominant than those of the posterior region, and the nonmotile female gametes are fertilized by the smaller, free-swimming male gametes inside the colony. The condition in which the sperm cell is motile and the egg cell is nonflagellated and nonmotile is called **oogamy**.

Colony development in ***Volvox*** is significantly more advanced than in any of the other green algae. Colonies in the Volvox genus are large, consisting of 500 to 50,000 cells. Cytoplasmic strands between the cells permit communication. Most of the cells are vegetative and do not function in reproduction. Scattered in the posterior half of the colony are a few large cells which are specialized for reproduction. Among these colonial forms we see an increase in the number of cells that make up the colonies, in the communication among cells, and in the specialization and differentiation of cells.

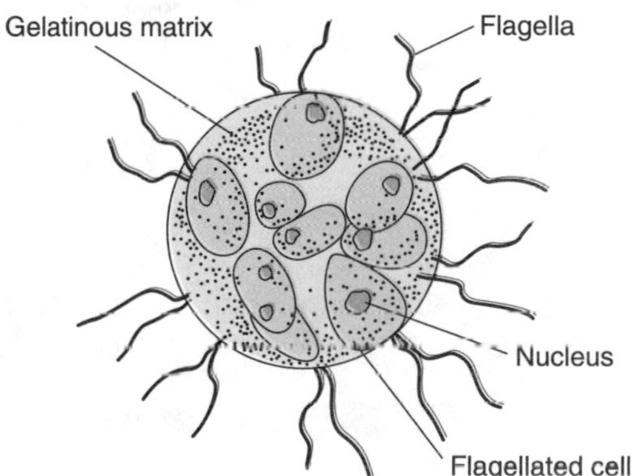

FIGURE 9.3 *Pandorina*. The cells are confined in an intercellular bed or matrix of gelatinous material.

Alternation of Generations

The sea lettuce ***Ulva*** is a green algae that lives in salt water. Its body has the form of a flat, leafy thallus and is two cell layers thick (Figure 9.4). The life cycle of ***Ulva*** is described as **alternation of generations** because one generation of *Ulva* is produced sexually by gametes while the next generation is produced asexually by zoospores. To the naked eye, there is no difference in the structures of the sea lettuce produced sexually and asexually.

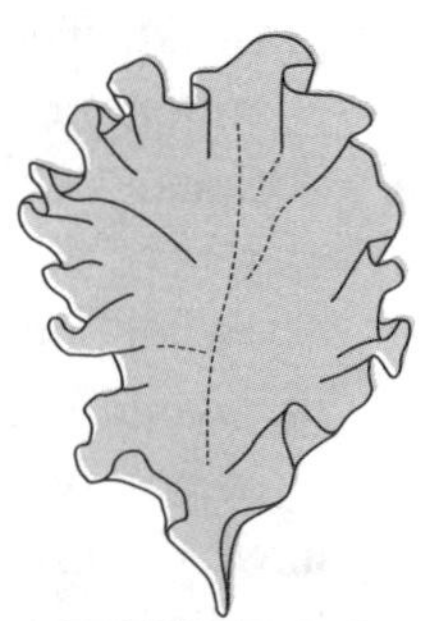

FIGURE 9.4
Sea lettuce

The **gametophyte generation** is a haploid thallus from which small, flagellated gametes are released into the water. A *thallus* is a flat sheet of photosynthetic cells. They pair off and fuse. Each fused pair of gametes forms a zygote that, after a short time, becomes a diploid thallus of a new generation of *Ulva*.

The diploid thallus is the **sporophyte generation**. It produces haploid zoospores that develop and grow into a haploid thallus, which is now the gametophyte (Figure 9.5).

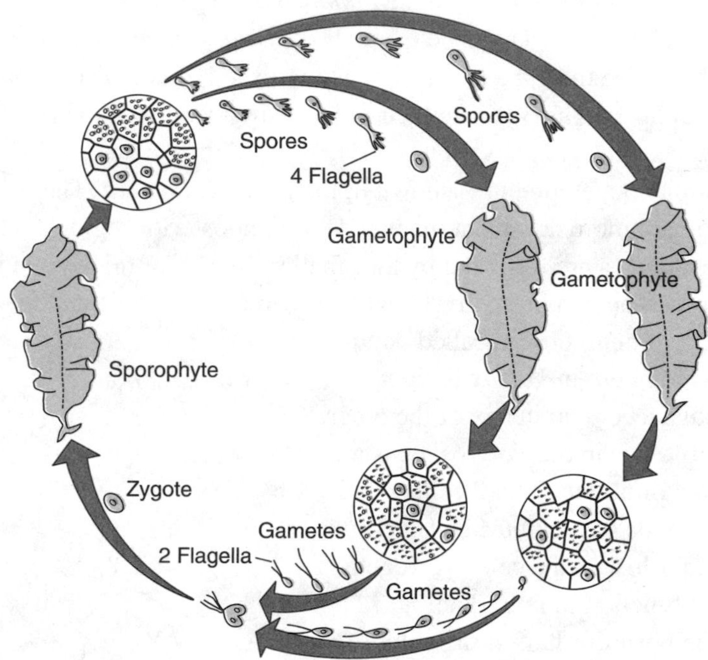

FIGURE 9.5 Alternation of generations in sea lettuce (*Ulva*)

PHAEOPHYTA—BROWN ALGAE

Most species of brown algae are marine, inhabiting the cooler ocean waters. They grow attached to rocks along the shallow waters of seacoasts. The brown algae are commonly called seaweed.

Species of the **Phaeophyta** are the largest of the algae and may reach lengths of 45 meters or more. **Kelps** are massive brown algae found usually on the Pacific Coast. They contain chlorophylls *a* and *c*, but lack chlorophyll *b*. The brownish pigment *fucoxanthin* gives them their characteristic color. In structure, the brown algae are quite complex, showing considerable differentiation among the cells. Most species have *holdfasts* that secure them to rock substrates. Some species have stem-like and leaf-like parts. Many species have **air bladders**, which give them buoyancy.

Fucus is a representative species of the Phaeophyta. Figure 9.6 shows its general structure, including the very prominent air bladders. *Fucus*, also known as bladder wrack and rock weed, lives on the rocky sea shores of temperate seas. The thallus of *Fucus* is unusual in that it has repeated double branches, with enlarged tips commonly called *conceptacles*. These conceptacles hold the sex organs. The **antheridia** contain the sperm, while the *oogonia* hold the egg cells. In some species the male and female sex organs are on separate plants; in other species they are produced in the same conceptacle. Gametes are released into the water through pores at the tips of the conceptacles. An egg cell fuses with a sperm cell to form a zygote. The zygote grows into a diploid plant. The haploid gametophyte stage is completely missing in *Fucus*.

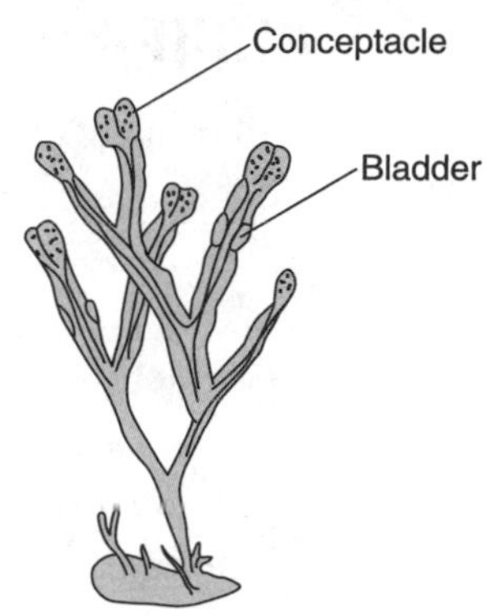

FIGURE 9.6 *Fucus*, the rockweed

RHODOPHYTA—RED ALGAE

The red algae are seaweeds inhabiting the warmer waters of the oceans and living at considerably greater depths than the brown algae. Very few species grow in fresh water. The leaf-like structures of a red alga are branched and filamentous, appearing rather feathery. Some species resemble gelatinous ribbons.

The life cycles of most species of **Rhodophyta** are quite complex. In general, they go through some sort of alternation of generations. During the asexual phase, the red algae reproduce by nonmotile spores. The heterogametes involved in the sexual generation are nonmotile also. The male gametes are carried by currents of water to the female sex organs known as **carpogonia**. For a short while, the sperm cell sticks to the elongated tip of the carpogonium. Then the sperm cell nucleus passes through the wall of the elongated tip and into the inner region of the carpogonium. There, the sperm nucleus fuses with the nucleus of the egg cell.

Some representative genera of red algae are ***Chondrus*** (also called Irish "moss"), ***Polysiphonia***, and ***Nemalion***. The cell walls of the red algae are made of a combination of cellulose and a gelatinous material. The reserve carbohydrate stored in the cells of red algae is known as *floridean starch*; it is not a true starch.

The Rhodophyta contain both chlorophyll *a* and chlorophyll *d*. The latter is a type of chlorophyll not found in any other species of plants. The characteristic color of the red algae is given by the pigment *phycoerythrin*.

The value of accessory pigments to the photosynthetic process is demonstrated by the red algae. Chlorophyll *a* is the pigment actively involved in trapping light energy for use in photosynthesis. However, chlorophyll a cannot trap light energy at the depths at which the red algae grow. The pigments *phycocyanins* and phycoerythrins can and they pass the energy along to chlorophyll *a*.

Red algae have a great deal of commercial value. They are dried and ground up to be used as agar for bacterial media. They also form colloids that are used as suspending materials in ice cream and binders in puddings and chocolate milk.

HIGHER PLANTS

Higher plants have developed a number of adaptations for life on land—structures and biochemical methods for conserving water that is necessary for all of the biochemical activities of the cell and thus for the maintenance of life. Chief among these adaptations was the evolution of sex organs that are protected by a surrounding layer of nonreproductive cells. **Antheridia** are male sex organs where sperm cells are produced. Egg cells are contained in organs called **archegonia**. Gametes enclosed in these organs are protected against drying out. Another adaptation for the prevention of water loss in the embryophytes is the cutin covering, a waxy substance impregnated in cell walls, which provides waterproofing to epidermal tissue and prevents evaporation of water. The higher plants also have special *vascular* or water-carrying tissues designed to distribute water efficiently throughout the plant body.

BRYOPHYTA—MOSSES, LIVERWORTS, AND HORNWORTS

The **bryophytes** are the first green land plants. They are primitive, small, and inconspicuous. Although multicellular, the tissue differentiation is quite simple. Bryophyte species have no tissues that are specialized for water-carrying and no cambium specialized for growing new cells. Bryophyte species do not have true stems, leaves, or roots. Simple root-like structures called *rhizoids* anchor the plants to the ground and absorb moisture from the soil.

Alternation of generations occurs in all bryophytes. The larger and more noticeable generation is the gametophyte, which usually supports a smaller (sometimes parasitic) sporophyte. Ciliated sperm cells, produced in the antheridia, swim to the archegonia and fertilize the egg cells held therein. The sporophyte generation begins with the zygote and is therefore diploid. Special cells in the sporophyte go through meiosis and produce haploid spores that begin the gametophyte generation.

There are 23,000 species of bryophytes among which are the mosses (Figure 9.7), the liverworts (Figure 9.8), and the hornworts, a group of plants structurally similar to the liverworts.

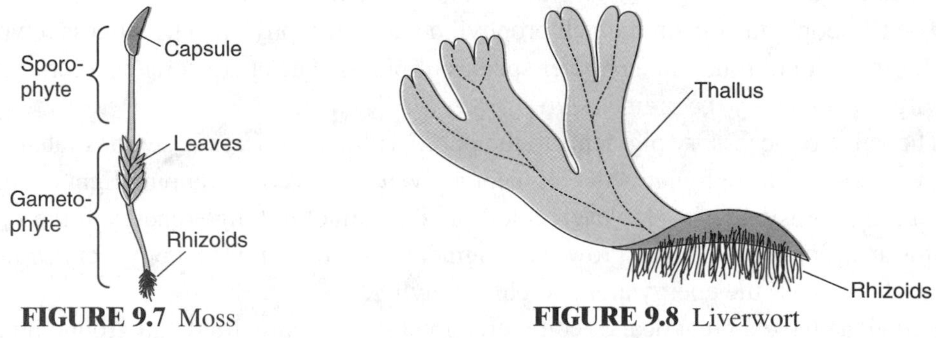

FIGURE 9.7 Moss

FIGURE 9.8 Liverwort

TRACHEOPHYTA—VASCULAR PLANTS

The **vascular** plants are truly land-dwelling plants. They have developed adaptations that permit them to live on land independent of bodies of water. The word *vascular* means that these plants have a water-carrying system. Water is conducted upward from the roots by **xylem** tubules. Fluid compounds are conducted downward from the leaves to lower plant organs by the **phloem** tubules.

The tracheophytes are divided into five subdivisions: psilopsids, club mosses, horsetails, ferns, and seed plants.

Psilopsida

There is disagreement among botanists as to whether there are two living genera—*Psilotum* and *Tmesiperteris*—of psilopsids or whether all members of this subdivision are extinct. The plants that may represent the psilopsids have rather simple bodies, with branched stems, but no roots. The leaves are absent or very small.

Lycopsida—Club Mosses

There are about 900 living species of club mosses. Plants in this group are usually one meter or less in height. Many are ground creepers. The club mosses have water-carrying (vascular) tissues and true roots, stems and leaves. The leaves are small and spirally arranged on the stems. The popular name of the Lycopsida is derived from the arrangement of the sporangia which are clustered on leaves formed into *cones* or *strobili*. The cones are positioned on the tips of stems as shown in Figure 9.9. Some club moss species bear one type of spore and one type of gametophyte; such species are described as being *homosporous*. Other species of club moss are *heterosporous* because they bear two types of spores and produce two types of gametophytes.

Sphenopsida—Horsetails

Only 25 living species of Sphenopsida remain. The horsetails are true land plants having a vascular system, true roots, stems, and leaves. Although the leaves are small and scale-like, they carry on photosynthesis. Horsetails grow well in both tropical and temperate climates, along river banks and in moist tracts of land (Figure 9.10).

The life cycle of the horsetails involves alternation of generations. The sporophyte generation is the conspicuous generation. Its haploid spores give rise to a small plant known as a **prothallus**. The prothallus represents the gametophyte generation. The zygote remains attached to the gametophyte and ultimately becomes the sporophyte.

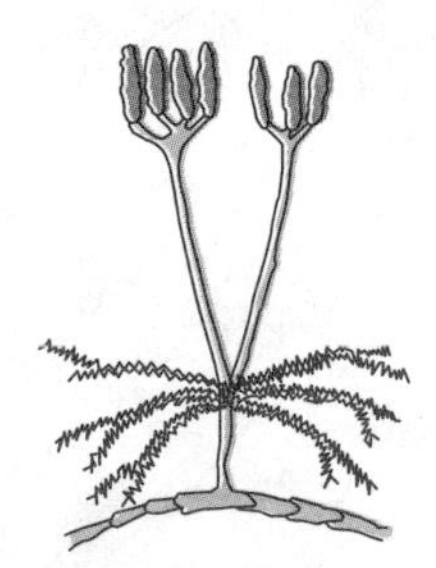

FIGURE 9.9 Club moss—*Lycopodium*

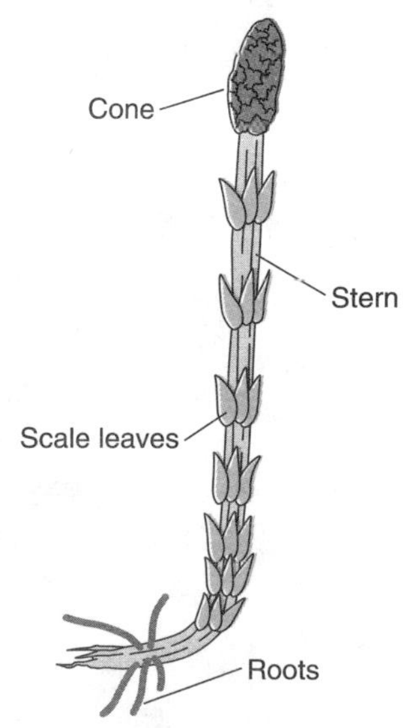

FIGURE 9.10 The horsetail—*Equisetum*

The cell walls in the leaves and stems of horsetails contain silica (sand), making the dried, ground stems useful as scouring powders.

Pteropsida—Ferns

The fern plant used in flower bouquets is the sporophyte generation. Remember that the sporophyte generation produces asexual spores. The mature fern has true roots, leaves, and stem. Ferns growing in temperate climates have an underground stem called a **rhizome**, which grows in a horizontal position. The rhizome not only stores food materials, but also gives rise to new fern plants that grow along its length. The stems of tropical species grow upright in a vertical position and serve as trunks of tree ferns.

Fern stems are composed of xylem and phloem vascular tissues but they do not have the growth layer of embryonic cells known as **cambium**. Growing from the lower side of the rhizome are fibrous roots, which absorb water and dissolved minerals from the soil. The leaves grow out from the top of the rhizome and break through the ground. Ferns have compound leaves composed of many leaflets.

The leaves of ferns serve a dual purpose. First, the green pigment chlorophyll enables them to carry on photosynthesis to produce food for the plant. Second, the undersides of certain green leaves are covered with small structures that resemble brown dots. Each dot, called a *sorus*, is really a cluster of spore cases (*sporangia*) (Figure 9.11). Inside the sporangia, haploid spores are produced. When the spores are ripe, they are discharged from the spore cases and carried by the wind to new soil habitats. A germinating spore grows into a small, inconspicuous, heart-shaped gametophyte plant. The fertilized egg cell becomes a zygote that develops into a young sporophyte plant that is nourished by the gametophyte. Many botanists believe that seed plants evolved from the ferns.

FIGURE 9.11 Fern leaf showing spore cases

Spermopsida—Seed Plants

The seed plants are divided into two groups: the *gymnosperms* and the *angiosperms*. The gymnosperms are known as the naked seed plants, while the angiosperms are referred to as the covered seed plants. Seed plants are the most successful land plants that have ever lived because they have reproductive mechanisms that do not require water. The plants that evolved before the seed plants produce sperm cells that can reach the egg cells only by swimming through a film of water.

Angiosperms are the most recently evolved major group of plants, dating from the early Cretaceous period, about 65 million years ago. Their success is measured in terms of the increase in number of species and their emergence as the dominant groups of plants, inhabiting varying environments throughout the world.

Gymnosperms Gymnosperms are cone-bearers. They are woody plants, chiefly evergreens, with needle-like or scale-like leaves. Cone-bearing plants grow in many parts of the world including tropical climates. However, most species are found in the cooler parts of temperate regions. Examples of gymnosperm species are pines, spruces, firs, cedars, yews, California redwoods, bald cypresses, and Douglas firs. Most biologists do not think that the gymnosperms evolved directly from ferns.

Reproduction in gymnosperms takes place on special structures called *cones*. The cones are specialized nongreen leaves where seeds are produced. Most gymnosperm species bear two different kinds of cones: seed cones, called *megasporophylls*, and pollen cones, called *microsporophylls*.

Seed cones are large and their woody "leaves" (sporophylls) bear on them an *ovule* (called also a *megasporangium*). Inside of the ovule is a **megaspore mother cell**. This mother cell goes through a meiotic division producing both an egg cell and a food storage cell.

Pollen cones are smaller in size than the seed cones. Their leaves are known as *microsporophylls*. These microsporophylls are really *stamens*, reproductive structures that produce pollen. Inside of each pollen grain is a *microspore* cell. When the pollen grain germinates, the microspore goes through a reduction division producing two sperm cells.

Pollen grains are carried by the wind from the pollen cones to the seed cones. When a pollen grain lands on an ovule, it begins to germinate (sprout). The germinating pollen grain commences to grow a pollen tube during its first summer and the following spring. As the pollen tube grows, changes take place in the "leaves" of the seed cone where the ovules are pollinated. These leaves grow together holding the ovules tightly inside of the cone. The pollen tube enters a pore (**micropyle**) in the ovule. The two sperm nuclei from the pollen grain travel through the pollen tube into the ovule. One sperm cell fertilizes the egg cell. The other sperm cell and the pollen tube cells disintegrate. The fertilized egg becomes a zygote and then develops into the embryo plant. Changes occur in the outer walls of the ovule where a tough seed coat develops.

The ripened ovule is now a **seed** containing both an embryo plant and food for the embryo plant. The stored food known as **endosperm** is derived from the female gametophyte. The embryo plant represents the new sporophyte generation. Seeds are carried to new locations by wind, water or animals. The embryo plant inside of a seed can live for long periods in a resting state. The outer coat of the seed protects the embryo from extremes of temperature, drying, chemical corrosion and even burning. The germinating pollen grain and the developing ovule represent the gametophyte generation. The large conspicuous plant is the sporophyte generation.

Angiosperms (Anthophyta)—Flowering Plants The reproductive structure of the angiosperm is the flower, which encloses the male and female sex organs. Figure 9.12 shows the parts of the flower. On the outside are the green, leaf-like *sepals*. The sepals protect the flower when in the bud stage. Collectively, sepals are known as the calyx. Just inside the calyx are colored *petals*, showy and conspicuous in flowers that are pollinated by insects or birds. All of the petals in a flower are known as the *corolla*. Look at Figure 9.12 below and locate the stamens. Notice that the top portion is called the **anther** and the stalk, the **filament**. The stamens are the male reproductive structures; pollen grains are produced in the anther. Locate the *pistil*, which is in the center of the flower. The top portion of the pistil is the *stigma*. The **style** is the long stalk that leads to the rounded portion of the pistil called the *ovary*. Inside the ovary are many *ovules*. The pistil and its many parts compose the female portion of the flower. The pistil is really a modified sporophyll which in the flowering plant is called the **carpel**.

When an ovule is ready for fertilization, its megaspore mother cell goes through a reduction division and two mitotic divisions yielding eight nuclei. One of these nuclei forms the *embryo sac*. A nucleus near the ovule micropyle (pore) becomes the egg cell (Figure 9.13).

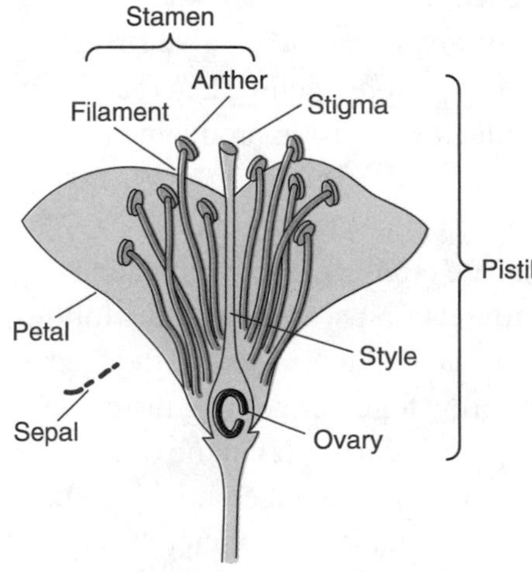

FIGURE 9.12 Parts of a flower

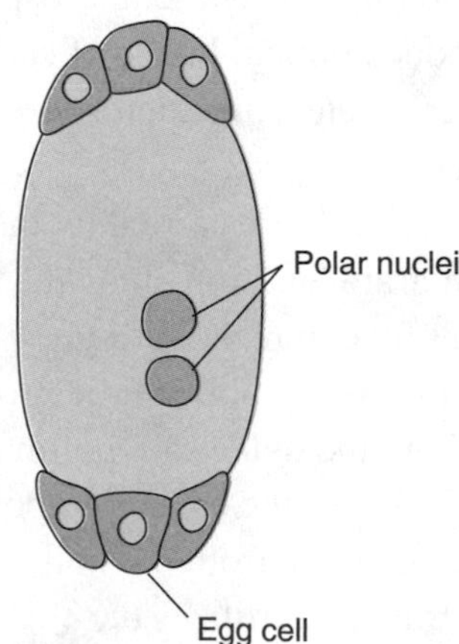

FIGURE 9.13 Maturation of the ovule

Each pollen mother cell in the anther divides by meiosis and produces four pollen grains, each with a haploid nucleus. When a pollen grain lands on the stigma, its cell divides by mitosis to form two nuclei. One, the tube nucleus, directs the growth of the pollen tube down to the micropyle in an ovule. The other nucleus (*generative nucleus*) divides and forms two sperm nuclei (Figure 9.14).

The sperm nuclei enter the ovule through the micropyle. One sperm fertilizes the egg cell and forms a diploid zygote. The other sperm cell fertilizes two polar nuclei in the embryo sac to form a triploid endosperm nucleus (Figure 9.15). The behavior of both sperm cells is described as a *double fertilization*. The zygote then undergoes a number of changes and develops into the embryo plant. The ovule coats increase in number and harden, changing into seed coats. After fertilization in the ovules, the ovary enlarges and usually becomes a **fruit**. The ovules enlarge, change shape and form *seeds*, each containing an embryo plant and stored food. All accessory nuclei disintegrate.

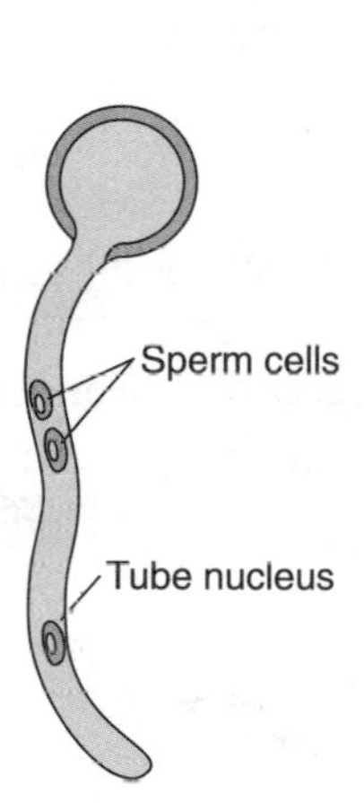

FIGURE 9.14 Maturation of the pollen grain

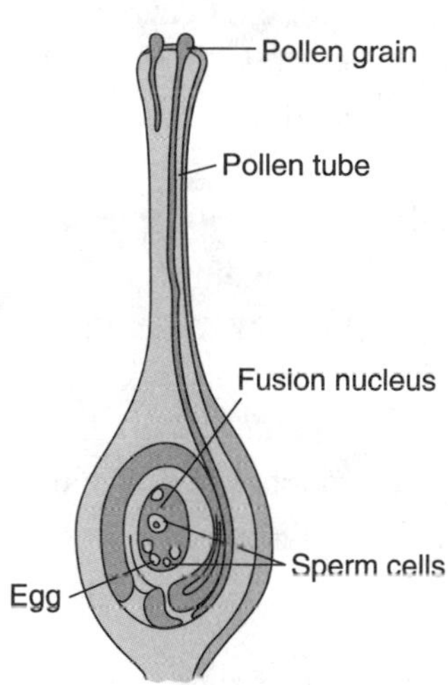

FIGURE 9.15 Fertilization in the flower

Parts of Higher Plants

The bodies of higher plants are made up of two organ systems, the root system below ground and the shoot system above ground. Organs of the root system include roots, tubers, and rhizomes. The shoot system contains organs such as the stem, leaves, fruit, and flowers. These organs are composed of three types of tissue. **Vascular** tissue includes xylem, phloem, cambium, and parenchyma cells; it functions in transporting water, food, minerals, and hormones within the plant. Epidermal cells make up **dermal** tissue, functioning in the prevention of water loss by producing a waxy surface cuticle. Most of a plant body is composed of **ground** tissue. Ground tissue includes parenchyma, collenchyma, and sclerenchyma cells. Versatile parynchyma cells are the most common ground tissue; they function in support, photosynthesis, gas exchange, secretion, growth, and wound repair. Collenchyma and sclerenchyma cells provide structural support for the plant.

ROOT

ROOT GROWTH AND FUNCTION

Roots anchor plants to the soil and absorb water and dissolved materials from the ground. The absorbed materials enter the root by way of root hairs, which are one-cell extensions of the epidermis. From the root hairs, dissolved materials pass through the cortex, endodermis, and pericycle into xylem cells. The xylem cells conduct the dissolved materials upward. The root cortex serves in the storage of food and water. A small amount of storage occurs in the parenchyma cells of the stele. Cells in the tips of the roots are responsible for the growth in length of the root. Growth in diameter of the root is controlled by the cambium between the xylem tissue and the phloem tissue. Asexual reproduction (vegetative propagation) occurs naturally in some plants when propagative buds arise spontaneously on roots and stems. Vegetative propagation can also be manipulated by humans. Many commonly cultivated plants are grown using horticultural technology rather than seeds. Apples, sugar cane, citrus fruits, and potatoes are just a few commonly encountered cultivated foods. Adventitious roots, which also originate through asexual reproduction, develop naturally above ground from the stumps, stems, branches, and leaves of many species of plants including willow trees, strawberry plants, and ivy.

GROSS STRUCTURE

There are two types of root systems: fibrous roots and taproots. The **fibrous root** has numerous slender main roots of equal size with many branch roots smaller in size. Examples of plants with fibrous roots are corn, wheat, and grasses (Figure 9.16). The **taproot** is the main root of the plant. It is longer and thicker than the smaller branch roots. Examples of taproot plants are carrots, beets, and dandelions (Figure 9.17).

MICROSCOPIC STRUCTURE

A longitudinal section of a young root shows four zones distinguished by the cell types within each region (Figure 9.18).

1. **Root cap.** This is a semicircular cap of cells forming the tip of the root and protecting the dividing cells just above it. Cells in the root cap are of moderate size and thick-walled. They protect the thinner-walled cells just above.
2. **Zone of cell division (meristemic region).** Here the cells are small and thin-walled with dense cytoplasm. These cells reproduce rapidly by mitosis and contribute to increased length of the root. This is the growth layer.

FIGURE 9.16 Fibrous root

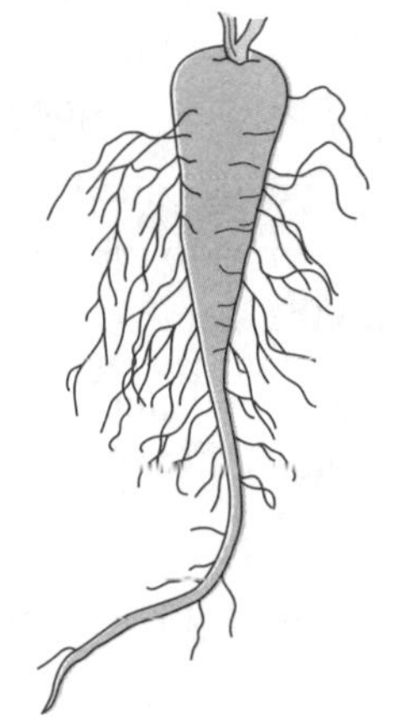

FIGURE 9.17 Taproot

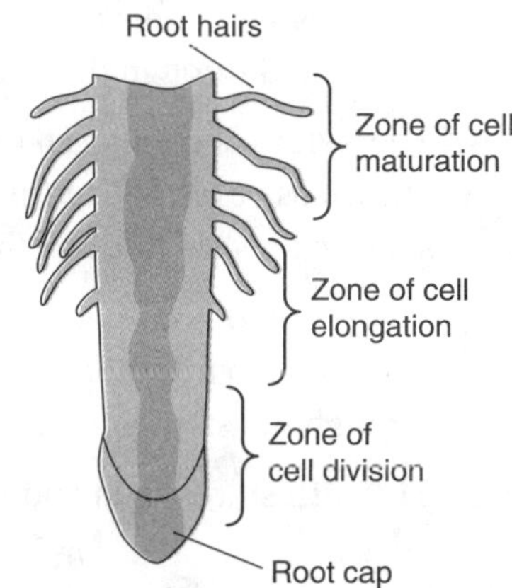

FIGURE 9.18 Longitudinal section of a root

3. **Zone of cell enlargement or elongation.** The cells in this zone were recently formed in the zone of cell division. These cells become elongated, produce new cytoplasm, and develop larger vacuoles.
4. **Zone of cell maturation.** In this zone the enlarged cells become *differentiated* into xylem, phloem, cambium, cortex and other tissues. Root hairs grow from the lower portion of the maturation zone.

A cross section of the root taken through the zone of maturation reveals several types of cell (Figure 9.19).

1. The **epidermis** is the outer layer of cells from which the root hairs develop. Epidermal cells are specialized for the absorption of water and minerals from the soil and for the protection of underlying tissues.
2. The **cortex** contains rather large, irregularly shaped parenchyma cells applied loosely to one another with a good deal of intercellular space. The function of this region is to store water and food. Water leakage from the cortex into the inner tissues is prevented by a band of thick-walled cells known as the *Casparian strip*. A waxy material, *suberin*, makes these cells waterproof and prevents leakage of water into the inner root tissues.

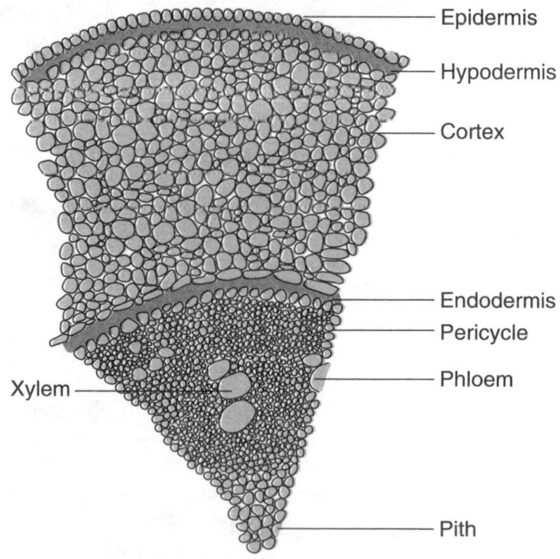

FIGURE 9.19 Cross section of a root

3. The **pericycle** is a layer of cells just inside the Casparian strip (endodermis) from which branch roots are produced. The cells of the branch roots grow through the cortex and through the epidermis and extend outside of the root into the soil.
4. The **xylem** is composed of conducting cells called *tracheids* and thin tubules known as *vessels*. The function of xylem is to conduct water and dissolved substances upward through the root into the stem.
5. The **phloem** is composed of companion cells and sieve tubes. The function of the phloem is to transport water with dissolved food downward from the leaves through the stem into the root.
6. The **pith** stores food and water and lends support to other tissues.

STEM

Stems have three major functions. First, they conduct water upward from the roots to the leaves and conduct dissolved food materials downward from the leaves to the roots. Second, stems produce and support leaves and flowers. Third, they provide the mechanisms for the storage of food.

Botanists call a stem with its leaves a **shoot**. The *shoot system* is the total of all of the stems, branches, and leaves of a plant.

Figure 9.20 shows a longitudinal section of a young stem.

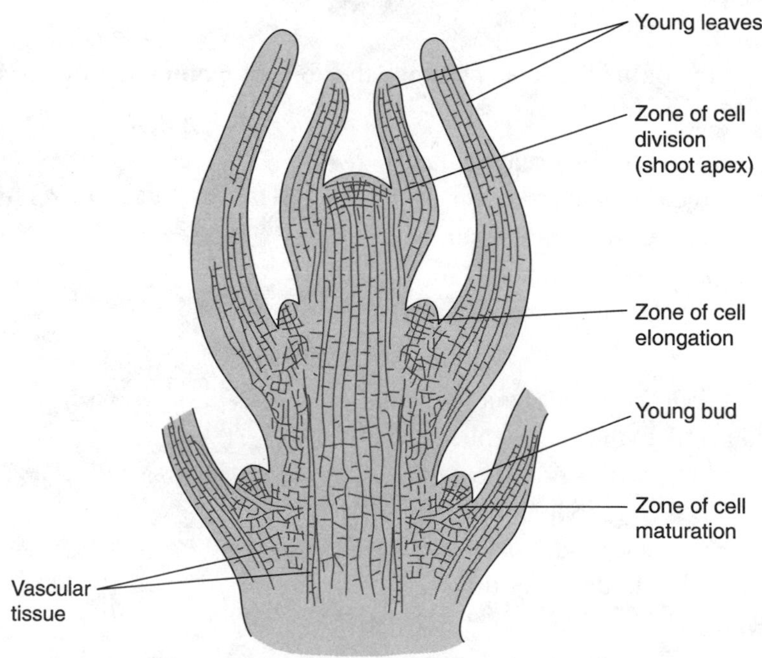

FIGURE 9.20 Longitudinal section of a young stem

There are two types of above ground (aerial) stems: woody stems and herbaceous stems. Major characteristics of each are shown in Table 9.1.

Herbaceous stems of dicotyledons have the following tissues: epidermis, schlerenchyma, cortical parenchyma, pericycle, phloem, cambium, xylem, and pith.

TABLE 9.1
TYPES OF AERIAL STEMS

Characteristic	Herbaceous Stems	Woody Stems
Texture	soft	tough
Color	green	nongreen
Growth	little in diameter	much growth in diameter
Tissues	primary	secondary
Life cycle	annual	perennial
Protective covering	epidermis	bark
Buds	naked	covered with scales
Examples	monocotyledons dicotyledons	all gymnosperms dicotyledons

Herbaceous stems of monocotyledons do not have cambium. Since no secondary growth takes place, the stems do not increase in size appreciably. Microscopic study of the stem cross section shows that the xylem and phloem are organized into vascular bundles. These are scattered throughout the parenchyma tissue which fills the stem.

Woody stems are composed of primary and secondary tissues. Primary tissues are those that develop from the meristems (embryonic tissue) of the buds on twigs during the first year of growth. After the first year, growth in the woody stem takes place in the secondary tissues. These are tissues that arise from the *cambium*. Figure 9.21 compares woody stems to the two forms of herbaceous stems—monocot and dicot.

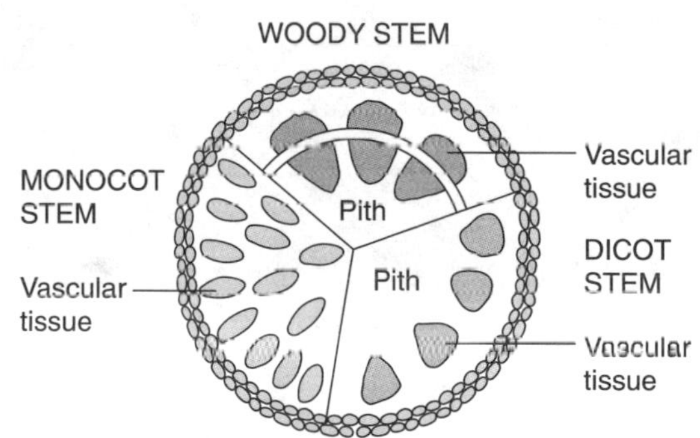

FIGURE 9.21 Comparison of monocot, dicot, and woody stems

Specialized stems have additional functions. The **tendrils** of grapevines function as climbing organs. *Thorns* on rose bushes offer protection against animal invaders. Desert cacti have stems specialized for the storage of water and food. *Runners* of strawberry and spider plants serve as organs of *vegetative propagation* in which new plants are produced at their nodes. **Rhizomes** are underground stems of ferns which serve the purpose of producing new plants. The underground stem of the white potato is called a *tuber*, and its function is to store carbohydrate in the form of starch. The **corms** of the crocus and gladiolus are underground storage stems consisting of fleshy leaves.

LEAF

The most important function of green leaves is to carry out photosynthesis, the food-making process in which inorganic raw materials are changed into organic nutrients.

A leaf consists of two parts: a stalk or *petiole* and the *blade*. The petiole attaches the blade to the stem. The blade is the place where photosynthesis takes place. Leaves vary greatly in shape (Figure 9.22).

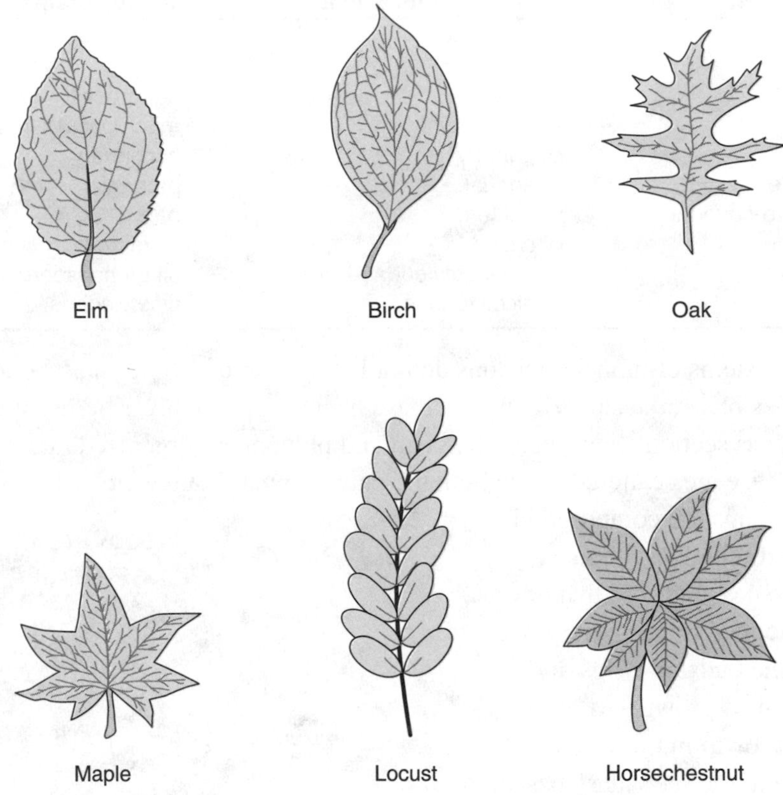

FIGURE 9.22 Some common leaf shapes

Study of a leaf cross section under the microscope reveals three types of tissue: upper and lower epidermis, mesophyll, and the vascular bundles (Figure 9.23).

The **epidermis** is a single layer of cells at the upper and lower surfaces of the leaf. The cells have thick walls made of cutin and lack chloroplasts. Their main function is to protect the underlying or overlying tissues from drying, bacterial invasion, and mechanical injury. On the underside of the leaf, the lower epidermis has pores known as **stomates**, the size openings of which are regulated by a pair of **guard cells**. The stomates serve as passageways for oxygen and carbon dioxide.

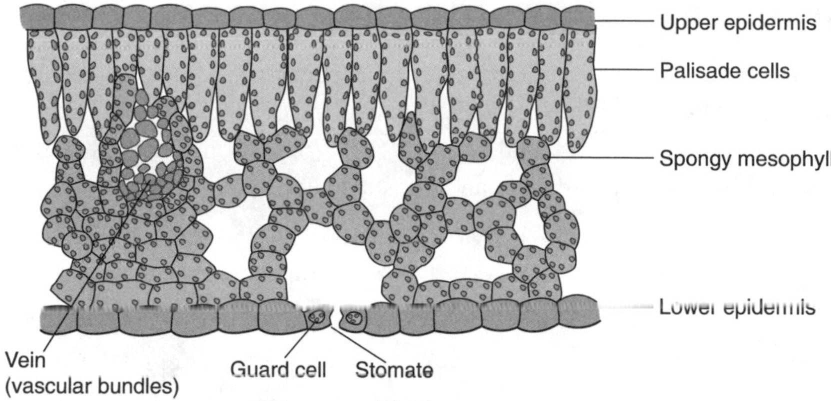

FIGURE 9.23 Leaf cross section

The **mesophyll** is positioned between the two epidermal layers. It is composed of two distinct tissues: the **palisade cells** and the **spongy cells**. Packed with chloroplasts, the cells of both types of mesophyll tissue are the sites for photosynthesis in the leaf.

The **vascular bundles** or veins consist of xylem cells (vessels and tracheids) and phloem cells (sieve tubes and companion cells). The xylem cells transport water and dissolved minerals upward into the mesophyll. The phloem conducts dissolved food materials downward to the lower parts of the plant.

Loss of water vapor through plant leaves is termed *transpiration*, a process which is responsible for the rise of sap in trees. *Guttation* is the loss of liquid water through the leaves of plants with short stems. Guttation is caused by the effect of root pressure on water flow and has hardly any physiological advantage to the plant.

Photosynthesis

Photosynthesis takes place in the chloroplasts of green leaves and stems. **Photosynthesis** is the food-making process of green plants made possible by the light-trapping pigment chlorophyll. An important event of photosynthesis is the changing of light energy into chemical energy, which is ultimately stored in sugar molecules. The raw materials of photosynthesis are carbon dioxide and water. The overall equation for photosynthesis is shown below:

$$CO_2 + 2H_2O + \text{light} \xrightarrow{\text{chlorophyll}} O_2 + (CH_2O) + H_2O$$

REDUCTION-OXIDATION REACTIONS (REDOX)

Reduction is the addition of an electron (e) to an acceptor molecule. Oxidation is the removal of an electron from a molecule. The addition of an electron (reduction) stores energy in a compound. The removal of an electron (oxidation) releases energy. Whenever one substance is reduced, another substance is oxidized.

Ae	+	B	→	A	+	Be
electron donor		electron acceptor		oxidized (loss of energy)		reduced (gain of energy)

In biological systems, removal or addition of an electron derived from hydrogen is the most frequent mechanism of reduction-oxidation reactions. Redox reactions play a major role in photosynthesis. For example, the synthesis of sugar from CO_2 is the reduction of CO_2. Hydrogen, obtained by splitting H_2O molecules, is added to CO_2 to form CH_2O units.

THE PROCESS OF PHOTOSYNTHESIS

Photosynthesis takes place inside **chloroplasts** (Figure 9.24), membranous structures within the cells of the leaf mesophyll. The chloroplasts have fine structures within—flattened membranous sacs called **thylakoids**. On the membranes of the thylakoids, chlorophyll and the accessory pigments are organized into functional groups known as **photosystems**. Each of these photosystems contains about 300 pigment molecules which are involved directly or indirectly in the process of photosynthesis.

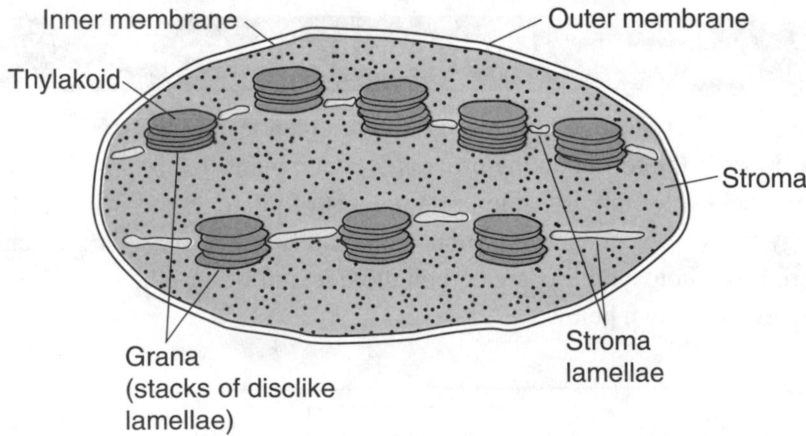

FIGURE 9.24 Structure of a chloroplast

Each of these photosystems has a *reaction center* or a *light trap* where a special chlorophyll *a* molecule traps light energy. There are two types of photosystems: Photosystem I and Photosystem II. In Photosystem I, the chlorophyll *a* molecule is named P700 because it absorbs light energy from the 700 nanometer wavelength. The chlorophyll molecule of Photosystem II is designated as P680 because this pigment molecule (chlorophyll *a*) absorbs light at the wavelength of 680 nanometers.

Photosynthesis involves four sets of biochemical events: photochemical reactions, electron transport, chemiosmosis and carbon fixation. The photochemical reactions and electron transport activities take place on the membranes of the thylakoids. The oval membranes of a thylakoid surround a vacuole or reservoir in which hydrogen ions are stored until needed in the Calvin cycle, or carbon fixation. Each thylakoid rests in the **stroma** or ground substance of the chloroplast. The stroma is the place where carbon fixation occurs.

THE EVENTS OF PHOTOSYSTEM I

Cyclic Phosphorylation—Electron Transport

Light energy strikes a photosystem. Pigment molecules absorb this energy and pass it on to the reaction center molecule. The energy level of an electron in P700 is raised to a higher level. This increased energy in the electron causes it to escape the confines of the P700 (chlorophyll *a*) molecule and to become temporarily attached to an acceptor molecule called *X*. In accepting the electron, molecule *X* is reduced. Molecule *X* passes the electron on to another acceptor molecule and becomes oxidized in the process. In a series of redox reactions, the electron is passed from one acceptor molecule to another. It finally returns to P700. Each step of these reduction-oxidation reactions is catalyzed by a specific enzyme.

The energy released as the electron is passed along the transport chain is used to synthesize ATP (adenosine triphosphate). Excess hydrogen ions released when ATP is formed are stored in the reservoir of the thylakoid. Inorganic phosphate from the fluid of the stroma is incorporated in the ATP molecule during photosynthetic phosphorylation. Photosynthesis requires the energy from ATP to synthesize carbohydrates.

Noncyclic Phosphorylation

During the events of Photosystem I, "excited" electrons may travel in another pathway different from that which builds ATP. Chlorophyll acts as an electron donor and later as an electron acceptor. It donates energy-rich (excited) electrons and accepts back energy-poor electrons.

Light energy strikes a chlorophyll *a* molecule. An electron in its reaction center molecule, P700, becomes elevated to a higher energy state. The electron passes from P700 to acceptor *X*. From acceptor *X*, the electron passes to ferridoxin (Fd), a compound containing iron. Fd passes the electron to an intermediate compound and then to nicotinamide adenine dinucleotide phosphate (NADP). Actually two P700 molecules release electrons along this pathway simultaneously. NADP accepts both electrons (2e) and becomes NADPH. NADPH keeps both electrons and does not pass them along. The energy from $NADPH_2$ will serve as an energy source when carbon dioxide is reduced to form sugar. By acquiring two extra electrons, the NADPH also attracts an H proton. Thus NADPH + H is written as $NADP_{re}$, where *re* means *reduced*.

Review: Photosystem I

1. Photons of light strike a chlorophyll *a* molecule.
2. The reaction center molecule (P700) absorbs the light.
3. One of its electrons is raised to a higher energy level.
4. The pathways followed by "excited" electrons are shown below.

Cyclic	e from P700 to *X* to acceptors to ATP
Noncyclic	2e from P700 to *X* to Fd to NADP → $NADP_{re}$

THE EVENTS OF PHOTOSYSTEM II

Photosystem II involves about 200 molecules in the reaction center of chlorophyll *a*, the light trapping pigment of green plants. In the blue-greens and in the bryophytes, the light trapping pigment is chlorophyll *b*; in the brown algae chlorophyll *c* and in the red algae chlorophyll *d*.

When light strikes the chlorophyll in Photosystem II, an electron in the reaction center molecule, P680, becomes "excited." The energized electron is passed to an electron "acceptor" molecule designated *Q*. Molecule *Q* passes the electron through a chain of acceptor molecules which pass the electrons to the holes in Photosystem I formed during the noncyclic synthesis of $NADP_{re}$. As electrons move along the chains of transport, they lose energy step by step. Some of the energy forms ATP. It is believed that P680 pulls replacement electrons from water, leaving behind free protons and molecular oxygen:

$$2H_2O \rightarrow \underset{\text{to P680}}{\underset{\downarrow}{4e}} + 4H^+ + O_2$$

The protons become associated with $NADP_{re}$.

A summary equation reviewing the electron pathways in Photosystem I and in Photosystem II follows:

> H_2O → 2e to Photo I to *Q* to transport chain to Photo II to *X* to transport chain to $NADP_{re}$ to Calvin cycle.

Figure 9.25 shows the pathways of photosynthesis.

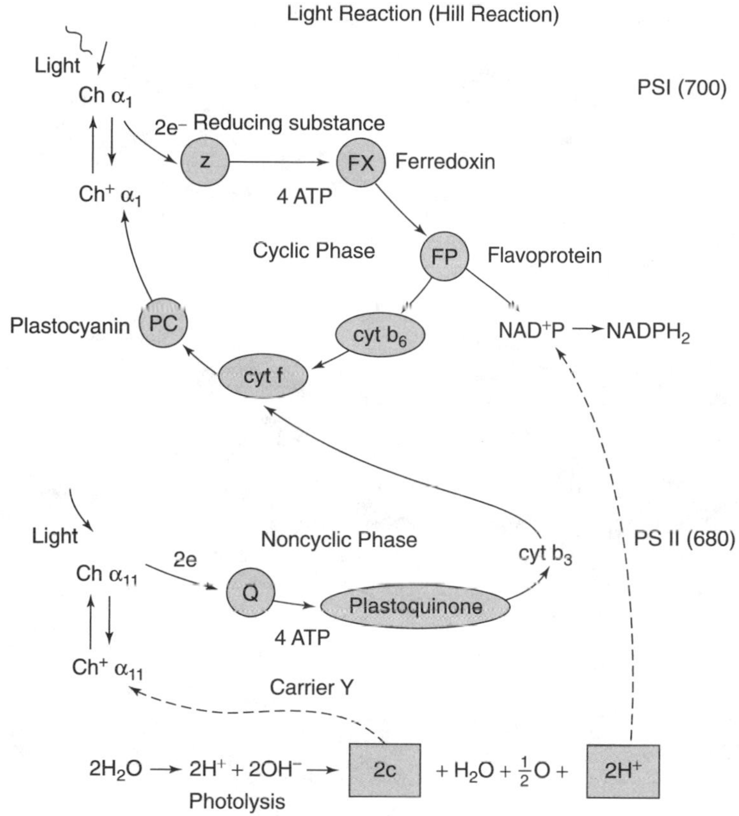

FIGURE 9.25 The pathways of photosynthesis involve many steps and many intermediate products and catalysts, including flavoprotein and cytochrome (cyt).

THE CALVIN CYCLE

The Calvin cycle is the series of events in photosynthesis during which carbon dioxide fixation takes place in the stroma of the chloroplast. $NADP_{re}$ and ATP that were produced during Photosystem I and Photosystem II are now used to attach CO_2 to a preexisting organic molecule. The enzymes that catalyze the events of the Calvin cycle are present in the stroma.

Carbon dioxide combines with the 5-carbon sugar ribulose biphosphate (RuBp), forming an unstable 6-carbon compound. This compound breaks into two molecules of a 3-carbon compound, phosphoglyceric acid (PGA). The two PGA molecules are reduced to two molecules of phosphoglyceraldehyde (PGAL) in two successive steps. First, the PGA molecules receive a high-energy phosphate from ATP. The high-energy phosphate bond is broken, and the phosphate is removed and replaced by a hydrogen atom from NADPH. Then, the joining together of two PGAL molecules results in the formation of a molecule of 6-carbon sugar. Some of the PGAL is used to replenish the store of ribulose biphosphate, the starting point of the Calvin cycle (Figure 9.26).

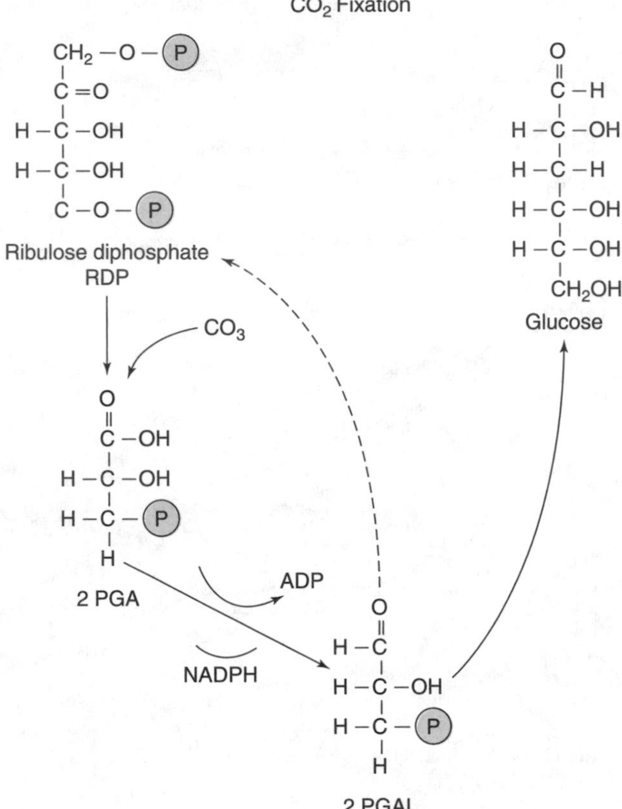

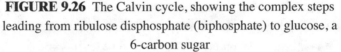

FIGURE 9.26 The Calvin cycle, showing the complex steps leading from ribulose disphosphate (biphosphate) to glucose, a 6-carbon sugar

PHOTORESPIRATION

Photorespiration is a rather strange series of events that occurs in the cells of green plants in the presence of sunlight. This process is considered strange because plant cells use energy in photorespiration rather than generating it. In the course of normal events, the enzyme ribulose biphosphate (RuBP) carboxylase joins a carboxyl group to ribulose biphosphate. The biochemical activity that follows was explained in the section entitled "The Calvin Cycle."

During photorespiration, oxygen, instead of carbon dioxide, binds to RuBP carboxylase. When RuBP carboxylase has oxygen bound to it, oxidation of ribulose biphosphate occurs. One molecule of PGA and a 2-carbon molecule are released. The PGA remains in the C_3 cycle, but the 2-carbon molecule leaves the chloroplast and enters into chemical reactions in a peroxisome and a mitochondrion. Some of the CO_2 produced in these reactions is released; the rest is returned to the chloroplast to take part in photosynthesis.

Photorespiration oxidizes organic compounds using oxygen and results in the release of carbon dioxide. This process does not utilize the electron transport system and therefore does not produce energy. Rather, it uses up energy, thereby appearing to be wasteful. To date, scientists do not really know how photorespiration benefits the cell's work in photosynthesis.

THE C_4 OR HATCH-SLACK PHOTOSYNTHETIC PATHWAY

In the late 1960s three research botanists (Kortschak, Hatch, and Slack) discovered another photosynthetic pathway, which has become known as the C_4 or Hatch-Slack photosynthetic pathway. Essentially, this is what happens. Carbon dioxide combines with a compound known as PEP (phosphoenolpyruvate), forming a 4-carbon compound, malate. Malate is transferred to bundle sheath cells in the leaf. This 4-carbon compound gives up carbon dioxide, which enters the C_3 or Calvin cycle in the photosynthetic bundle sheath cells.

Plants that carry on C_4 photosynthesis exhibit special arrangement in their leaf tissues. In this special arrangement, termed **Kranz anatomy**, the *bundle sheath cells* are positioned in a circle around the vascular bundles (made up of phloem and xylem tubes). The mesophyl cells comprise the rest of the leaf's interior. The air spaces are very small (Figure 9.27). Plants in tropical and desert regions with especially high rates of photosynthesis are C_4 plants; among them are crabgrass, sugarcane, millet, and sorghum. Interestingly enough, corn, a temperate-region plant, also carries on C_4 photosynthesis.

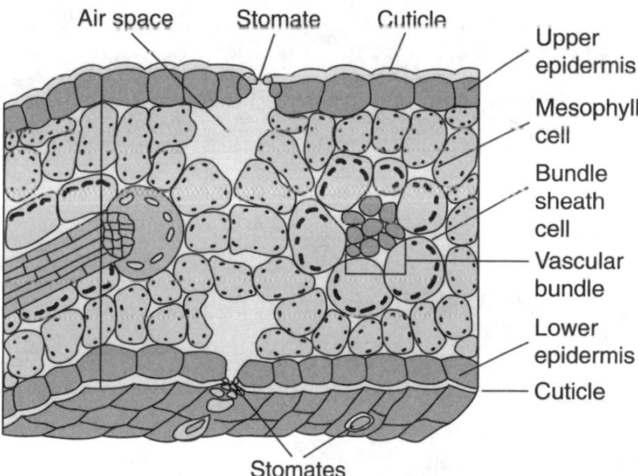

FIGURE 9.27 Kranz anatomy. Bundle sheath cells surround vascular bundles.

Plant Hormones

Hormones, which are chemical messengers produced by tissues in small concentrations, effect change in the organism. In general, plant hormones affect plant growth and responses to the environment. Plant hormones differ from the hormones produced by animals in a number of ways. Animal hormones are produced by endocrine glands and affect target organs. Plant hormones, on the other hand, are produced by different tissues at different stages and affect virtually every tissue in the plant. Unlike animal hormones, plant hormones do not affect *homeostasis* (steady-state control of physiological activity). A summary of plant hormones and their effects follows.

Auxins (indoleacetic acid or IAA) cause plant stems to bend (grow) toward the light by promoting enlargement of the stem cells. Plant growth response to light is known as **phototropism**. **Apical dominance** is the control exercised by the growing tip of the stem over lower stem structures. Auxin accumulates in the growing stem tip and restrains the development of side stems or branches.

Gibberellins, isolated from the fungus *Gibberella fujikuroi*, belongs to a family of 57 closely related compounds. Gibberellic acid, produced in the young leaves of plants, controls stem growth, leaf growth, and root elongation. The gibberellins have another important function. The embryos of germinating cereal grains, such as corn and barley, release gibberellins, which, in turn, stimulate the release of food from the endosperm.

Cytokinins, produced in the roots, are distributed upward through the xylem. Together with auxins, the cytokinins control differentiation of plant tissue, stimulate cell division and the production of fruits and seeds, and slow the process of aging in picked leaves. Cytokinins also prevent the onset of dormancy in plants.

Abscisic acid, produced in mature leaves and transported to other plant tissues and organs, is a growth inhibitor, causing dormancy in trees. Roots grow downward in a positive response to gravity known as *geotropism*. It is believed that abscisic acid controls root growth response.

Ethylene, produced as a gas by plants, controls the ripening of fruit and plays an important role in the growth of seedlings upward through the soil.

Oligosaccharines are sugar compounds that protect the cell wall against insect invasion by stimulating cells to produce antibiotics known as *phytoalexins*. Phytoalexins inhibit the synthesis of protein-digesting enzymes of insects and thus prevent predation.

Photoperiodicity

Each species of plant has its own *photoperiod* for flowering. Although plants flower in response to length of darkness, photoperiodicity is described in terms of *day length*. Some plants belong in the category of *long-day* plants because they flower in response to a shortening night. Plants that flower in response to lengthening nights are called *short-day* plants. *Day-neutral* plants seem not to be affected by the ratio of daylight to darkness.

Phytochrome has been identified as the pigment that controls the "dark clock" of flowering. Cocklebur, Maryland mammoth (a species of tobacco), and Biloxi soybeans are short-day plants. Spinach, radish, and barley are long-day plants requiring shorter hours of darkness. Tomatoes and cucumbers are day-neutral and flower without regard to length of night.

REVIEW EXERCISES FOR CHAPTER 9

WORD-STUDY CONNECTION

abscisic acid
alternation of generations
angiosperm
anther
antheridia
apical
archegonia
auxin
Calvin cycle
calyx
cambium
carpel
carpogonia
conceptacle
cone
corolla
cutin
cytokinins
division of labor
embryo sac
ethylene
endosperm
fibrous root
filament
flower
fucoxanthin
gametophyte
gibberellin
gymnosperm
heterogamy
dominance holdfasts
hormones
isogametes
kelps
Kranz anatomy
megasporophyll
micropyle
microspore
microsporophyll
multicellular
oligosaccharine
oogamy
oogonia
ovary
ovule
petal
phloem
photoperiod
photorespiration
photosystem
phototropism
phytochrome
pistil
Plantae
polar bodies
pollen
prothallus
pyrenoid body
rhizoids
rhizome
sepal
sorus
specialization
sporophyte
stamen
stigma
strobilus
style
taproot
thallophytes
thallus
vascular
xylem

SELF-TEST CONNECTION

PART A. Completion. *Write in the word that correctly completes each statement.*

1. Single-cell plants include free-living forms, cells that live in colonies, and _____ forms.
2. The chromosome number of vegetative *Chlamydomonas* cells is best described as being _____.
3. Primitive gametes that are similar in appearance are known as _____.
4. *Gonium* species live as a simple _____ of cells.
5. The *Volvox* series of cells demonstrates an evolutionary trend toward _____.
6. Chlorophyll _____ is found only in the red algae.
7. The water-carrying tissues of green plants are known collectively as _____ tissues.
8. The cycle of a sexual generation following an asexual generation is known as _____.
9. Fluid compounds are conducted downward in plants by _____ tubules.
10. A sorus (plural, sori) can be found on the leaves of _____ plants.
11. The dominant generation in the ferns is the _____ generation.
12. The gymnosperms are best described as _____ bearers.
13. The prefix *-mega* means _____.
14. The microsporophylls are _____ cones.
15. In gymnosperms, the agent of pollination is _____.
16. In gymnosperms, the megasporangium produces the _____.
17. Petals are known collectively as the _____.
18. The tube nucleus develops from the germinating _____.
19. A fruit is a ripened _____.
20. Growth in length of roots occurs at the _____.

21. A stem with its leaves is called a _____.

22. In the plant, the Casparian strip is located in the _____.

23. Secondary plant tissues arise from the _____.

24. The most effective light-trapping pigment is _____.

25. Carotene and phycoerythrin function as _____ pigments during the light driven stages of photosynthesis.

26. Carbon fixation takes place in the _____ of the chloroplast.

27. The response that plants make to light is called _____.

28. Indoleacetic acid belongs to a group of hormones known by the general name _____.

29. A plant's photoperiod is described in terms of _____.

30. The group of hormones produced in plant roots and distributed upwards are known as _____.

PART B. Multiple Choice. *Circle the letter of the item that correctly completes each statement.*

1. *Chlamydomonas* resembles the protists because it lives in water and swims by means of
   (a) cilia
   (b) pseudopods
   (c) tentacles
   (d) flagella

2. The function of the pyrenoid body is to
   (a) synthesize glucose
   (b) store starch
   (c) resorb flagella
   (d) secrete cellulose

3. Sea lettuce is best classified as a (an)
   (a) protist
   (b) bryophyte
   (c) euglenoid
   (d) alga

4. A true statement about *Volvox* is
   (a) Most of its cells do not function in reproduction.
   (b) All of its cells are specialized for reproduction.
   (c) There is no communication between its cells.
   (d) The egg cells have paired flagella.

5. The sporophyte generation of sea lettuce is
   (a) haploid
   (b) monoploid
   (c) diploid
   (d) tetraploid

6. *Fucus* does not have
   (a) air bladders
   (b) an asexual cycle
   (c) a sexual cycle
   (d) a branching thallus

7. The greatest hazard to land-dwelling plants is
   (a) drying out
   (b) insect infestation
   (c) loss of sperm cells
   (d) nonmotile egg cells

8. Waterproofing of land cells is made possible by cell walls composed of
   (a) chitin
   (b) cutin
   (c) cellulose
   (d) glycoprotein

9. Examples of bryophytes are
   (a) moss, *Volvox, Fucus*
   (b) *Ulva*, hornwort, *Pandorina*
   (c) *Euglena*, moss, red algae
   (d) hornwort, liverwort, moss

10. Species that bear two types of spores are described as being
   (a) homosporous
   (b) heterosporous
   (c) homozygous
   (d) heterozygous

11. Dried, ground stems of the horsetails are used in scouring powder because their cell walls contain
    (a) cutin
    (b) cellulose
    (c) silica
    (d) suberin

12. An underground stem is called a
    (a) prothallus
    (b) root
    (c) strobilus
    (d) rhizome

13. Ferns are successful land plants because they have well developed
    (a) vascular systems
    (b) digestive systems
    (c) reproductive systems
    (d) excretory systems

14. In the gymnosperms, the megasporophyll is a (an)
    (a) large spore
    (b) seed cone
    (c) egg cell
    (d) large ovule

15. The embryo plant is protected inside of the
    (a) ovule
    (b) egg
    (c) pollen
    (d) seed

16. The production of two types of gametes is known as
    (a) heterospory
    (b) heterogamy
    (c) isogamy
    (d) isospory

17. A true statement about the gymnosperms is that they do not produce
    (a) flagellated sperm
    (b) seed cones
    (c) nonmotile eggs
    (d) pollen cones

18. In the flowering plants, the male reproductive structures are the
    (a) calyx
    (b) corolla
    (c) pistils
    (d) stamens

19. The chromosome number of the fertilized endosperm nucleus is best described as
    (a) haploid
    (b) monoploid
    (c) diploid
    (d) triploid

20. The part of the young root that pushes its way through the soil is the
    (a) root hair
    (b) tap root
    (c) root cap
    (d) root zone

21. The word *meristem* refers to
    (a) rapidly growing cells
    (b) old cells
    (c) cells specialized for reproduction
    (d) differentiated cells

22. The primary function of the root cortex is to
    (a) absorb minerals from the soil
    (b) differentiate into other cells
    (c) store water and food
    (d) give rise to sperm cells

23. Secondary tissues in roots arise from the
    (a) meristem
    (b) embryo
    (c) cambium
    (d) phloem

24. A tendril is a specialized
    (a) stem
    (b) root
    (c) flower
    (d) leaf

25. Sap rises in trees due to the force created by
    (a) evaporation
    (b) transportation
    (c) transpiration
    (d) guttation

26. Photosystems are functional pigment groups located on the
    (a) proteins of the plasma membrane
    (b) membranes of the thylakoids
    (c) in the stroma of the chloroplasts
    (d) in fluids of vacuoles

27. As an outcome of cyclic phosphorylation
    (a) P700 is destroyed
    (b) ATP is formed
    (c) NADPH is synthesized
    (d) carbon is fixed

28. Ribulose biphosphate
    (a) begins the Calvin cycle
    (b) functions as an enzyme
    (c) breaks into two equal parts
    (d) is a 6-carbon sugar

29. During photorespiration, O_2 binds to
    (a) phosphoglyceric acid
    (b) RuBP carboxylase
    (c) malate
    (d) carbon dioxide

30. A plant that has Kranz anatomy in the arrangement of its leaf tissues is
    (a) *Volvox*
    (b) white oak
    (c) corn
    (d) wheat

PART C. Modified True-False. *If a statement is true, write "true" for your answer. If a statement is incorrect, change the* underlined *expression to one that will make the statement true.*

1. *Chlamydomonas* is classified as a protist.

2. Zoospores represent sexual reproduction.

3. If the original parent colony of *Gonium* contained 8 cells, each new colony produced will contain <u>32</u> cells.

4. Oogamy refers to conditions in which the egg cell is <u>motile</u>.

5. The sporophyte generation is produced by <u>spores</u>.

6. Brown algae are called <u>seaweed</u>.

7. Water is carried upward in plants by special tubules known as <u>phloem</u>.

8. A strobilus is a <u>stem</u>.

9. The fern plant is the <u>gametophyte</u> generation.

10. A germinating fern spore develops into a <u>gametophyte</u> plant.

11. The reproductive mechanisms of seed plants do not require <u>gametes</u>.

12. A megaspore mother cell is found inside of an <u>egg</u>.

13. Reproduction in the gymnosperms takes place on special structures called <u>stamens</u>.

14. In gymnosperms, the egg cell develops from the <u>microspore</u> cell.

15. The flowering plants are called <u>gymnosperms</u>.

16. Green leaves that protect flower buds are called <u>petals</u>.

17. The stigma and the style are best associated with the <u>filament</u>.

18. The ovules are found inside the <u>ovary</u>.

19. A seed is a ripened <u>ovary</u>.

20. Growth in diameter of roots takes place in the <u>epidermis</u>.

21. Two types of root systems are fibrous roots and <u>root hairs</u>.

22. The phloem is composed of companion cells and <u>sieve</u> tubes.

23. The underground stem of the white potato is a <u>rhizome</u>.

24. Runners of strawberry plants are specialized for <u>climbing</u>.

25. All gymnosperms have <u>herbaceous</u> stems.

26. The two types of cells that compose the leaf mesophyll are the spongy cells and meristemic cells.

27. Oxidation occurs by loss of an electron.

28. An "excited" electron escapes to a lower energy level.

29. During photorespiration, a two-carbon molecule leaves the chloroplast and enters the reactions of the mitochondrion and the lysosome.

30. The compound PEP is actively involved during C_3 photosynthesis.

CONNECTING TO CONCEPTS

1. Distinguish between heterotroph and autotroph.

2. Why is chlorophyll important to the life process of green algae?

3. How do cells that live in colonies differ from cells that live independently?

4. Would life on Earth be possible without the existence of green plants? Explain.

5. For each of the following, list the functions that would be lost to green plants if their cells could not synthesize these products: auxins, hormones, phytochrome.

ANSWERS TO SELF-TEST CONNECTION

PART A

1. filamentous
2. haploid
3. isogametes
4. colony
5. multicellularity (specialization)
6. d
7. vascular
8. alternation of generations
9. phloem
10. fern
11. sporophyte
12. cone
13. large
14. pollen
15. wind
16. ovule
17. corolla
18. pollen grains
19. ovary
20. tips
21. shoot
22. root
23. cambium
24. chlorophyll *a*
25. accessory
26. stroma
27. phototropism
28. auxins
29. daylength
30. cytokinins

PART B

1. **(d)**	6. **(b)**	11. **(c)**	16. **(b)**	21. **(a)**	26. **(b)**
2. **(b)**	7. **(a)**	12. **(d)**	17. **(a)**	22. **(c)**	27. **(b)**
3. **(d)**	8. **(b)**	13. **(a)**	18. **(d)**	23. **(c)**	28. **(a)**
4. **(a)**	9. **(d)**	14. **(b)**	19. **(d)**	24. **(a)**	29. **(b)**
5. **(c)**	10. **(b)**	15. **(d)**	20. **(c)**	25. **(c)**	30. **(c)**

PART C

1. green alga
2. asexual
3. 8
4. nonmotile
5. gametes
6. true
7. xylem
8. cone
9. sporophyte
10. true
11. water
12. ovule
13. cones
14. megaspore mother cell
15. angiosperms
16. sepals
17. pistil
18. true
19. ovule
20. cambium
21. taproots
22. true
23. tuber
24. reproduction
25. woody
26. palisade
27. true
28. higher
29. peroxisome
30. C_4

CONNECTING TO LIFE/JOB SKILLS

If you like working outdoors in association with plants, the careers listed below may be of interest to you.

- A **tree surgeon** takes care of trees, cutting off dead and diseased branches. You must be strong and able to withstand heights.
- A **horticulturist** is a flowering plant specialist who plans and cares for gardens. A **horticulturist technician** works under the direction of the horticulturist.
- A **park ranger** works in national and state parks overseeing the environment of plants and animals.
- A **plant physiologist** is a scientist who specializes in the structures and functions of plants, both in health and in disease.

All of these careers have specific educational and/or training requirements. The libraries of your local colleges are good places to start your research. For additional information, you may wish to contact the following:

U.S. Department of the Interior
National Park Service
1849 C Street, N.W.
Washington, D.C. 20240

American Society for Horticultural Science
710 North Saint Asaph Street
Alexandria, VA 22314

Chronology of Famous Names in Biology

1679 **Marcello Malpighi** (Italy)—determined the functions of xylem and phloem by conducting a series of girdling experiments.

1729 **Stephen Hales** (England)—demonstrated that transpiration can pull sap up through the xylem and phloem.

1772 **Joseph Priestley** (England)—demonstrated that the air is replenished by green plants.

1782 **Jean Senebier** (France)—showed that photosynthesis depends on "fixed air," now known as carbon dioxide.

1796 **Jan Ingen-Housz** (Netherlands)—concluded that plants use carbon dioxide in photosynthesis.

1804 **Nicolas Theodore De Saussure** (France)—demonstrated that water is necessary for photosynthesis.

1879 **James Clerk Maxwell** (England)—discovered that light travels in waves.

1880 **Charles Darwin** and **Francis Darwin** (England)—were the first to propose the existence of a plant hormone.

1883 **T. W. Engelmann** (Germany)—provided evidence that chlorophyll plays a role in photosynthesis and demonstrated that red light is most effective in photosynthesis.

1904 **B. Haberland** (Germany)—discovered that plants native to tropical climates have a different arrangement of the bundle sheath cells.

1905 **F. F. Blackman** (England)—was the first to present evidence that photosynthesis has a light-driven stage and a stage not requiring light.

1905 **Albert Einstein** (United States)—proposed that light energy travels in packets called *photons*.

1910 **Max Planck** (Germany)—established that the energy of radiation is contained in packets called *quanta*.

1926 **Frits W. Went** (Netherlands)—gave the name "auxin" to the substance in plant stems that controls their elongation.

1934 **C. B. van Niel** (United States)—proposed that water is the source of the oxygen in photosynthesis.

1941 **Samuel Ruben, Merle Randall, Martin Kamen,** and **James L. Hyde** (United States)—reported that oxygen liberated in photosynthesis comes from water.

1950s **George Wald** (United States)—was the greatest living expert on light and life.

1950 **Daniel Arnon** (United States)—identified the reactions that take place in light reactions of photosynthesis.

1954 **Hugo Kortschak** (United States)—discovered that Kranz anatomy plants begin carbon dioxide fixation with a 4-carbon compound.

1960 **Haraguro Yomo** (Japan)—discovered that cereal grain embryos release gibberellin, which dissolves stored food in the endosperm.

1960 **H. P. Kortschak** (United States), **M. D. Hatch**, and **C. R. Slack** (Australia)—elucidated photosynthetic pathways in C_4 plants such as sugarcane, corn, and sorghum.

1961 **Melvin Calvin** (United States)—discovered the pathway of events by which green plants incorporate carbon dioxide into carbohydrate molecules.

1994 **Elliott Meyerowitz** (United States)—published an account of his ongoing research in the genetics of flower development, using the species *Arabidopsis thaliana*, commonly known as mouse ear cress.

10

Invertebrates: Sponges to Mollusks

WHAT YOU WILL LEARN

In this chapter you will review the major differences between animals and plants. One of these differences is the method by which animals obtain their food.

SECTIONS IN THIS CHAPTER

- Basic Organization of the Animal Body
- General Characteristics of Invertebrates
- Phylum Porifera—Sponges
- Phylum Cnidaria (Coelenterates)—Hydrozoa, Jellyfish, Sea Anemones, and Their Relatives
- Phylum Platyhelminthes—Flatworms
- Phylum Nematoda—Roundworms
- Phylum Annelida—Segmented Worms
- Phylum Mollusca—Clams and Their Relatives
- Review Exercises for Chapter 10
- Connecting to Life/Job Skills
- Chronology of Famous Names in Biology

OVERVIEW

All animals belong to the kingdom Animalia, a grouping of 35 phyla (sing., phylum). Modern taxonomists support the theory that the lineage of all animal phyla can be traced back to a common multicellular ancestor. Over billions of years phyla evolved, one from another in a branching pattern, each more complex in body structure and systems than its most recent ancestor. The chart that follows presents six invertebrate phyla. Notice the increasing complexity in body structure.

KINGDOM: INVERTEBRATES

Invertebrates are animals without backbones. They are multicellular, composed of cells that lack walls and are the most diverse group of the kingdom Animalia.

Phylum	Characteristics
Porifera: sponges	Body structure: two layers of cells not considered tissues, asymmetrical
Cnidaria: *Hydra*, *Obelia*	Body structure: central digestive cavity, simple nervous system, radial symmetry, stinging cells, two tissue layers—ectoderm and endoderm
Platyhelminthes: flatworms	Three primary germ layers—ectoderm, mesoderm, endoderm; marine, fresh water, land; no body cavity; bilateral symmetry
Nematoda: roundworms	Body structure: endoderm, body cavity partly lined by mesoderm, ectoderm; bilateral symmetry; unsegmented body
Annelida: segmented worms	Three tissue layers; digestive system with specialized regions; bilateral symmetry
Mollusca: mollusks; snails, slugs, squids	Muscular foot; organs in body cavity; mantle cavity; bilateral symmetry

Basic Organization of the Animal Body

The organization of the animal body is different from that of the plant and requires a different set of terms for accurate descriptions.

SYMMETRY

The animal body form is often described in terms of **symmetry**, the relative positions of parts on opposite sides of a dividing line. Symmetry helps to define the degree of similarity between two species, or between parts of the same animal. **Spherical symmetry** describes the symmetry of a ball. Any section through the center of a ball-shaped organism divides the animal into equal and symmetrical halves (Figure 10.1).

Radial symmetry describes the symmetry of a wheel, in which the parts are arranged in a circle around a central hub or axis. A vertical cut through the central axis divides a wheel-shaped organism into equal (symmetrical) halves (Figure 10.2). Radial symmetry is a characteristic of animals that are sessile. However, radially symmetrical animals that have some head development are capable of rather quick movements. **Bilateral symmetry** is characteristic of animals that have a head end, a tail end, and right and left sides. The body of a dog or a horse is a good example of bilateral symmetry. A line drawn through the center of the body lengthwise divides the body into halves that are mirror images of each other (Figure 10.3).

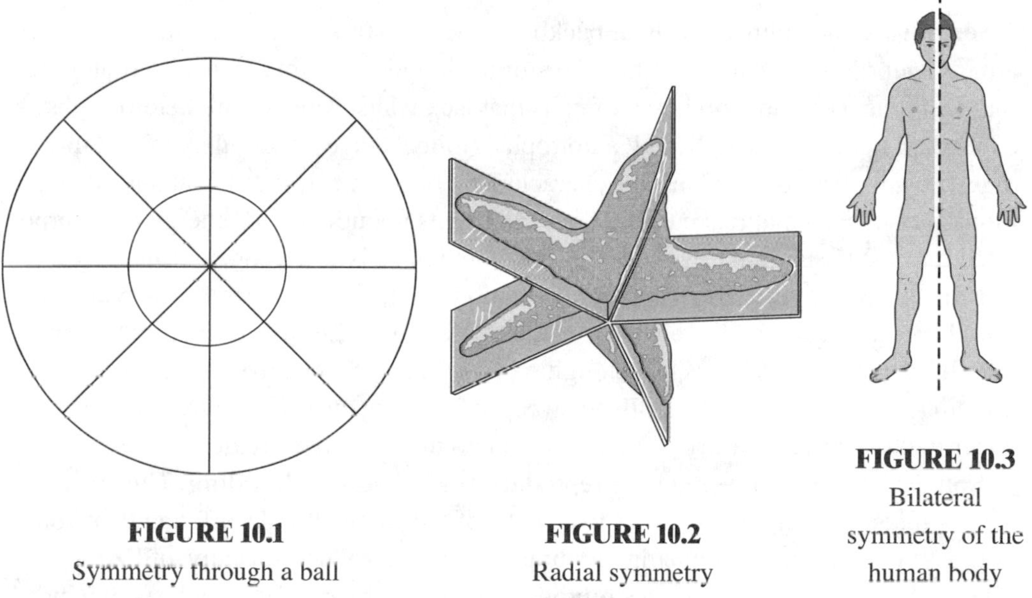

FIGURE 10.1 Symmetry through a ball

FIGURE 10.2 Radial symmetry

FIGURE 10.3 Bilateral symmetry of the human body

EMBRYONIC CELL LAYERS

All of the tissues, organs, and systems of animals develop from two or three *embryonic cell layers*, often referred to as **primary germ layers**. These three cell layers are the **ectoderm** (meaning outside skin), the **mesoderm** (middle skin), and the **endoderm** (inside skin). Animals with simple body structures may develop from only ectoderm and endoderm because the middle tissue layer remains undifferentiated. The tissues of animals with more complex body forms develop from the three primary germ layers. In these animals there is a body cavity or **coelom** (see-lom) which is lined with endoderm.

BODY DIRECTIONS

The kinds of directions used to locate body structures permit precise description. The forward end of an animal is the *anterior* end. The *posterior* end is opposite to the anterior region. For example, the head end of a dog is the anterior end; its tail represents the posterior end. *Dorsal* refers to the back of an animal. In all animals except humans (and some primates) the dorsal side faces upward. The underside or bellyside of an animal is the *ventral* side. The sides of an animal, such as the right-hand and left-hand sides of a human or the right and left fore and hind limbs of a horse, are the **lateral** sides of the body. The point of attachment of a structure, such as the wing of an insect, is the *proximal* end. The free, unattached end is known as the **distal** end.

General Characteristics of Invertebrates

Invertebrates are animals without backbones. About 90 percent of all animal species are invertebrates. Like all other members of the kingdom Animalia, invertebrates are multicellular. They are composed of cells that lack walls. Animals are heterotrophs, dependent upon food supplied by autotrophs. Animals ingest food; they take it in, digest it, and then, by some means characteristic of the species, distribute nutrient molecules to cells that make up the body. Nutrients not utilized for energy or incorporated into the tissue-building processes are stored in cells as glycogen or fat.

Most invertebrates are capable of locomotion and have specialized cells with contractile proteins that facilitate movement. However, the adult forms of some lower invertebrate species are **sessile**, belonging to a group of **filter feeders**. These animals use cilia, flagella, tentacles or gills to sweep smaller organisms from the currents of water that flow over or through their bodies into the digestive cavities.

Some of the lower invertebrates reproduce vegetatively by **budding**. This means simply that a new organism grows from cells of the parent, breaks off, and then continues its own existence. Other invertebrate species reproduce sexually, utilizing sperm and egg. Still other species reproduce asexually by **parthenogenesis** in which an unfertilized egg develops into a complete individual. Some invertebrates have marvelous powers of **regeneration**, the growing back of lost parts or the production of a new individual from an aggregate of cells or from a piece broken off from the parent organism.

Phylum Porifera–Sponges

In a practical sense, you are familiar with the word *sponge*. You know that a sponge is able to soak up large amounts of water because of the many pores and spaces that form its structure. Living sponges are pore-bearing animals and, therefore, were given the scientific name of *Porifera*. These are the simplest of the multicellular animals. About 15,000 sponge species live attached to rocks bathed by ocean waters. Only a few species are freshwater dwellers.

STRUCTURE

Sponges are **asymmetrical**; they are irregular in shape and do not have a defined or predictable body arrangement. Figure 10.4 illustrates the structure of a simple sponge. Notice that the sponge has a cylindrical shape resembling a vase. At the anterior end is the opening of the central cavity, which extends through the Porifera body. Sponges are composed of undifferentiated cells not organized into specialized tissues. The cells show a division of labor much like that of the cells in an algal colony. The cells of the sponge are arranged around the central cavity. In some species this cavity is divided into compartments; in others, it is a continuous space. The central cavity is brought into contact with the surrounding water by means of pores that lead into a system of canals.

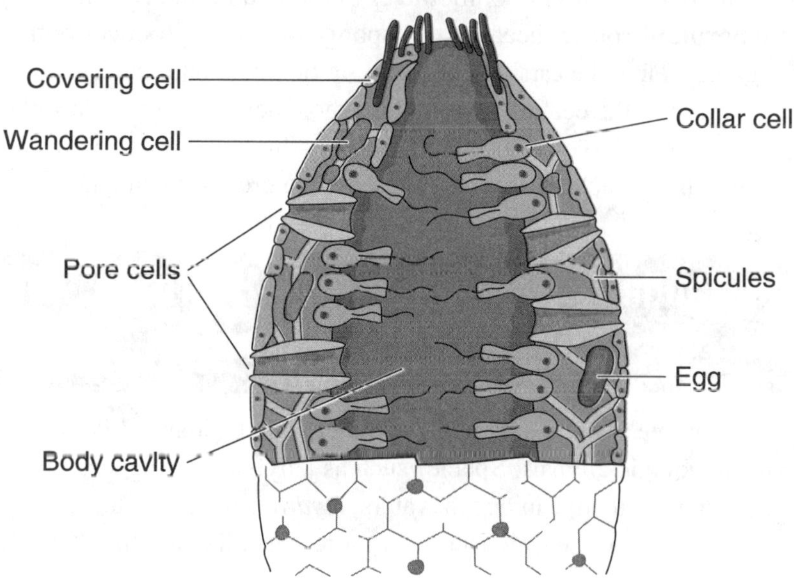

FIGURE 10.4 Internal structure of the sponge

If you have ever held a natural sponge, you are aware that it is not flimsy. Sponges have an internal skeleton. In some species the skeleton is made of needle-like crystals called **spicules**, which are made of lime or silica (glass). In other species, a fibrous protein called *spongin* forms the skeleton, making the sponge firm, but pliable, wettable and absorbent.

Sponges are filter-feeders. The internal cavity is lined with *collar cells* equipped with flagella which capture microorganisms from water as it flows through the central cavity. Epithelial cells cover the outer body wall of the sponge. Between the outer wall and the cells that line the inner cavity is a jelly-like layer known as the *mesoglea*. Here amoeboid wandering cells known as *mesenchyme cells* help to transport digested molecules from the collar cells to cells in other parts of the body. The mesenchyme cells have the ability to change their form and function. They can change into collar cells, epithelial cells, or cells secreting the materials that make spicules for the skeleton.

REPRODUCTION AND REGENERATION

Sponges have exceptional powers of regeneration. If a sponge is sieved through a fine silk mesh and the cells left undisturbed, the cells will come together to reform a complete sponge. Porifera reproduce asexually by **budding**. A new individual grows from cells on the parent body, breaks off, and grows into an adult sponge. Sponges also reproduce sexually by means of nonmotile egg cells and motile flagellated sperm. Sponges are **hermaphrodites** because one sponge organism produces both egg and sperm. The gametes may be carried out to sea by flowing water currents or fertilization may take place in the central cavity. The zygote develops into an *amphiblastula* or early embryo. The embryo escapes from the cavity through a pore, swims about for a short while, and then attaches itself, settling down to grow into an adult sponge.

Phylum Cnidaria (Coelenterates)–Hydrozoa, Jellyfish, Sea Anemones, and Their Relatives

The next group on the evolutionary scale is the phylum Cnidaria. Like the sponges, the Cnidaria are aquatic animals. Species such as jellyfish, Portuguese man-of-war, sea anemones, and corals live in ocean waters; *Hydra* is a freshwater genus. The cells that compose the body of the Cnidaria show more specialization than the body cells of sponges and are grouped together into simple tissues.

GENERAL CHARACTERISTICS

In general, the Cnidaria body is shaped like a hollow sac composed of two tissue layers. The **ectoderm** covers the outer surface of the body; the **endoderm** lines the inner body surfaces. Between these two tissues is a gelatinous mass of undifferentiated material known as the **mesoglea**.

A unique feature of the Cnidaria is the *medusa* and *polyp* forms shown by representative classes in this phylum. Figure 10.5 illustrates the structure of a jellyfish. Note that its body shows radial symmetry. Notice also that the outer surface of the body curves outward very much like a bowl or a bell. The under surface curves inward and is best described as being concave. An animal with this type of body shape is known as a *medusa* (plur., medusae). Some cnidarians have the medusa shape all of their lives. Other species exhibit the medusa shape in part of the life cycle. Figure 10.6 shows a simplified diagram of a medusa. Notice the *gastrovascular cavity*; extending downward from it is the centrally located *manubrium*, where the mouth is positioned. Tentacles hang from the outer edges of the jellylike bowl. The gonads are suspended under the radial canals and open inward in some species and outward in others.

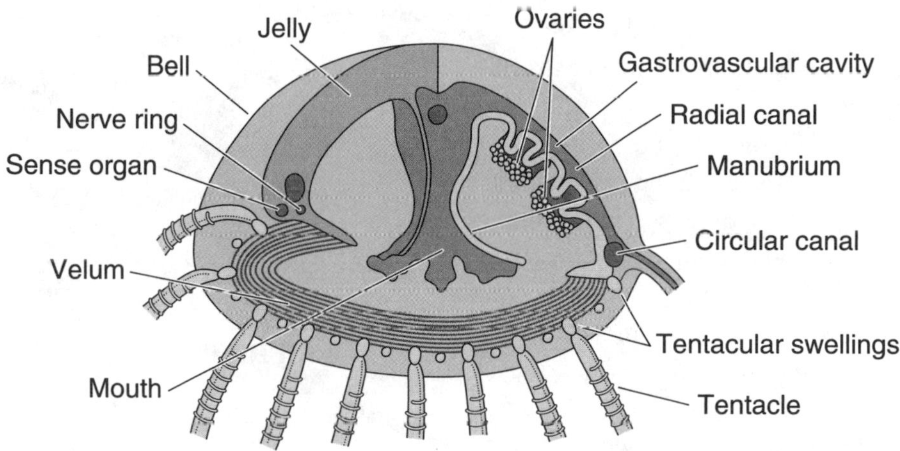

FIGURE 10.5 The structure of a jellyfish

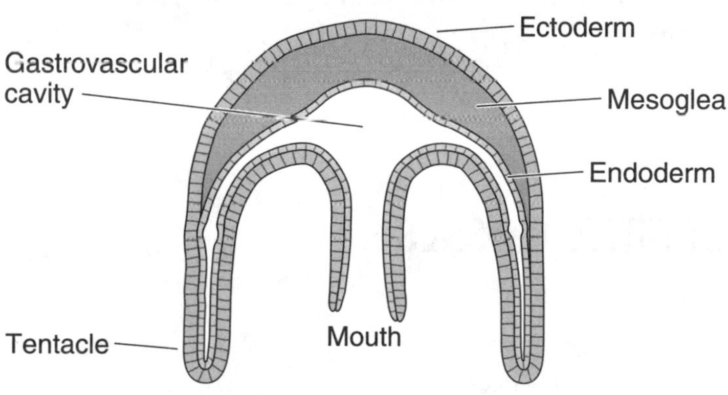

FIGURE 10.6 Diagram of medusa

A number of Cnidaria species have a body shape resembling a vase or cylinder. This type of body is known as a *polyp*. For some species the polyp is a stage in the life cycle; in others, the polyp is the adult body form. Figure 10.7 provides a simple diagram of a polyp. Notice that the proximal end is attached. The mouth, surrounded by tentacles, is at the distal end of the animal.

The gastrovascular cavity, where digestion takes place, is a distinctive feature of the animals in this phylum. Enzymes are secreted into the cavity, where food is partially digested extracellularly (outside of cells). The partially digested food molecules are then engulfed by cells lining the cavity; these cells complete the digestion intracellularly (inside of cells). In an evolutionary sense, the cnidarian method of digestion is a signpost pointing toward increased specialization of body tissues and organs for digestion. Particles not digested are expelled through the single *mouth-anus*.

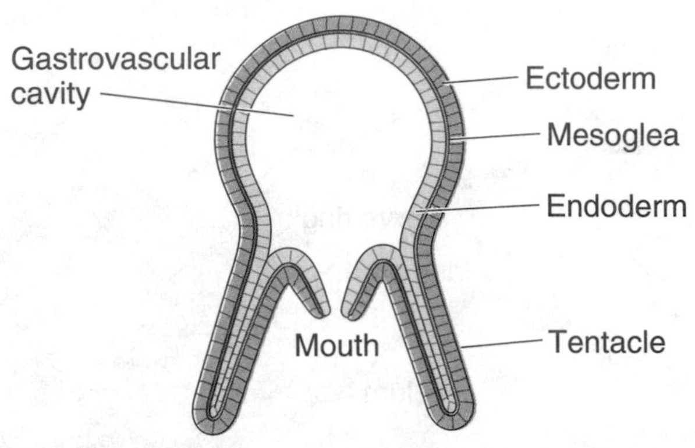

FIGURE 10.7 A diagram of a polyp

Biologists prefer to use the name Cnidaria for this phylum, instead of the former Coelenterata, because of the *cnidocytes* or stinging cells that are in the tentacles of Cnidaria. These animals feed on live prey. They capture smaller animals by immobilizing them with toxins secreted by the stinging cells. Each cnidocyte contains a thread capsule called a **nematocyst**, which discharges a thread in response to touch or chemical stimulation of the *cnidocil*, a trigger-like device. The thread may be barbed or coated with toxin. The nematocyst thread either hooks the prey, ensnares it, paralyzes it with toxin, or does all three of these. Once immobilized, the small animal is swept by the tentacles into the mouth-anus.

REPRESENTATIVE CLASSES

CLASS HYDROZOA

The Hydrozoa are the most primitive of the Cnidaria. Many of the Hydrozoa species are colonial, with the members of a colony showing a division of labor. Some of the hydrozoans go through an alternation of generations in which an **asexual polyp generation** alternates with a **sexual medusa generation**. In this Cnidaria class, the polyp form is dominant. However, the polyps of some species reproduce medusae by the asexual means of budding. The medusae produce gametes (eggs and sperm) and thus begin the sexual generation. Species in the class Hydrozoa include *Hydra, Obelia, Gonionemus*, and the Portuguese man-of-war.

Hydra is a freshwater cnidarian that is representative of a genus of the same name. *Hydra* is a polyp and has no medusa form in its life history. In length, this hydrozoan is about 12 millimeters and has about eight tentacles that surround the mouth-anus (Figure 10.8). Hydras move about by somersaulting, end over end. The animal's locomotion is made possible by cells that have locomotor and sensory functions. Reproduction in *Hydra* is sexual and asexual. A single organism produces both egg and sperm, which are discharged into the water, where fertilization takes place. Asexual reproduction occurs by budding.

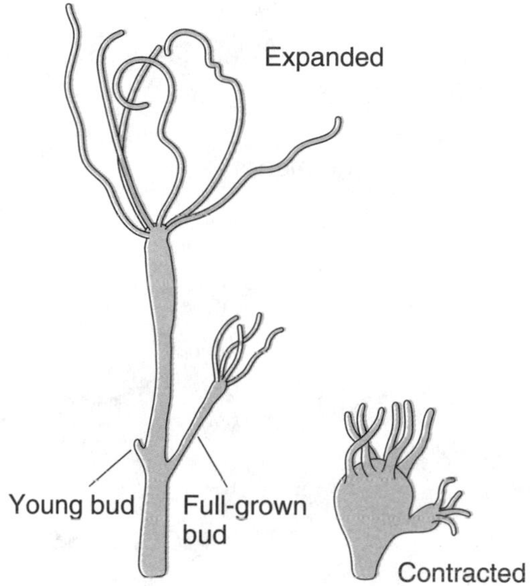

FIGURE 10.8 *Hydra*, expanded and contracted

Nervous response in *Hydra* is controlled by a simple nervous system. Slim, pointed *sensory cells* scattered throughout the endoderm and ectoderm layers receive stimuli. From the sensory cells, the sensory impulses are passed to nerve cells which form a *nerve net* spread throughout the ectoderm. The nerve net coordinates *Hydra*'s activities, enabling the animal to respond to chemical and tactile stimuli in the environment. The nerve net is a very primitive nervous system in which nervous impulses travel in either direction.

Obelia is a marine hydroid—a colony of subindividuals. The size of the colony may range from 2.5 to 10 centimeters in height. It is sessile, attached to substrate rocks along the shoreline. Look at Figure 10.9. You can see that a number of polyps are attached to a stalk. These polyps are specialized for carrying out specific life functions. Some polyps are vegetative and attend to the feeding needs of the group; others have a reproductive function. The *hydrotheca* is a transparent covering that surrounds the vegetative polyps; the *gonotheca* covers the reproductive polyps.

The reproductive polyps do not have tentacles and consequently cannot feed. They produce new polyp individuals by budding. The reproductive polyps produce another kind of bud that resembles a saucer. This bud escapes the polyp through an opening in the gonotheca, becomes free swimming and develops into a medusa, resembling a jellyfish. The medusa is the sexual stage producing egg and sperm. Fertilization takes place in the water. The zygote develops into a ball of cells which changes into a ciliated larva and then becomes a young polyp. The polyp attaches itself and settles down. On maturity, the polyp begins colony building by budding and the cycle repeats. Thus, there is an alternation of an asexual generation with a sexual generation.

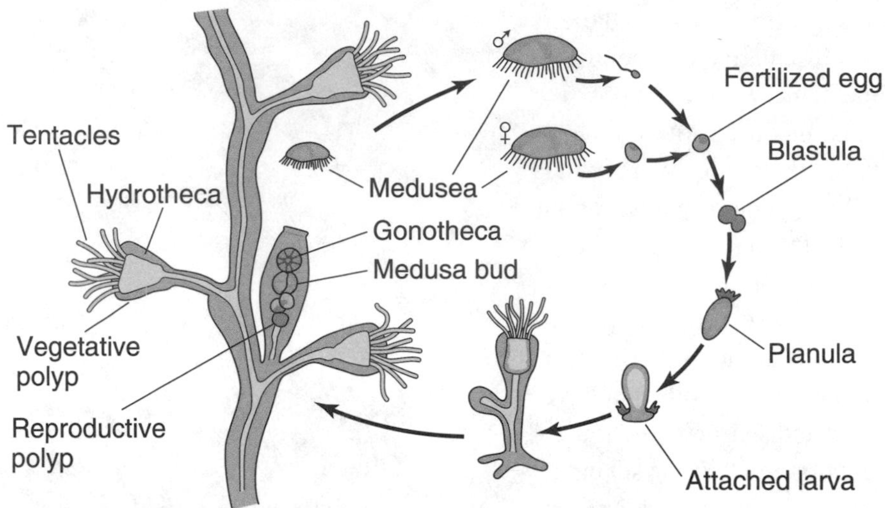

FIGURE 10.9 Life cycle of *Obelia*

CLASS SCYPHOZOA—JELLYFISH

The Scyphozoa represent the true jellyfish. These cnidarians are free-swimming and inhabit marine waters. In size, a jellyfish may be as small as 2 centimeters in diameter or may be an enormous 4 meters across with trailing tentacles about 10 meters in length. As shown in Figure 10.5, a jellyfish has the medusa body form with notches in the margin of the bell. Four *oral arms* extend from the mouth opening. Surrounding these arms are four *gastric pouches* where digestion occurs. A complex of canals radiate through the medusa. Attached to the membranes of the canals and encircling the gastric pouches are the gonads.

In some species of jellyfish a polyp stage occurs, but it is subordinate and inconspicuous compared to the medusa. Reproduction by the polyp is asexual, accomplished by a kind of terminal budding known as *strobilation*. Reproduction by the medusa is sexual.

Jellyfish show greater cell specialization than *Hydra* or *Obelia*. Underlying the ectoderm are true muscle cells that propel the animal through water by regular contractions. Sensory nerves connect with the fibers of the nerve net and serve as the nerve supply for the tentacles, the muscle cells, and the sense receptors. Now for the first time we see the emergence of true sense organs: **statocysts** and **ocelli**. The statocyst is a sense organ specialized for receiving and coordinating information that enables an organism to orient itself in respect to gravity. Statocysts are spaced around the margin of the bell. Each statocyst is composed of a circle of *hair cells*, which surround a central hardened crystal of calcium carbonate known as the **statolith**. In response to movement of the statolith, the hair cells send impulses to the nerve fibers. The organism can adjust to an up or down position in the water as indicated by the statocyst. The ocelli are groups of light receptor cells located at the base of the tentacles.

CLASS ANTHOZOA—SEA ANEMONES

This class of Cnidaria includes the sea anemones, sometimes called "animalflowers," because of their brightly colored tentacles (Figure 10.10), and the coral-building animals. Both of these groups are polyps and have no medusa stage. They are sessile, attached to substrate rocks at the edge of the sea. Anthozoans have numerous tentacles that surround an elongated mouth which opens into a tube (stomodaeum) that extends into the gastrovascular cavity. Reproduction may be asexual by budding or sexual involving eggs and sperm.

FIGURE 10.10
Sea anenome

The epithelial cells of the coral-building cnidarians secrete calcium carbonate walls in which the living polyps hide themselves. The compounds secreted by these polyps build limestone coral reefs.

Phylum Platyhelminthes–Flatworms

The tissues and the organs of the flatworm are developed from the three primary germ layers: ectoderm, mesoderm, and endoderm. The simplest of the flatworms demonstrate bilateral symmetry. This phylum represents a step up the evolutionary scale showing definite development of excretory, nervous and reproductive systems. Most flatworms are hermaphrodites, and most are parasites, including several serious parasites of humans and other animals. Among the flatworms are the planaria, flukes, and tapeworms.

THE PARASITIC WAY OF LIFE

By definition, a **parasite** is an organism that lives on or inside of the body of a plant or animal of another species and does harm to the host. Parasites offer physical discomfort to the host and tend to kill slowly, meanwhile having had time to reproduce themselves for several generations. **Ectoparasites** live on the host's body: body lice, dog fleas, ticks. **Endoparasites** live within the hosts's body and exhibit several adaptations for life in an intestine or in muscle or in the blood.

REPRESENTATIVE CLASSES

CLASS TURBELLARIA—PLANARIA

Planaria is a genus of small, freshwater, free-living flatworms that are studied extensively in biology classes. Figure 10.11 illustrates the general external structure of a planarian. The digestive system consists of a ventral mouth-anus, a pharynx that can be pushed outward, a digestive cavity and an intestine. A true body cavity (coelom) is not present in *Planaria* as indicated by lack of an anus separate from the mouth. The excretory system of *Planaria* consists of a network of branching tubes. The outer ends of these tubes open to the outside through an excretory pore. The inner portion of the tube connects to ciliated cells called **flame cells**, which remove excess water from spaces around the cells.

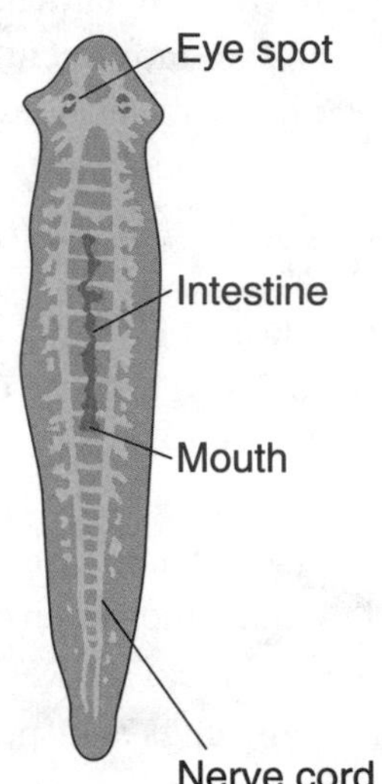

FIGURE 10.11
Planaria

A consequence of bilateral symmetry is the refinement of body systems. The nervous system of *Planaria* represents an evolutionary advancement. The nerve net of *Hydra* is replaced by two lateral nerve cords extending longitudinally from the anterior end to the posterior end of the body. The presence of anterior **ganglia**, or groups of nerve cells, to sort and coordinate nerve impulses is a signpost to **cephalization**, the development of the head region. The eyespots or *ocelli* can distinguish light from dark and can also discern the direction from which the light comes. The head region has many **chemoreceptors**, which aid in the locating of food.

Of special interest to students of biology is the regenerative powers of *Planaria*. Figure 10.12 illustrates the regeneration process. Some turbellarians are parthenogenic, producing individuals from nonfertilized eggs.

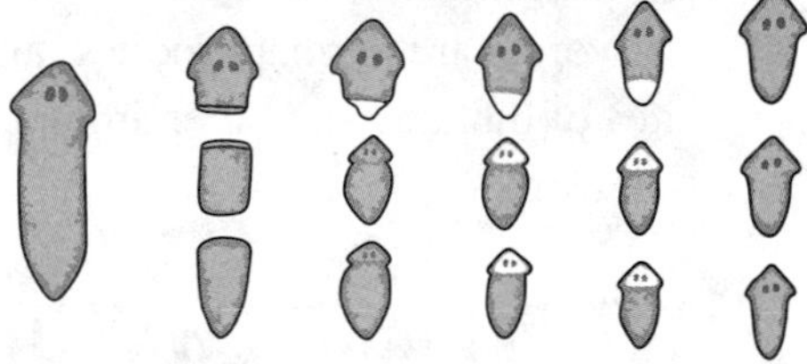

FIGURE 10.12 Regeneration in *planaria*

CLASS TREMATODA—FLUKES

Opistorchis sinensis is the species name of the Chinese liver *fluke*, a parasite for which the life cycle involves three hosts: snails, fish, and humans. Aquatic (water-dwelling) snails eat fluke eggs that have been discharged into the water in contaminated human feces. Once inside the snail, the eggs reproduce asexually, resulting in hundreds of offspring. Three generations are nonmotile. The fourth generation of offspring, free-swimming larvae, is discharged into the water. These **cercaria** larvae burrow under the scales of fish, where they encyst in the muscles. When the infected fish are eaten by humans, the encysted cercariae become activated, mature into adult flukes, and move into the bile ducts and livers of their human hosts.

The body plan of the fluke is adapted for parasitism (Figure 10.13). Most flukes range in length from 1 to 5 centimeters. A sucker at the head end and one on the underside enable them to adhere to host organs. The internal body consists of a muscular throat (pharynx) that sucks body fluids in through the mouth and passes them into a two-branched intestine. The greater portion of the internal body is filled with reproductive organs. Flukes are **hermaphroditic**, possessing the reproductive machinery for producing both eggs and sperm.

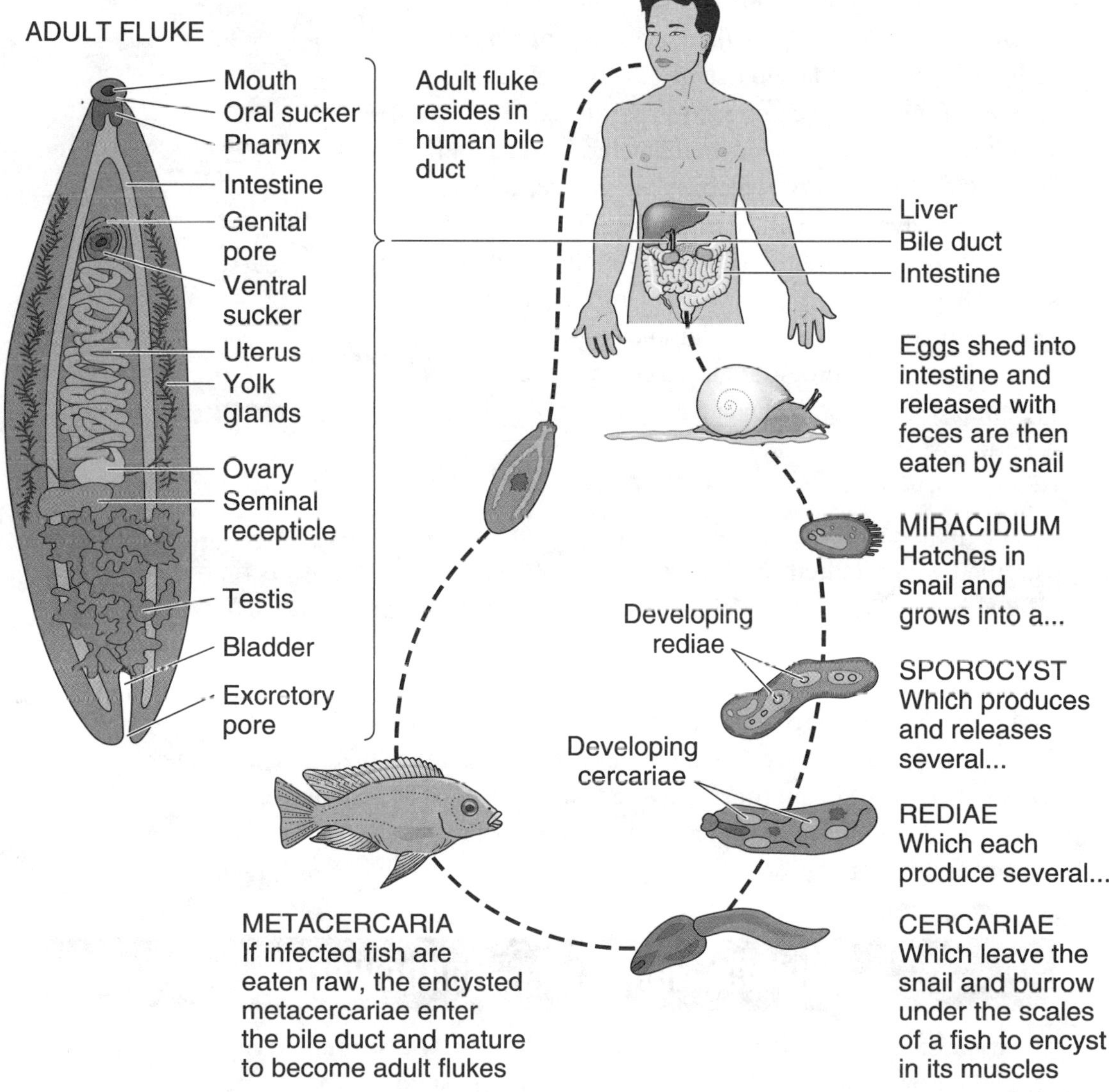

FIGURE 10.13 Life history of the Chinese liver fluke

Blood flukes of the genus *Shistosoma* infect more than 200 million people in 70 nations of the tropics. The human host is robbed of nutrition by the fluke invaders, which cause a wasting of the limbs, distended abdomen, intestinal malfunction, and urinary disorders. At this time the newer, more effective drugs now available seem to be the best means of diminishing the effects of fluke infection.

CLASS CESTODA—TAPEWORM

Taenia solium is the species name of the tapeworm that infects pigs and people. Having no mouth or digestive system, the adult worm obtains nourishment by absorption through its body wall from the human intestine where it lives. The body of *Taenia* consists of a head, called the **scolex**, attached to a neck which is followed by a series of body segments called **proglottids**. The scolex, which is about 2 millimeters in diameter, is fitted with hooks and four suckers, which enable the worm to attach itself to the intestine wall. The proglottids bud from the neck and become progressively more mature and larger as the chain of segments moves toward the posterior end of the animal. A chain of proglottids may be two to three meters long. Each proglottid has a complete reproductive system producing both egg and sperm. The nervous and excretory systems extend through the chain of proglottids (Figure 10.14).

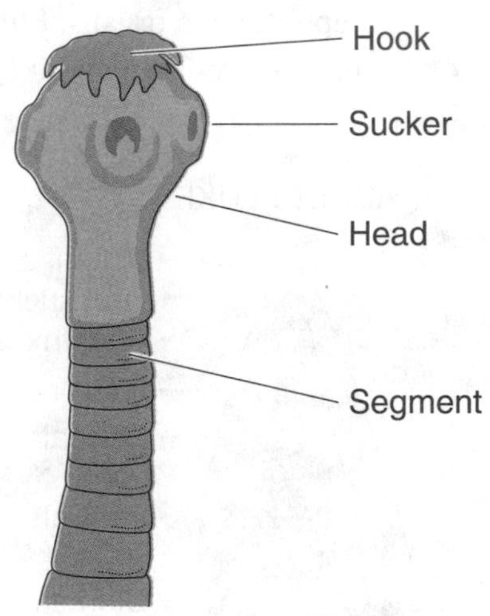

FIGURE 10.14 Tapeworm

The life cycle of the tapeworm involves two hosts. A mature proglottid contains a sac filled with hundreds of fertilized eggs. When the proglottid walls rupture, the ground becomes infected with fertilized eggs. If these eggs are ingested by a pig, the protective walls surrounding each egg are digested, releasing developing embryos of the tapeworm into the digestive system of the pig. These embryos bore into the pig's capillaries and are carried by the blood to the muscles where the scolex forms a cyst. The worm now remains encysted in the muscles (meat) of the pig. If the butchered pig (now called pork) is improperly cooked and eaten by a human, the encysted worm becomes activated. Its head begins to bud proglottids and the cycle of infection repeats.

Phylum Nematoda–Roundworms

The roundworms, also called thread worms, are widely distributed, living in the mud of salt and fresh water and in soil. Roundworms are small, tapered at both ends; the elongated body is covered with an enzyme-resistant cuticle. The free-living species have well-developed sense organs, including eyespots and complex mouthparts; parasitic forms are simpler in structure. Anterior ganglia connect with dorsal and ventral nerve cords, articulating with smaller branching nerve fibers. Male and female gametes are produced by separate sexes.

Ascaris lumbricoides is a roundworm parasite that lives in the human intestinal tract. It ranges in length from 15 to 40 centimeters. The eggs escape through the feces of an infected individual and contaminate soil or water. People become infected with *Ascaris* by eating contaminated food.

Hookworms are small nematodes, measuring about seven centimeters in length. Well-developed hooks surround the male genitalia and are present in the mouths of both sexes. Embryos in contaminated soil bore through the skin of human feet and travel through the blood vessels to the lungs. Then they bore through the lung tissue into the bronchi and the windpipe into the throat. They are swallowed and pass into the intestines where they become attached to the intestinal wall. Hookworms drain blood from the host and often cause severe anemia.

Phylum Annelida–Segmented Worms

An important indicator of evolutionary advancement is the development of the body cavity known in technical language as the **coelom**. Species in the phylum Annelida possess a true coelom. The coelom is a body cavity that has two openings, beginning with an anterior *mouth* and terminating in a posterior *anus*. It is lined with **mesoderm** and separates the internal body organ systems from the muscles of the body wall. The coelom allows space for the development of complex and specialized systems: circulatory, digestive, reproductive and excretory.

GENERAL CHARACTERISTICS

The annelids are segmented worms that live in soil, fresh water, or the sea. Most of the annelids are free-living, although some of the marine forms burrow in tubes and some species (class Myzostoma) are parasites on echinoderms.

The body of an annelid is divided into a series of similar segments and is said to be **metamerically segmented**. Most annelids have a closed **circulatory system** where the blood is contained in vessels. Enlarged muscular blood vessels function as hearts and pump the blood through the system of vessels. Annelids may be **dioecious** (have separate sexes) or hermaphroditic. Most annelid species go through a ciliated larval stage known as the *trochophore* larva. This is a larva of evolutionary importance because the same type appears in several phyla.

Table 10.1 summarizes the important characteristics of three classes of annelids. The first two classes are discussed in more detail.

REPRESENTATIVE CLASSES

CLASS POLYCHAETA—SANDWORM

Nereis inhabits burrows in sand and rocks at the edge of the sea (Figure 10.15). The body of the worm is markedly segmented. The head end is well-developed. It has a muscular pharynx that can be extended outward to capture food. The pharynx is equipped with a pair of hard curved jaws which are designed for grasping. The head bears four simple eyes. four short tentacles, and two longer ones. Each body segment following the head has fleshy, protruding appendages called *parapodia* from which grow bristles, or *setae* (sing., seta).

The digestive system of *Nereis* is a straight tube consisting of a pharynx, esophagus, and a stomach-intestine, where the major part of digestion takes place. Undigested food is eliminated through the anus. Oxygen is taken into the body through the skin covering the parapodia, and it diffuses through the thin walls of the blood vessels. The blood of *Nereis* contains the red pigment hemoglobin and is enclosed in muscular vessels. A pair of coiled **nephridia** in each segment filter out waste materials. Well-developed ganglia in the dorsal part of the head relay nerve signals from the sense organs to the ventral nerve cord. Lateral branching nerves from the ventral nerve innervate the body organs. The sexes are separate. Mature gametes are discharged into the water where fertilization takes place. The zygote develops into a trochophore larva.

TABLE 10.1
PHYLUM ANNELIDA

Class	Examples	Characteristics
Polychaeta	sandworm, sea mouse, fanworm, lugworm	Marine; segmented worms; setae, parapodia; tube-dwelling, free-living; sexes separate
Oligochaeta	earthworm, giant Australian worm	Freshwater and land (*Tubifex*) dwelling; segmented worms; setae, no parapodia; hermaphroditic
Hirudinea	leeches	Mostly freshwater, some marine, few terrestrial; body flattened; reduced segmentation and body cavity; no circulatory system or setae; ectoparasites, predators, scavengers

CLASS OLIGOCHAETA—EARTHWORM

Lumbricus terrestis is representative of the earthworms and is studied extensively in biology classrooms. The earthworm burrows in moist soil, feeding on organic materials in the earth (Figure 10.16). The body is segmented and may have over one hundred metameres. Its under-developed head without eyes and tentacles is well suited to a burrowing way of life.

FIGURE 10.15 Nereis, the sandworm

The digestive system consists of a mouth, a pharynx, a long narrow esophagus, a thin-walled crop, a muscular gizzard, an intestine, and an anus. The intestine has a dorsal infolding called the **typhlosole**, an adaptation for increasing surface area necessary for absorptive purposes. The respiratory system of *Lumbricus* is similar to that of *Nereis* except that in the earthworm oxygen diffuses through moist body skin, not through parapodia. Paired nephridia are present in each body segment and are used to filter out wastes from the coelomic fluid. Excretion of these wastes takes place through the **nephrostome**, a ventral excretory pore. In the skin, there are sense organs that function as light and touch receptors. The ventral nerve cord with its branching nerve fibers and the anterior ganglia are not unlike that of the sandworm.

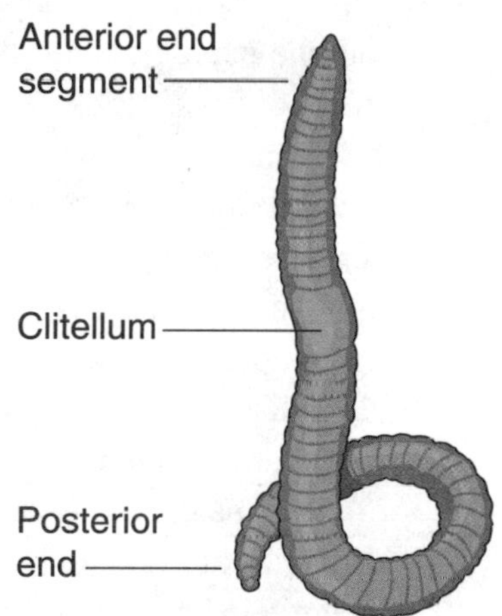

FIGURE 10.16 Earthworm

Lumbricus is a hermaphrodite. However, earthworms do not self-fertilize; they exchange sperm during copulation. The **clitellum**, a smooth circle of tissue on the outside of the body, is the place where copulating worms attach to each other.

Phylum Mollusca–Clams and Their Relatives

The mollusks include the chitons, snails, clams, scallops, squids, and octopuses. This is one of the largest animal phyla and includes about 1,000 species.

GENERAL CHARACTERISTICS

Mollusks are generally soft-bodied, nonsegmented, and usually enclosed within a calcium carbonate shell. They are most abundant in marine waters, although some species inhabit fresh water and others live on land. All mollusks have a *mantle*, a flattened piece of tissue that covers the body and which may secrete the calcareous shell. The body of the mollusk is described as being a **head-foot**, a muscular mass having different shapes and functions in the various classes. Between the body and the mantle is the **mantle cavity**, which functions in respiration. A large *visceral mass* contains most of the body organs. Water enters through an *incurrent siphon* and is expelled through an *excurrent siphon*.

The mollusks are the most highly developed of the nonsegmented animals and are considered to be the most advanced invertebrates. They have well-developed digestive, respiratory, excretory, and reproductive systems. The circulatory system includes a two-chambered heart equipped with an auricle and a ventricle. Oxygen-laden blood is pumped both anteriorly and posteriorly through two arteries to parts of the body. Blood carrying respiratory wastes is collected through a vein that is closely applied to the nephridia. The deoxygenated blood is transported back to the gills. The excretory system has a pair of nephridia that excrete filtered wastes through an excretory pore into the mantle cavity. The excurrent siphon removes the dissolved wastes out of the shell.

The nervous system contains three pairs of ganglia positioned near the esophagus, in the foot, and at the end of the visceral mass. These ganglia are connected by transverse nerve fibers. Simple sense receptors in the form of sensory cells sensitive to light and touch are positioned in the margin of the mantle.

In most mollusks, the sexes are separate. The reproductive system is a mass of gland tissue that lies in the muscular foot near the coiled intestine. In some species sperm are conducted through the excurrent siphon of the male into the mantle cavity of the female by way of the incurrent siphon, and fertilization takes place in the mantle cavity of the female. In other species, eggs and sperm are released into the water where fertilization takes place. Some mollusks—oysters, scallops, and primitive worm-like forms—are hermaphroditic.

The fertilized eggs of clams and other bivalves develop into a larval stage known as a *glochidium*. The glochidium is discharged into the water and attaches as a parasite to the gills of fish, where it stays until reaching maturity.

Table 10.2 summarizes the important characteristics of the major classes of mollusks. One class—the bivalves—is discussed in more detail.

A REPRESENTATIVE CLASS–PELECYPODA (BIVALVIA)

The clam is an excellent representative of this class, usually known as the bivalves because of the presence of two valves or shells. The right and left valves are hinged on the dorsal side and held tightly together by muscles attached to the inner surfaces of the shells. Lining these inner surfaces is the membranous mantle. The cavity inside of the shells is the mantle cavity.

The clam has a large muscular foot, which can be thrust between the opening of the shells and which contracts and expands during locomotion due to the action of protractor and retractor muscles attached to the inner faces of the shells. A large visceral mass at the base of the foot contains most of the animal's organs.

Four parallel gill plates are adjacent to the visceral mass. These gills function in respiration. Oxygen from the water in the mantle cavity diffuses into the gills, which are surrounded by capillaries. Oxygen diffuses into the capillaries and then into the bloodstream. Carbon dioxide diffuses out into the surrounding water by way of the capillaries and the gills.

The digestive system of the clam is complete. A mouth occupies the anterior end of the visceral mass. Leading from the mouth is a short esophagus followed by a stomach and a coiled intestine extending partly into the foot. At the end of the intestine is an anus. A bilobed digestive gland is on either side of the stomach (Figure 10.17).

TABLE 10.2
PHYLUM MOLLUSCA

Class	Examples	Characteristics
Amphineura	chiton	Marine; bilaterally symmetrical; ventral foot; shell of eight calcareous plates; gills in mantle cavity.
Pelecypoda (Bivalvia)	clam, mussel, oyster, scallop, shipworm	Marine and freshwater; body enclosed in right and left shells; head reduced; filter feeders; compressed foot; sexes separate or hermaphroditic.
Gastropods	snail, whelk, limpet, slug	Marine, freshwater, land; most with spiral, single shells; well-developed head having tentacles and eyes; foot for locomotion; shell absent in slugs; trochophore larva.
Cephalopods	squid, octopus, nautilus	Marine; well-developed head with large eyes; shell absent or present; tentacles; siphon used for locomotion; ink gland; sexes separate.

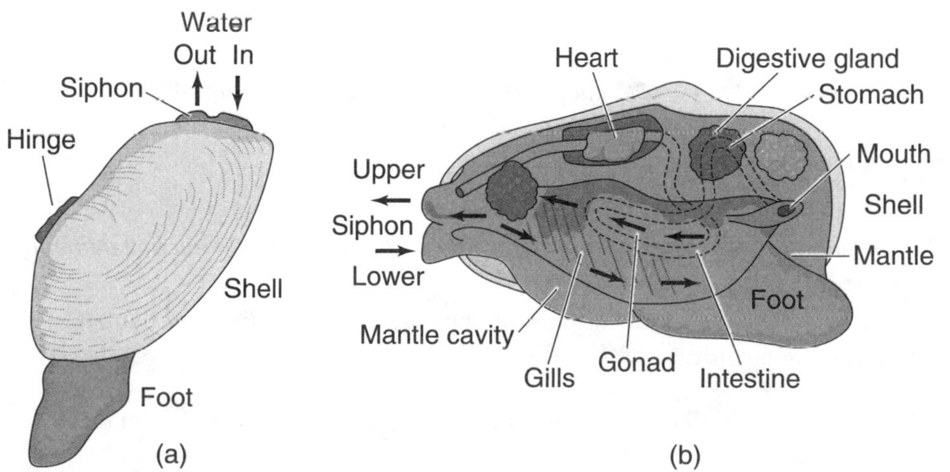

FIGURE 10.17 Structure of the clam—(a) external view; (b) internal structures

REVIEW EXERCISES FOR CHAPTER 10

WORD-STUDY CONNECTION

amphiblastula
anterior
antimeres
blastula
budding
cephalization
cercaria
chemoreceptors
clitellum
cnidarian
cnidoblast
cnidocil
cnidocyte
coelom
collar cells
dioecious
distal
dorsal
ectoderm
ectoparasite
endoderm
endoparasite
filter feeder
flame cells
fluke
ganglia
gastrovascular cavity
glochidium
gonotheca
hair cells
hermaphrodite
hydrotheca
larva
lateral
mantle
manubrium
medusa
mesenchyme cells
mesoderm
mesoglea
metameres
myoneme
nematocyst
nephridia
nephrostome
ocellus
parapodia
parasite
parthenogenesis
polyp
posterior
primary germ layer
proglottids
proximal
regeneration
scolex
setae
spicules
spongin
statocyst
statolith
strobilation
symmetry
trochophore larva
typhlosole
ventral

SELF-TEST CONNECTION

PART A. Completion. *Write in the word that correctly completes each statement.*

1. Relative positions of body organs on opposite sides of a dividing line are described in terms of _____.

2. The portion of the fish fin that is attached to the body is the _____ end.

3. Invertebrates are animals without _____.

4. Budding is a form of _____ reproduction.

5. The name Porifera means _____ bearing.

6. Needle-like crystals that make up the skeleton of a sponge are called _____.

7. An organism that produces both egg and sperm is known as an _____.

8. Cnidocytes are _____ cells.

9. A thread capsule found in Cnidaria tentacles is the _____.

10. *Hydra* lives in _____ water.

11. The polyps in a colony of *Obelia* reproduce asexually by _____.

12. Orientation to gravity is controlled by jellyfish sense organs known as _____.

13. *Taenia* is the genus of the _____.

14. *Ascaris* is a parasite belonging to phylum _____.

15. Infection by hookworm can be avoided by wearing _____.

16. The true coelom first appears in the _____.

17. Animals in which the sexes are separate are said to be _____.

18. The body of all mollusks is covered by a membranous _____.

19. The incurrent and excurrent siphons are characteristic of species in the phylum _____.

20. The head-foot body is characteristic of the phylum _____.

PART B. Multiple Choice. *Circle the letter of the item that correctly completes each statement.*

1. Bilateral symmetry is characteristic of animals that have a
   (a) shell and plates
   (b) head and tail
   (c) foot and coelom
   (d) spines and tube feet

2. *Distal* is a body directional term that refers to
   (a) a forward end
   (b) a backward end
   (c) an attached end
   (d) a free end

3. In animal cells, excess nutrients are stored as
   (a) glucose and glycogen
   (b) glucagon and fatty acid
   (c) glycogen and fat
   (d) glycogen and protein

4. The growing back of a lost part is known as
   (a) regeneration
   (b) budding
   (c) vegetative propagation
   (d) parthenogenesis

5. A true statement about sponges is
   (a) The cells are organized into tissues.
   (b) They move quite rapidly.
   (c) The digestive system is well developed.
   (d) The cells show a division of labor.

6. Types of sponge cells that can change their form and function are
   (a) collar cells
   (b) flame cells
   (c) mesenchyme cells
   (d) epithelial cells

7. In the Cnidaria the body is composed of
   (a) only undifferentiated cells
   (b) one tissue layer
   (c) two tissue layers
   (d) three tissue layers

8. In jellyfish, digestion takes place in the
   (a) velum
   (b) tentacle
   (c) manubrium
   (d) gastrovascular cavity

9. Medusae reproduce by
   (a) budding
   (b) eggs and sperm
   (c) regeneration
   (d) parthenogenesis

10. Alternation of generations is demonstrated in the life cycle of
    (a) *Hydra*
    (b) sea lily
    (c) pelecypods
    (d) *Obelia*

11. Coral reefs are formed by
    (a) lime-secreting polyps
    (b) the bodies of sea anemomes
    (c) the shells of foraminifera
    (d) jellyfish statocysts

12. Flatworm embryos differentiate into
    (a) ectoderm, mesoglea, mesoderm
    (b) ectoderm, mesoglea, endoderm
    (c) ectoderm, mesoderm, endoderm
    (d) ectoderm, mesenchyme, endoderm

13. Bilateral symmetry brings with it
    (a) refinement of body systems
    (b) the development of the nerve net
    (c) increased reproduction
    (d) alternation of generations

14. Each tapeworm segment has a complete
    (a) digestive system
    (b) nervous system
    (c) reproductive system
    (d) locomotor organ

15. A true statement about *Planaria* is
    (a) *Planaria* are parasites.
    (b) *Planaria* lack powers of regeneration.
    (c) *Planaria* show the first signs of cephalization.
    (d) *Planaria* have segmented bodies.

16. The circulatory system of the annelid is
    (a) open
    (b) closed
    (c) partly closed
    (d) varying

17. The term *metamerism* refers to
    (a) male and female gonads in the same animal
    (b) a parasitic infestation of humans
    (c) an evolutionary trend in roundworms
    (d) a series of identical segments

18. The trochophore larva
    (a) is a fossil remnant
    (b) shows relationships between phyla
    (c) is part of the roundworm life cycle
    (d) has free-flowing pseudopods

19. A true statement about the earthworm is
    (a) Earthworms are self-fertilized.
    (b) The earthworm sexes are separate.
    (c) Earthworms copulate and exchange sperm.
    (d) Earthworms copulate and exchange eggs.

20. Cercaria larvae are best associated with
    (a) hydras
    (b) earthworms
    (c) flukes
    (d) jellyfish

PART C. Modified True-False. *If a statement is true, write "true" for your answer. If a statement is incorrect, change the <u>underlined</u> expression to one that will make the statement true.*

1. A true body cavity is lined with <u>mesoderm</u>.

2. The dorsal region of the body is the <u>under</u> surface.

3. Contractile proteins are associated with <u>feeding</u>.

4. The Cnidaria are the first group to show <u>tissue</u> organization.

5. Sponges have an <u>external</u> skeleton.

6. Spongin is a fibrous <u>carbohydrate</u>.

7. The bowl shape of a jellyfish is known as an <u>umbrella</u>.

8. The body form of *Hydra* is a <u>polyp</u>.

9. The nerve net is associated with <u>sponges</u>.

10. Hair cells are part of the ocelli.

11. "Animal-flowers" refer to jellyfish.

12. The function of flame cells is related to digestion.

13. Proglottids are segments in Planaria.

14. The head of a tapeworm is known as the proglottid.

15. *Taenia* lives in two hosts: human and the snail.

16. Annelids are round worms.

17. *Nereis* is the genus name of the tapeworm.

18. The sandworm has a well developed head.

19. The most advanced phylum of invertebrates is Annelida.

20. The life cycle of the Chinese liver fluke includes four host organisms.

CONNECTING TO CONCEPTS

1. What purpose is served by describing the animal body in terms of symmetry?
2. How do powers of regeneration affect the mortality of the sponge?
3. How has the body of the tapeworm been adapted for a life of parsitism?
4. Why are the mollusks considered to be the most advanced of the invertebrates?

ANSWERS TO SELF-TEST CONNECTION

PART A

1. symmetry
2. proximal
3. backbones
4. asexual
5. pore
6. spicules
7. hermaphrodite
8. stinging
9. nematocyst
10. fresh
11. budding
12. statocysts
13. tapeworm
14. Nematoda
15. shoes
16. Annelids
17. dioecius
18. mantle
19. Mollusca
20. Mollusca

PART B

1. **(b)**	6. **(c)**	11. **(a)**	16. **(b)**
2. **(d)**	7. **(c)**	12. **(c)**	17. **(d)**
3. **(c)**	8. **(d)**	13. **(a)**	18. **(b)**
4. **(a)**	9. **(b)**	14. **(c)**	19. **(c)**
5. **(d)**	10. **(d)**	15. **(c)**	20. **(c)**

PART C

1. true	6. protein	11. sea anenomes	16. segmented
2. back	7. medusa	12. excretion	17. sandworm
3. movement	8. true	13. tapeworms	18. true
4. true	9. *Hydra*	14. scolex	19. Mollusca
5. internal	10. statolith	15. pig	20. three

CONNECTING TO LIFE/JOB SKILLS

Marine science is the study of the oceans and the plants and animals that live therein. A number of careers are related to marine science, such as **marine biologist, taxonomist, marine archeologist,** and **fisheries biologist**. All of these careers require a B.S. and an M.S. in biology or marine biology. A Ph.D. is required for research opportunities. Other related careers worth looking into are **underwater film maker, oceanographer, aquarist** (a scientist who heads an aquarium), and **marine science teacher**. Use the Internet to find more information about these careers. Your school or local library will be of help to you also.

Chronology of Famous Names in Biology

1907 **H. V. Wilson** (United States)—discovered that individual cells of the sponge, if left undisturbed in a culture dish, will reform sponge aggregates.

1929 **Charles M.Yonge** (Scotland)—wrote a treatise on the biology of coral reefs in which he elucidated the patterns of behavior of coral reef invertebrates.

1939 **Ernest E. Just** (United States)—developed basic research techniques to study eggs of marine animals.

1940 **Libbie Hyman** (United States)—was a noted authority on the invertebrates; demonstrated metabolic gradients in *Planaria*.

1965 **T. L. Lentz** and **R. J. Barrnett** (United States)—elucidated the fine structure of the *Hydra* nervous system; demonstrated how the feeding response is stimulated in *Hydra*.

1967 **Donald Kennedy** (United States)—discovered the effects of the abdominal ganglia on reflex behavior in the slug.

1971 **A.O.D. Williams** (United States)—discovered the function of brain cells in mollusks as these cells relate to behavior.

1975 **Charles M. Yonge** (Scotland)—elucidated the life cycle of the giant clam *Tridocna gigas*.

1980s **Martin Wells** (England)—carried out major studies on octopus behavior and the sensory functions of the arms.

11

Invertebrates: The Arthropoda

WHAT YOU WILL LEARN

In this chapter you will learn about one of the most successful groups of animals on Earth. The arthropods comprise a large number of adaptable animal species living on land, swimming in fresh water and in the oceans, and flying in the air.

SECTIONS IN THIS CHAPTER

- Major Representative Classes
- Importance of Arthropoda to Humans
- Review Exercises for Chapter 11
- Connecting to Life/Job Skills
- Chronology of Famous Names in Biology

OVERVIEW

The Arthropoda is the largest animal phylum in numbers of individuals, encompassing approximately 800,000 species, more than all of the other animal species combined. It includes spiders, ticks, mites, lobsters, crabs, insects, centipedes, and millipedes (Figure 11.1). Arthropods are widely distributed, occupying habitats in marine, freshwater, and terrestrial environments.

The name ***arthropod*** in literal translation means "jointed foot," a distinctive characteristic of this group, expressed traditionally as "*jointed appendages*." The arthropods are segmented animals protected by an exoskeleton made of protein and the flexible but tough carbohydrate chitin. The chitinous exoskeleton is fashioned in articulating plates held together by hinges covering both the body and the appendages, and attached to muscles that make possible quick and unencumbered movements.

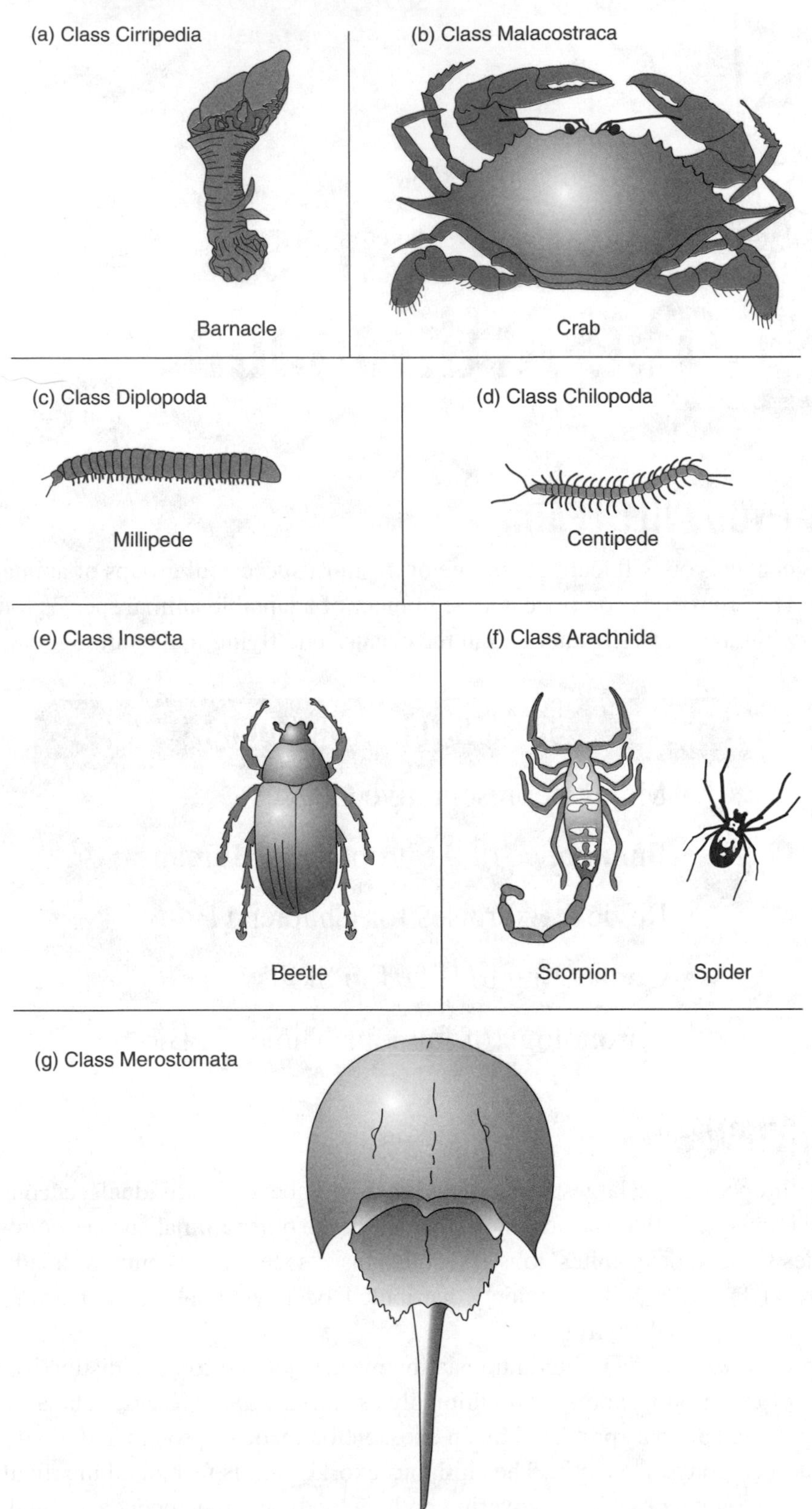

FIGURE 11.1 Some representative classes of arthropods

The body cavity (coelom) is reduced in size and in its place is a *hemocoel* (blood cavity), a part of the **open circulatory system** in which a pulsating section functions as a dorsal heart. **Hemolymph** bathes the body organs directly because it is not confined to vessels. Respiration in many arthropod species makes use of **tracheal tubes**, which communicate to the external environment by **spiracles**, pores that appear one pair to a segment. Some arthropods use gills as organs of respiration. Nervous impulses are carried on a ventral nerve cord occupying a mid-position in the body. The sexes are separate. Some of the insects reproduce parthenogenetically, a process in which eggs develop without fertilization.

Major Representative Classes

The phylum Arthropoda is divided into several classes. The Arachnida, Merostomata, Malacostraca, Insecta, Diploda, and Chilopoda are the most important. Table 11.1 summarizes the characteristics of these classes, and a more detailed discussion of three classes follows.

TABLE 11.1
PHYLUM ARTHROPODA

Class	Examples	Characteristics
Arachnida	spider, tick, mite, scorpion, harvestmen	Mostly terrestrial; small-to moderate size; body divided into cephalothorax and abdomen; 6 pairs of appendages; 4 pairs of walking legs; no antennae; simple eyes; book lungs or trachae.
Merostomata	horseshoe crab	Marine; cephalothorax covered with chitinous exoskeleton; 5 walking legs; 5 pairs of chelicerae adapted for walking.
Malacostraca	crab, shrimp, lobster	Mostly aquatic; marine; gill breathers; exoskeleton of chitin; chewing mouthparts; 2 pairs of antennae; 3 or more pairs of legs.
Insecta	crayfish, water flea sow bug grasshopper, dragonfly, cockroach, termite, louse, flea, fly, butterfly, bee, beetle, mosquito	Freshwater Terrestrial Mostly terrestrial, freshwater, no marine forms; body divided into head, thorax, abdomen; mouthparts for biting, sucking, lapping; 2 pairs of wings; 3 pairs of legs; breathing by tracheae; excretion by Malpighian tubules; 1 pair of antennae; 2 compound eyes.
Diplopoda	millipede	Terrestrial; herbivores; head with antennae and chewing mouthparts; body segmented; 2 pairs of walking legs ventral on each segment; 2 eyes; breathing by trachea; 2 pairs of spiracles on each segment.

Class	Examples	Characteristics
Chilopoda	centipede	Terrestrial; carnivorous; body segmented; 1 pair of walking legs lateral on each segement; 1 pair of long antennae; poison fangs on first body segment; chewing mouthparts.

CLASS ARACHNIDA–SPIDERS

By careful examination of a spider, you would be able to note the major external characteristics. The body is divided into two regions: the **cephalothorax** and the **abdomen**. The cephalothorax, an arthropod characteristic, is the fusion of the head and the thorax. The cephalothorax of the spider supports six pairs of jointed appendages. The first appendage has been modified into jaws called **chelicerae**. The second are the **palps**, sense receptors and grasping organs, leg-like in females but bulbous in males. The remaining appendages are four pairs of walking legs characteristic of arachnids.

In spiders, the posterior end of the abdomen lying underneath (ventral to) the anus contains several pairs of rounded projections called the **spinnerets**. They contain a group of flexible tubules through which silk secreted by the silk glands leaves the body when the spider spins a web.

On the anterior dorsal surface of the cephalothorax are eight simple eyes. These are the external parts of specialized systems that control the physiological processes of the spider. Nerve fibers connect the eight eyes with a nerve mass that surrounds the esophagus, where the ventral and dorsal ganglia unite. Branching nerve fibers service the various parts of the body.

The food of the spider consists of body juices of other animals. These juices are drawn through the mouth and the esophagus by action of a sucking stomach, which then passes the food on to another stomach where digestion takes place. Digestion is aided by five pairs of pouched glands. The digestive tract ends in a sac-like rectum and an anus.

Most spiders breathe by means of **book lungs**. These are chambers or cavities containing many thin, hollow membranous plates. The hollow spaces are connected to the outside by fine tracheal tubes. Oxygen is distributed by the blood circulating around the hollow spaces (sinuses). The heart is in the dorsal part of the abdomen fitted into the *pericardial cavity*. Blood enters the heart through a pair of valve-like openings and is pumped out through vessels which empty into the body spaces.

Excretion in the spider is controlled by a pair of **Malpighian tubules**. These are long slender tubules attached at one end to the digestive tract. Nitrogen-containing wastes in the body fluid are changed into uric acid which is then moved through the Malpighian tubule to the end of the digestive tract where it is ultimately excreted as dry crystals (Figure 11.2).

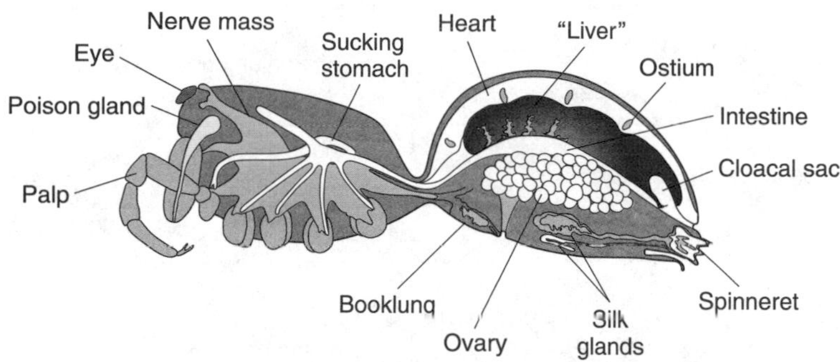

FIGURE 11.2 Internal structure of the spider

The sexes in the spider are separate. The sex organs are positioned in the ventral part of the abdomen and communicate with the outside through a ventral opening. During copulation, the male uses modified palps to transfer sperm to the female, effecting internal fertilization. The fertilized eggs are laid in a silk cocoon by some species of spiders. In other species, the eggs are carried around by the female until they are hatched.

CLASS MALACOSTRACA–LOBSTERS AND THEIR RELATIVES

Malacostraca is a class of the phylum Crustacea, mandibulate arthropods. Derived from the Latin *cursta*, meaning "crust," the name Crustacea describes the lobsters and their relatives aptly. The crustacean body is covered by a tough exoskeleton arrangement in the form of arched plates that thin out at the joints to permit maximum movement. The lobster is representative of this class.

EXTERNAL CHARACTERISTICS

The segmented body is divided into a cephalothorax and an abdomen. Six segments of the head and eight segments of the thorax are fused into a single cephalothorax; seven segments compose the abdomen. Pairs of jointed appendages are outgrowths of each of the body segments except the first and the last (Figure 11.3).

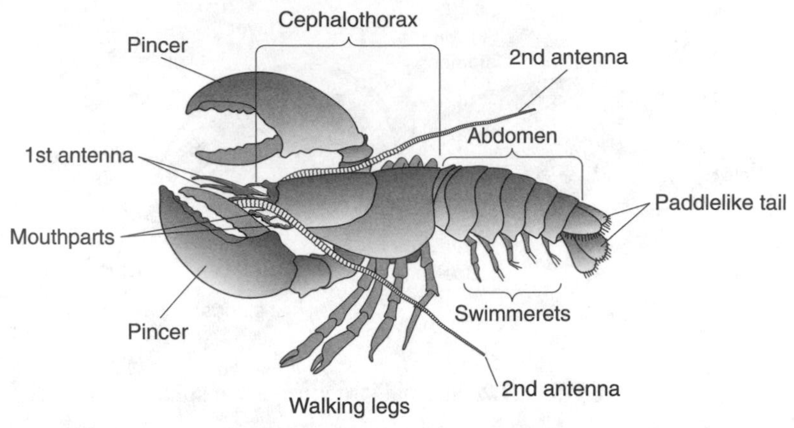

FIGURE 11.3 External structure of the lobster

The first body segment of the cephalothorax has a pair of **compound eyes** (Figure 11.4) positioned at the ends of long flexible stalks that can be extended or retracted. The compound eye is an arthropod characteristic that represents an evolutionary development in light receptors. The compound eye is a collection of thousands of light-gathering units, the **ommatidia**, each with its own lens and light-sensitive cells. Each unit makes its own picture of part of an object. These separate pictures are put together in a pattern resembling a collage so that the compound eye perceives thousands of pictures of the same object but at slightly different angles. The compound eye does not render a clear picture of an object but is a very efficient device for spotting movement of prey.

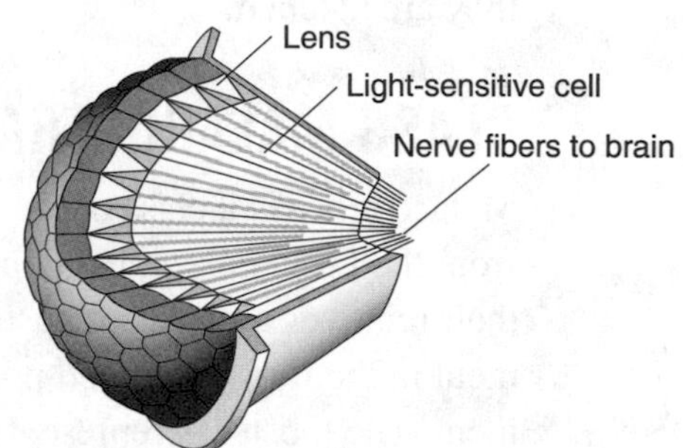

FIGURE 11.4 A diagram of the arthropod compound eye

Segments 2 and 3 on the cephalothorax bear pairs of **antennae**, sense receptors for touch and chemical stimuli. On segments 4–9 are appendages that have been modified into biting and chewing mouthparts. Segment 10 bears two large pincers (claws) used to fight off enemies and to capture food. Segments 11–14 have four pairs of walking legs. Segments 15–19 are abdominal segments; they bear appendages called **swimmerets** that function mainly in circulating water over the gills. There are two paddle-like appendages on segment 20 which constitute the tail, an appendage modified for swimming.

INTERNAL STRUCTURE

The respiratory system consists of feathery **gills**, outgrowths of the body wall that lie on both sides of the body. The gills, surrounded by blood vessels, are bathed by water in the **gill chamber**, a space under the anterior part of the exoskeleton. The gills absorb oxygen from the water and release carbon dioxide into the water.

The heart is located dorsally. Blood enters the heart through three pairs of valve-like openings. It is pumped out through several arteries which branch to all parts of the body, where it delivers oxygen and food to the cells. Blood is emptied into the body sinuses which drain into the capillaries of the gills. Here blood exchanges carbon dioxide, a waste product of respiration, for oxygen.

The excretory system consists of a pair of **green glands** that are located in the head. These glands filter out the nitrogenous wastes produced during protein metabolism and pass them into a small **bladder**. The bladder empties out through a small pore located at the base of the antennae (Figure 11.5).

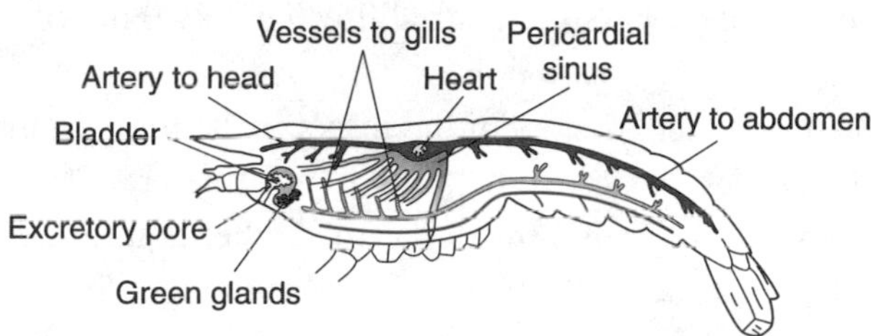

FIGURE 11.5 Excretory system of the lobster

The digestive system consists of a mouth, a short esophagus, a stomach divided into cardiac and pyloric chambers, an intestine, and an anus. The cardiac portion of the stomach contains special grinding organs known as the *gastric mill*. Finely ground food enters the pyloric chamber of the stomach where most of the digestion takes place and where useful nutrients are absorbed into the blood.

The nervous system includes a dorsal "brain," which is really a large ganglion situated in the cephalothorax above the digestive system. These rings of tissue connect with a ventral chain of ganglia from which a double nerve cord extends into the tail. Nerves branch out from the dorsal brain and the ventral cord (Figure 11.6).

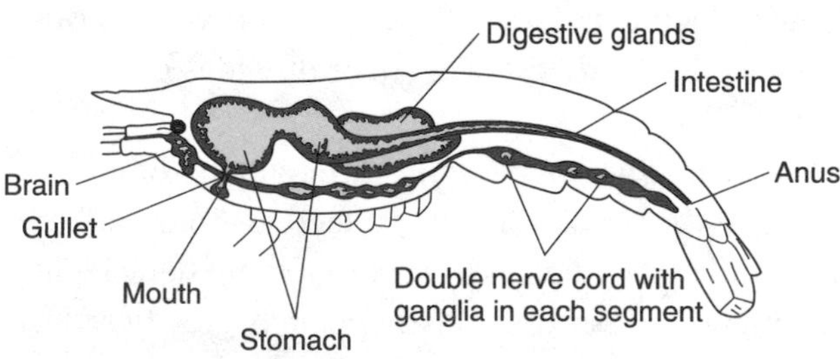

FIGURE 11.6 Digestive and nervous systems of the lobster

Sexes in the lobster are separate. Bilobed gonads are located ventrally in the cephalothorax. Sperm cells leave the testis by way of *sperm ducts* that open into pores at the base of the fifth pair of walking legs. *Oviducts* in the female conduct eggs from the ovary to openings at the base of the third pair of walking legs. The early embryos develop while they are attached to the swimmerets of the female.

CLASS INSECTA

The Insecta represent the most advanced class of the modern arthropods. Scientists estimate that the number of species in this class is probably as high as several million. To date, more than one-half million have been described in scientific literature. It is known that there are more kinds of insects than all of the other animals combined. The insects are by far the dominant form of terrestrial life. The diversity of the insects is unmatched.

There are many reasons for the success of the insects. The modern insects are relatively small organisms ranging in size from 1.5 to 50 millimeters. Their small body size and consequently small food requirement has made them quite successful in the struggle for survival. Another reason for their survival is the fact that they have wings and can fly away from predators or toward food sources. (In fact the insects are the only invertebrates that can fly.) That insects have developed adaptations for survival can be attested to by their wide distribution in terrestrial and aquatic environments. Insect species occupy habitats from the equator to the arctic and from sea level to the snow fields of the highest mountains. Many live in fresh water during the larval stages and several species spend their adult lives near the water.

GENERAL CHARACTERISTICS

There are several characteristics that all insects have in common. The insect body is divided into a well-defined head, thorax and abdomen. Segments 1 to 6 are fused to form the head on which there are a pair of compound eyes and one or more simple eyes. Segment 2 carries a single pair of antennae. The mouthparts vary according to the species of insect. The grasshopper, for example, has biting mouthparts, whereas in the butterfly the mouth is structured for sucking and in the mosquito the mouth serves as a piercing device. Segments 7–9 make up the thorax. Each of these segments bears a pair of walking legs. Most insects bear a pair of wings on the last two segments of the thorax.

The composition and structure of wings differ among orders of insects. For example, the front wings of the beetle serve as protective armor and are generously impregnated with chitin, while the wings of the dragonfly are delicate and membranous. The second wing pair in mosquitoes and grasshoppers is greatly reduced in size and these stubs are used as balancing organs rather than for flight purposes.

The number of segments that make up the insect abdomen varies from a maximum of 11 downward. Although the abdomen bears no appendages, it does have a number of breathing pores (spiracles). The spiracles open into air tubes called tracheae that conduct oxygen to all of the body cells.

The life history of an insect may involve several stages in which the young form of the insect does not resemble the parent. Such a life cycle in which there is change of body form is known as *metamorphosis*.

METAMORPHOSIS

Metamorphosis usually involves distinct stages. The life cycle of the butterfly provides an excellent example of four stages: (1) the egg or embryonic stage (2) the *larva* or feeding stage (3) the *pupa* or cocoon stage and (4) the adult stage. The larva is an active stage of life when the feeding organism is entirely different in body form from the parent. The caterpillar, for example, does not resemble the adult butterfly. The pupa is a quiescent nonfeeding stage in the life cycle (Figure 11.7).

FIGURE 11.7 Complete metamorphosis in the butterfly

The process of metamorphosis in insects is under control of **hormones** secreted by cells in the brain. Hormones are protein molecules that are secreted by **endocrine** (ductless) gland cells into the blood for transport to the sites of action. Hormones are specific and stimulate certain *target* cells or organs to grow or to carry out general metabolic activities. *Brain hormone*, secreted by certain cells in the insect's brain, stimulates the *prothoracic gland*, which, in turn, secretes **ecdysone**, a growth and differentiation hormone. Ecdysone controls the molting of the larva and the change into the pupa stage. Another endocrine gland, the *corpus allatum*, lies near the brain in the larva (caterpillar) and secretes a hormone called **juvenile hormone**. This hormone encourages larval growth and molting (shedding of skin), but prevents the larva from changing body form. After the corpus allatum stops secreting juvenile hormone, metamorphosis to the adult body form occurs.

REPRESENTATIVE ORDERS

The large number of insect species are classified into more than 15 orders. The members of these orders vary in size, feeding habits, habitat, life cycle, and behavior, with some showing complex patterns of social behavior. A detailed discussion of three orders follows. The grasshopper illustrates the structure of a typical insect; the honeybee, the life style of a social insect. The mosquito is important in public health.

Order Orthoptera—Grasshopper

External Characteristics Figure 11.8 shows parts of the grasshopper. The head of the grasshopper is clearly defined. There are two lateral compound eyes separated by three simple eyes or ocelli. The single pair of antennae have sensory functions. The grasshopper has biting mouthparts, which include a pair of chewing jaws, a pair of flap-like structures for manipulating food, a lower lip-like structure, and a tonguelike organ used for the mastication and manipulation of food. The forewings are hard, leathery, and opaque. The hind wings are thin, membranous, and transparent.

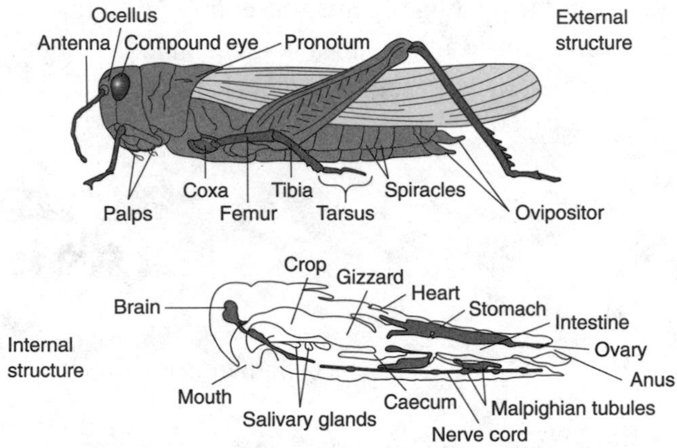

FIGURE 11.8 Parts of the grasshopper

The thorax of the adult grasshopper is divided into three clearly seen parts: the prothorax (with a saddle-like covering called the *pronotum*), mesothorax, and metathorax. The end of the abdomen is modified to form the sex organs. In males, the posterior tip of the abdomen is rounded and the projection used in copulation is present. The posterior end of the female's abdomen is a forked structure, the *ovipositor*, used to dig holes into which the eggs are deposited.

The grasshopper has three pairs of legs; one pair each is on the prothorax, mesothorax, and metathorax. The third pair of legs are strong and elongated, adapted for jumping.

Internal Structure The digestive system consists of a mouth that leads into a narrow esophagus, which connects with a thick-walled grinding organ, the **gizzard**, which is followed by a thin-walled stomach. The narrow intestine ends with the anus. The cavity of the mouth gets secretions from two salivary glands. Six doubleblind sacs surround the gizzard and stomach and empty enzymes into the stomach.

On the first segment of the abdomen, one of each side of the animal, are two **tympanic** membranes. These structures respond to vibrations and sound waves. On two thoracic segments and on eight segments of the abdomen are the spiracles, the external openings of the tracheal tubes and air sacs that branch (ramify) throughout the body. The small branches of the tracheal tubes are in contact with the body cells conducting oxygen to them and removing carbon dioxide wastes from them.

The circulatory system includes a dorsal heart. Blood enters the heart through five pairs of valve-like openings called *ostia* and is pumped into the body sinuses. The hemocoel receives blood from all parts of the body and returns it to the heart. The blood of the grasshopper is merely a circulating medium for food and wastes.

The excretory system consists of numerous fine **Malpighian tubules** which reabsorb water and excrete protein wastes in the form of dry uric acid crystals.

The sexes in the grasshopper are separate. The male reproductive system consists of two **testes** located near the intestine, a sperm duct, and a copulatory organ situated at the tip of the abdomen. In the female, **eggs** are formed in two ovaries and travel through an oviduct to the vagina. During copulation, the seminal receptacle of the female receives and stores sperm. When eggs pass through the vagina, they are fertilized by the stored sperm. The female deposits the eggs in the soil by digging a hole with the ovipositor.

The zygote develops quite rapidly into an embryo, but then growth and development stop for a while. This rest period in embryonic growth is called *diapause*, and, in the case of the grasshopper, lasts over the winter. Development resumes in the spring and by early summer young grasshoppers, called **nymphs**, emerge. The nymphs are wingless but, after a series of molts, grow to adult size.

Order Hymenoptera—Honeybee

The honeybee exhibits a specialized social structure called **polymorphism**, a condition in which individuals of the same species are genetically specialized for different functions (Figure 11.9). In a honeybee colony, three classes of individuals arise: fertile haploid males called **drones** (n); fertile diploid females or **queens** ($2n$); and sterile diploid females or **workers** ($2n$). The workers have a special concave surface on the second pair of walking legs called a *pollen basket*, which is used to carry pollen.

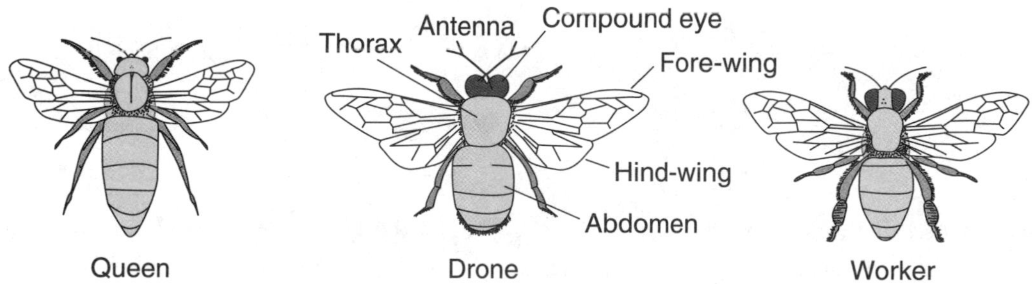

FIGURE 11.9 Polymorphism in the honeybee

The queen bee receives sperm from the drone once during her lifetime. The sperm are stored in a special organ called the **spermatotheca**, in which they may live for years. Fertilized eggs give rise to females, most of which remain workers. A special female may be selected by the colony and fed a diet of "royal jelly," which causes her to grow larger than the others and become fertile. This fertile female will become a queen, and either take over the existing colony or start a colony of her own. Drones develop from unfertilized eggs by the process of parthenogenesis.

Order Diptera—Mosquito

Characteristics A large number of insects are classified as dipterans because they exhibit similar structural characteristics. Houseflies, fruit flies, gnats, and mosquitoes each have only one pair of functional wings. A second pair are reduced to small **halteres**, rounded bodies representing incompletely developed (vestigial) posterior wings. The larvae undergo a complete metamorphosis.

The body of the female mosquito is elongated and covered with scales. The legs are long and fragile looking. The mouthparts are structured for piercing and shaped like an elongated **proboscis**. The antennae (feelers) of the male are bushier than those of the female. Both male and female feed on plant juices, which they obtain through the piercing proboscis. Female mosquitoes, however require a blood meal in order for their eggs to mature. Blood is obtained by biting animal organisms that have blood.

The eggs are laid on the surface of still water, where they develop into larvae called **wrigglers**, a term descriptive of the way in which they swim with jerky, wriggling movements. Mosquito larvae feed on algae and organic debris. In most insects the pupae are quiet, but mosquito pupae, called, **tumblers**, are active, breathing through structures on the thorax.

Mosquito species are identified by the resting position of the adult and the shape of the larva. Figure 11.10 shows the adult resting position and larva of the *Culex* mosquito in the resting position. Figure 11.11 is a similar illustration for the *Anopheles* mosquito.

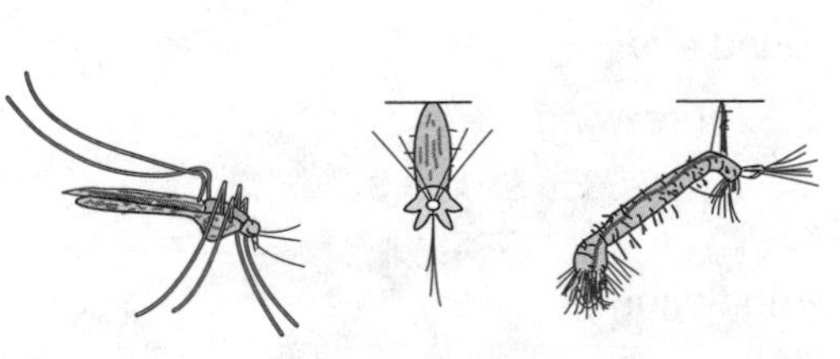

FIGURE 11.10 *Culex* mosquito: adult in the resting position and larva

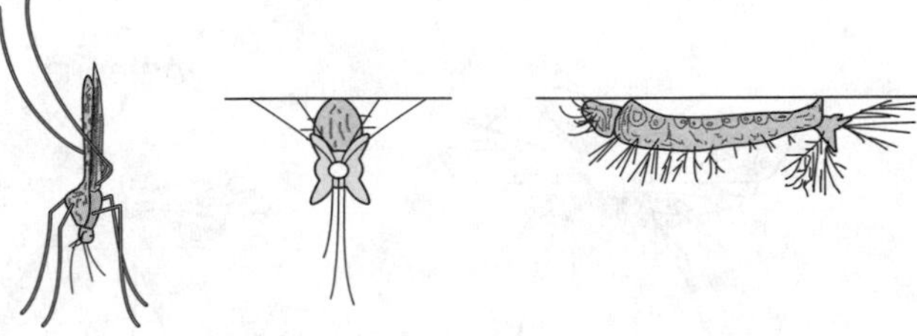

FIGURE 11.11 *Anopheles* adult in the resting position and larva

The Mosquito as a Disease Carrier Globally, mosquito species number about 2,500. Mosquitoes are a public health concern because several species transmit diseases that disable and even kill humans. In Chapter 7 you read about plasmodium, the mass of protist cells that causes **malaria**. Plasmodium is transmitted (carried) by the *Anopheles* mosquito. About 200,000 people a year (usually in Third World countries) contract malaria, and many suffer and die from the disease. Occasional cases of malaria show up in the southern part of the United States.

In northeast America, the *Culex* mosquito is the carrier of the virus that causes **encephalitis**, a disease in which inflammation of the brain often induces coma and death. More recently, the mosquito species *Culex pipiens*, identified as the carrier of the West Nile virus, has aroused much concern. The virus is named after the West Nile district in Uganda, where it first appeared. Through the bite of *C. pipiens*, 60 people were sickened and 8 died from encephalitis. All of these infected persons resided in Queens, New York. In other locales, the virus infection has resulted in meningitis.

Public health officials have expressed alarm as they recover the West Nile virus from dead birds and mosquitoes. It is not understood at this time why some people infected with the virus died of high fever and irreversible brain swelling, while others exhibited relatively mild symptoms of low-grade fever and achy joints.

Reproduction Mosquitoes are prolific breeders. *Culex pipiens* breeds in any stagnant pool of water, depositing at a single laying 100 or more eggs. Held together by a gelatinous substance, the *C. pipiens* eggs form a floating egg raft. The *Culex* life cycle lasts from 10 to 14 days.

Each species of mosquito, however, demonstrates a different breeding pattern. The *Aedes* mosquito, carrier of yellow fever, dengue (pronounced den-gay) fever, and encephalitis, breeds in flood waters, rain pools, and salt marshes. Long periods of dryness do not affect *Aedes* eggs. The life cycle of this mosquito may be as short as 10 days or as long as several months.

Mosquito—Control Measures Although yellow fever is no longer a threat, there have been increasing numbers of cases of dengue fever and encephalitis. Modern travel has brought the world's countries closer, and therefore diseases no longer remain isolated. Insects move from one country to another via airplane, and so do infected people.

The West Nile virus has been found in New York, Connecticut, Egypt, Israel, South Africa, and Romania. It has been suggested that the virus may have entered the New York area through mosquitoes, viremic humans, and the importation of infected birds.

In New York City, control measures have been put into place to eliminate the breeding sites of the *Culex* strain that carries the West Nile virus. These methods include massive aerial spraying of vegetation and marshy areas to eliminate mosquito breeding sites. Other control measures include application of oil on stagnant pools to clog the breathing tubes of wrigglers, and the use of chemical larvicides.

Identification of the Virus Genome Dr Ian Lipkin of the University of California at Irvine and his research team have identified a complete genome of the West Nile virus. A **genome** is the total number of genes in a virus particle. This is the first time any scientist has cloned a viral genome in the human brain. The technique used by Dr. Lipkin and his colleagues may help identify causes of encephalitis in other outbreaks.

Importance of Arthropoda to Humans

The arthropods have very definite effects on human life. The arachnids influence the quality of human life in diverse ways. Spiders are predators on insects that may offer discomfort or harm to humans, and some feed on decomposed organic matter and serve as decomposers. On the other hand, ticks cause Rocky Mountain spotted fever and mites are spoilers of grain.

The crustaceans for the most part are of great value to humans. Crabs, lobsters, and shrimp serve the cause of human nutrition deliciously.

Insects cannot be avoided. Some insects are helpful and others are harmful. Honeybees provide honey and pollinate flowers. Ladybird beetles destroy other insects that are crop destroyers. Many insects such as flies, fleas, and mosquitoes are carriers of disease; others such as Japanese beetles, tent caterpillars, and grasshoppers cause serious damage to foliage and food crops. A greater-than-billiondollar industry has been built for exterminating insects that infest our homes and gardens. Insects help in the balance of nature and also in certain situations help to unbalance natural communities. This very successful species is human's greatest competitor for food on Earth.

The recent outbreak of encephalitis in countries around the world, including the northeastern United States, has public health scientists concerned. At a time when it was thought that infectious diseases had been conquered, incidences of mosquito-borne viruses have aroused concern in both health officials and the public. It is quite evident that insects that transmit diseases to humans are still a problem for society.

REVIEW EXERCISES FOR CHAPTER 11

WORD-STUDY CONNECTION

abdomen
Anopheles
arthropod
book lungs
cephalothorax
chelicerae
compound eyes
corpus allatum
crustacean
Culex
diapause
drones
ecdysone
encephalitis
gastric mill
gill
gizzard
green glands
hemocoel
hemolymph
hormone
jointed appendages
juvenile hormone
larva
Malpighian tubules
metamorphosis
mosquito
nymphs
ommatidia
open circulatory system
ostia
oviduct
ovipositor
palp
pericardial cavity
polymorphism
proboscis
pronotum
prothoracic gland
pupa
queens
spermatotheca
sperm duct
spinneret
spiracles
swimmerets
tracheal tubes
tumbler
tympanic
West Nile virus
workers
wriggler

SELF-TEST CONNECTION

PART A. Completion. *Write in the word that correctly completes each statement.*

1. The animal phylum that is largest in number of species and number of individuals is the _____.

2. The development of eggs without fertilization is known as _____.

3. The body of a spider is divided into _____ and an abdomen.

4. Simple eyes are known as _____.

5. Green glands have a (an) _____ function.

6. The most advanced class of the modern anthropods are the _____.

7. The grasshopper has _____ mouthparts.

8. Wing stubs in mosquitoes are used for the purpose of _____.

9. A hormone is secreted by _____ or ductless glands.

10. A membrane specialized for gathering sound vibrations is the _____ membrane.

11. The honeybee exhibits a specialized social structure known as _____.

12. The *Aedes* mosquito is historically known as the carrier of the virus that causes _____ fever.

13. Wingless young grasshoppers reach adult size after undergoing a series of _____.

14. The spermatotheca is correctly associated with the type of honeybee known as the _____.

15. In mosquitoes, halteres are vestigial or underdeveloped _____.

PART B. Multiple Choice. *Circle the letter of the item that correctly completes each statement.*

1. The outstanding characteristic of the anthropods is
   (a) membranous wings
   (b) bilobed antennae
   (c) a simple eye
   (d) jointed appendages

2. Book lungs are the breathing mechanisms of
   (a) arachnids
   (b) termites
   (c) lobsters
   (d) grasshoppers

3. Examples of Malacostraca are
   (a) barnacles and beetles
   (b) lobsters and sow bugs
   (c) horseshoe crab and shrimp
   (d) water flea and mite

4. Ommatidia are structural and functional units of the
   (a) thorax
   (b) flame cell
   (c) compound eye
   (d) brain

5. Lobster embryos
   (a) are free-swimming
   (b) fall to the bottom of the sea
   (c) are maintained in the male green gland
   (d) attach to the female swimmerets

6. The group that represents the dominant form of terrestrial life is the
   (a) insect
   (b) human
   (c) scorpion
   (d) spider

7. The insect body is divided into the
   (a) cephalothorax and abdomen
   (b) wing, abdomen, head
   (c) head, tail, wing
   (d) head, thorax, abdomen

8. The breathing holes on the abdomen of the grasshopper connect to
   (a) green glands
   (b) nephridia
   (c) Malpighian tubules
   (d) tracheal tubules

9. A nymph refers to a young
   (a) maggot
   (b) grasshopper
   (c) honeybee
   (d) blowfly

10. The growth and differentiation that take place in insect larva is controlled by the hormone known as
   (a) juvenile
   (b) brain
   (c) ecdysone
   (d) prothoracic

11. If an outbreak of encephalitis should occur in a specific geographic locale, public health investigators would most likely search for organisms belonging to the order
   (a) Hymenoptera
   (b) Diptera
   (c) Orthoptera
   (d) Isoptera

12. *Culex pipiens* is best described as a
    (a) pathogen
    (b) virus
    (c) carrier
    (d) wriggler

13. Plasmodium, a pathogenic protist, is most closely associated with the insect genus
    (a) Hymenoptera
    (b) *Aedes*
    (c) *Culex*
    (d) *Anopheles*

14. The mouthparts of mosquitoes are structured for
    (a) biting
    (b) chewing
    (c) stinging
    (d) piercing

15. In order for their eggs to mature, female mosquitoes require a meal of
    (a) blood
    (b) algae
    (c) honey
    (d) organic debris

PART C. Modified True-False. *If a statement is true, write "true" for your answer. If a statement is incorrect, change the underlined expression to one that will make the statement true.*

1. The body cavity of the arthropods is replaced by the coelom.

2. Arachnids are ants.

3. Malpighian tubules change nitrogenous wastes into dry crystals of bile.

4. Spinnerets circulate water over the gills of the lobster.

5. The lobster has a (an) closed circulatory system.

6. The only flying invertebrates are the waterfleas.

7. Butterfly mouth parts are structured for biting.

8. The feeding stage of the developing insect is the pupa.

9. Change in body form of the butterfly is known as cocoon.

10. Juvenile hormone is secreted by the corpus allatum.

11. Humans infected with the West Nile virus most likely will contract the disease enteritis.

12. Mosquito larvae are inactive while breathing through tubes.

13. The total number of genes in a virus particle is known as a genome.

14. Species of mosquitoes are identified by their flying positions.

15. Houseflies and gnats each have two pairs of wings.

CONNECTING TO CONCEPTS

1. There are 80,000 species of Arthropoda. Why is this phylum so successful?

2. How does metamorphosis extend the life of the butterfly?

3. How does the compound eye of the lobster work?

4. Describe polymorphism in the honeybee.

ANSWERS TO SELF-TEST CONNECTION

PART A

1. Arthropoda
2. parthenogenesis
3. cephalothorax
4. ocelli
5. excretory
6. insects
7. biting
8. balancing
9. endocrine
10. tympanic
11. polymorphism
12. yellow
13. molts
14. queen
15. wings

PART B

1. **(d)**
2. **(a)**
3. **(b)**
4. **(c)**
5. **(d)**
6. **(a)**
7. **(d)**
8. **(d)**
9. **(b)**
10. **(c)**
11. **(b)**
12. **(c)**
13. **(d)**
14. **(d)**
15. **(a)**

PART C

1. hemocoel
2. spiders
3. uric acid
4. swimmerets
5. open
6. insects
7. sucking
8. larva
9. metamorphosis
10. true
11. encephalitis
12. active
13. true
14. resting
15. one pair

CONNECTING TO LIFE/JOB SKILLS

Have you ever tried to organize people in your neighborhood to carry out a community project? A very worthwhile project would be to involve the folks in your community to carry out an on-going search for stagnant pools of water or other places where mosquitoes are likely to breed. Consult your local health department on the best way to eliminate areas that pose a threat to public health.

Chronology of Famous Names in Biology

1669 **Marcello Malpighi** (Italy)—first to describe metamorphosis in the silkworm; discovered the excretory tubules of insects.

1737 **Jan Swammerdam** (Netherlands)—prepared a detailed monograph on insect structure, including many fine illustrations.

1824 **Straus-Durckheim** (Germany)—produced an illustrated work on insect anatomy; a detailed study of the European beetle, the cockchafer.

1834 **Leon Dufour** (France)—published monographs describing the anatomy of several insect families.

1841 **George Newport** (England)—elucidated the embryology of insect phyla.

1864 **Franz Leydig** (Germany)—was the first to study tissues of insects under the microscope.

1959 **V. B. Wigglesworth** (England)—elucidated the physiological process that controls metamorphosis in insects.

1963 **G. G. Johnson** (England)—determined the means by which flight carries insects over large distances.

1964 **Wolfgang Beerman** and **Ulrich Clever** (Germany)—discovered that chromosome puffs seen in insect species are active genes.

1965 **Miriam Rothschild** (England)—described the role of hormones in the life cycle of the rabbit flea.

1967 **Suzanne Batra** and **Lekh Batra** (United States)—studied the mutualistic relationship between some insects and fungi.

1977 **G. Adrian Horridge** (England)—explained the workings of the ommatidia of the insect compound eye.

1978 **Lorus Milne** and **Margery Milne** (United States)—presented a research study on the four types of insects that live on the surface of quiet waters.

1980 **William G. Eberhard** (United States)—discovered the function of the horns in the horned beetle as organs of lifting.

1982 **Thomas D. Seeley** (United States)—discovered how honeybees locate a site for a hive.

1999 **Ian Lipkin** (United States) and colleagues—identified 100 percent of the genome of the West Nile virus that caused the encephalitis outbreak in New York. The colleagues are **Xi-Yu Jie, Thomas Brieze, Ingo Jordan, Andrew Rambadt, Hen Chang Chi, John S. Mackenzie, Roy A. Hall,** and **Jacqui Scharret**.

12

Deuterostomes, Chordates, Mammals

WHAT YOU WILL LEARN

In this chapter you will explore how animals carry out their life functions. All animals share certain basic characteristics. Animal characteristics become more defined with increasing complexity in body structure.

SECTIONS IN THIS CHAPTER

- The Deuterostomes
- Phylum Chordata—the Chordates
- Subphylum Vertebrata—the Vertebrates
- Review Exercises for Chapter 12
- Connecting to Life/Job Skills
- Chronology of Famous Names in Biology

EVOLUTIONARY LINKS

Coelomates (see-loh-mates) are animals that have a true body cavity completely lined by mesoderm. In evolutionary consideration, the development of the **coelom** (body cavity) enabled the formation of separate and more efficient body systems. The coelomates include two distinct evolutionary lines consisting of the deuterostomes (represented by the echinoderms) and the chordates.

The Deuterostomes

PHYLUM ECHINODERMATA– SEA STARS AND THEIR RELATIVES

Echinoderms are spiny-skinned invertebrates that include the sea stars, brittle stars, sand dollars, sea urchins, and sea cucumbers. Although they do not look very much like vertebrate animals, the development of the echinoderm embryo strongly resembles that of the chordates in the early stages. The larval stage is free-swimming and shows bilateral symmetry.

GENERAL CHARACTERISTICS

All echinoderms live in the sea. Most species are capable of a very slow, creeping locomotion. The only group of sessile echinoderms is the sea lilies.

The name *echinoderm* means "spiny skin," a distinctive feature of the phylum members. Just under the skin, calcareous spines and plates form a skeleton. Another distinctive characteristic of the echinoderms is pentaradial symmetry: the body is built on a plan of five **antimeres** radiating from a central disc in which the mouth is in the middle. The digestive system is complete, although the anus does not function. The echinoderms have no head and no excretory and respiratory systems. They do, however, have a **water vascular system** composed of a series of fluid-filled tubes that are used in locomotion. Changes of pressure in this system enable an echinoderm to extend and retract **tube feet**. The tube feet are used in locomotion and in some species they are used to capture prey. In the echinoderms, the sexes are separate.

Table 12.1 summarizes the important characteristics of the classes of echinoderms. One class—that of the sea star—is discussed in more detail.

TABLE 12.1
ECHINODERMS

Class	Examples	Characteristics
Crinoidea	sea lily, feather	Sessile, attached by a stalk; branched arms; ciliated tube feet used for feeding; some species are free swimming; more abundant during Paleozoic era.
Asteroidea	sea star	Free moving by means of tube feet; arms branching from a central disc.
Ophiuroidea	brittle stars, serpent stars, basket stars	Free moving; thin flexible arms marked off from disc; tube feet used as sensory organs and for feeding.

Class	Examples	Characteristics
Echinoidea	sand dollar, sea biscuit, sea urchin	Free moving; body fused plates or flattened disc, without free rays, covered with calcareous plates; some species covered with spines.
Holothurioidea	sea cucumber	Free moving; elongated flexible body with mouth at one end; sometimes with tentacles; skeletal elements of the skin reduced.

A REPRESENTATIVE CLASS—ASTEROIDEA (SEA STARS)

The **sea star** is an excellent representative of the echinoderms (Figure 12.1). (Formerly, sea stars were known as starfish.) A sea star has all of the distinguishing characteristics of echinoderms: radial symmetry, spiny skin, tube feet controlled by a water vascular system, no head, and no excretory or respiratory system. Protruding from the wall of the coelom, and extending out between the calcareous plates into the sea water, are the **papulae**, sac-like structures that function as respiratory and excretory organs.

FIGURE 12.1 Some representative echinoderms

The mouth is located in the center of the disc on the underside of the body. The mouth side of the body is called the oral side. The aboral side is the upper surface without the mouth. A short esophagus leads from the mouth to the cardiac portion of the stomach. A constriction in the stomach wall separates the cardiac portion of the stomach from the pyloric part. The cardiac stomach is turned inside out and pushed through the mouth when the sea star is eating. The stomach engulfs the food, usually mollusks or crustaceans, and digests it before pulling the stomach back to the inside. The intestine and the anus of the sea star are practically nonfunctional.

On the aboral side of the sea star is a colored plate known as the **madreporite**. Water enters the sea star through minute openings in this plate. Water is drawn by ciliary action down into the stone canal (made rigid by calcareous rings) to the ring canal that encircles the central disc. The ring canal has five radiating canals that extend into the arms of the sea star. Short side branches connect the radial canals with many pairs of tube feet, which contract and expand in response to the water pressure in the **ampulla**, a muscular sac at the upper end of the tube feet (Figure 12.2).

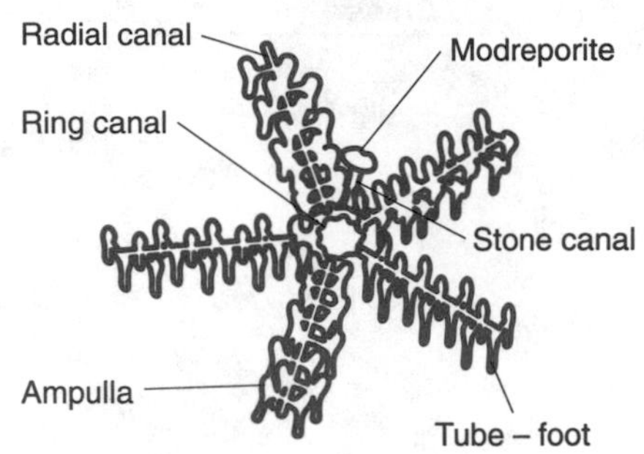

FIGURE 12.2 Water vascular system of the sea star

The nervous system is composed of a nerve ring located in the disc from which a ventral and radial nerve branch into each arm. The radial nerves have finer branches that extend throughout the body. At the tip of each arm is a light sensitive eyespot that is innervated by the radial nerve.

Sea star sexes are separate. Paired gonads are to be found in each ray. The eggs of the female and sperm of the male escape through pores on the aboral surface of the sea star. Fertilization takes place in the water. During embryonic development, sea stars pass through several larval stages.

Sea stars have remarkable powers of **madreporite**. If an arm breaks off, another arm grows back. Should a piece of the central disc be attached to the amputated arm, a new individual will grow from the dismembered part. Sea stars prey on oysters. At one time, oyster "farmers" would clear the oyster beds of sea stars, cut them up, and throw the cut pieces back into the water. What they accomplished was an increase in the numbers of sea stars. In effect, they aided the process of sea star regeneration.

SUBPHYLUM HEMICHORDATA (ENTEROPNEUSTA)

The hemichordates, worm-like animals, are entirely marine. Some species live near the shore, while others burrow in sediments at the bottom of shallow seas. Their bodies are divided into three regions: the *proboscis*, covered with cilia; a short *collar*; and a long *trunk*. The proboscis, used for burrowing, resembles an acorn—thus the name "acorn worms" (Figure 12.3). The cilia that cover the proboscis sweep water, food, and mucus into the mouth. The flattened body contains gill slits, through which excess water is pushed out. A digestive tract extends through the entire length of the body. Food is carried into the pharynx and filtered into the digestive tract.

At one time biologists classified the hemichordates as chordates, believing that the connective tissue in the collar was hollow notochord. Newer evidence shows, however, that this is not so. Hemichordates are considered to be the evolutionary link between the echinoderms and the chordates. The ciliated larva of the hemichords resembles the larval stage of some echinoderms. The pharangeal gill slits are a major chordate characteristic found in all chordates but nowhere else in the kingdom Animalia.

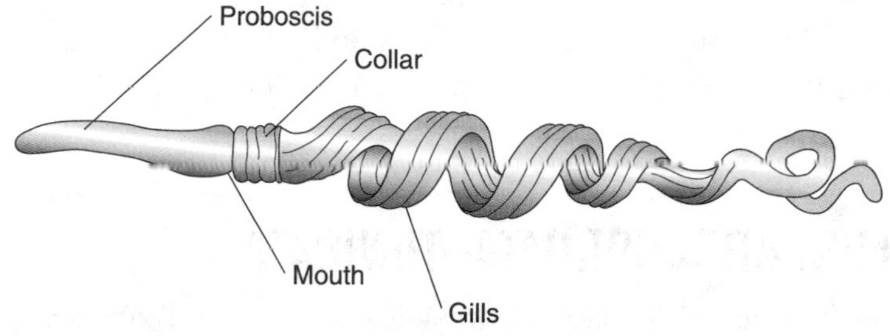

FIGURE 12.3 *Balanoglossus*, the acorn worm

Phylum Chordata–the Chordates

CHARACTERISTICS OF THE CHORDATES

The chordates are animals that have a notochord at some time in their life cycle. The **notochord** is a living, internal skeletal axis in the form of a compact cellular rod-like structure that extends the length of the body. It lies dorsal to the digestive tract and is not to be confused with the backbone. In lower chordates (invertebrate chordates), the notochord prevents the body from shortening when muscles contract. During the embryonic stages of the vertebrate chordates, the notochord is replaced by a column of bones, the **vertebrae**, which form the backbone.

Chordates are set apart from lower animals by several distinguishing characteristics in addition to having a notochord.

1. All chordate embryos have the **three primary germ layers** from which all specialized tissues and organs develop.
2. Chordates are **bilaterally symmetrical** animals with anterior-posterior differentiation (Figure 12.4).
3. The body has a **true coelom** and a digestive tract that begins with a mouth and ends with an anus.

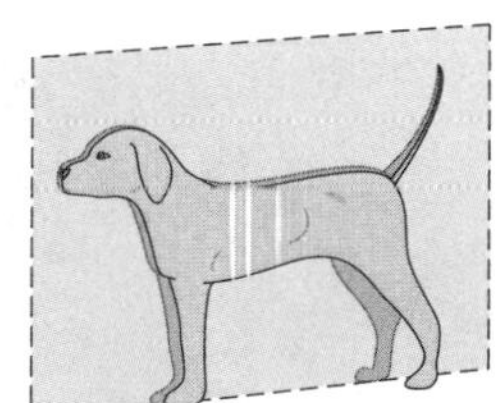

FIGURE 12.4 Bilateral symmetry. There is only one way that the dog could be divided into two equal halves.

4. Other characteristics that differentiate the chordates from other animals are the presence of pharyngeal gill slits and the dorsal hollow nerve cord. The pharyngeal gill slits are paired vertical slits in the wall of the *pharynx* (throat) positioned directly behind the mouth and connecting to the outside of the body. The gill slits may be present only in the embryo stage, or they may persist and be used in breathing. The dorsal hollow nerve cord is a hollow cylindrical tube, usually expanded into a brain at the anterior end. It lies dorsal to the notochord. The hollow space is called the *neurocoel*.

SUBPHYLUM UROCHORDATA–TUNICATA

The adult tunicates are sessile, sac-like animals, often referred to as **sea squirts**. They are filter-feeders, taking in water which passes through the gill slits into the atrial chamber and out by way of the *atriopore*. The tunicates reproduce asexually by budding and also sexually by eggs and sperm. The larval forms have a notochord, a nerve cord, a pharynx with gill slits, and an *endostyle*. An endostyle is aciliated or glandular outpocketing from the wall of the throat of the urochords. Cilia and mucus from the gland sweep food backwards into the gullet. The adult forms lose the chordate characteristics. Examples of the tunicates are *Cynthia, Salpa,* and sea pork.

SUBPHYLUM CEPHALOCHORDATA–LANCELETS

The representative organism for the group is *Amphioxus*, the lancelet. It burrows in the sand at the shoreline of tropical or temperate waters. The body is small, elongated, and shaped like a fish without paired fins. It has gill slits, a well-developed notochord and a dorsal hollow nerve cord (Figure 12.5).

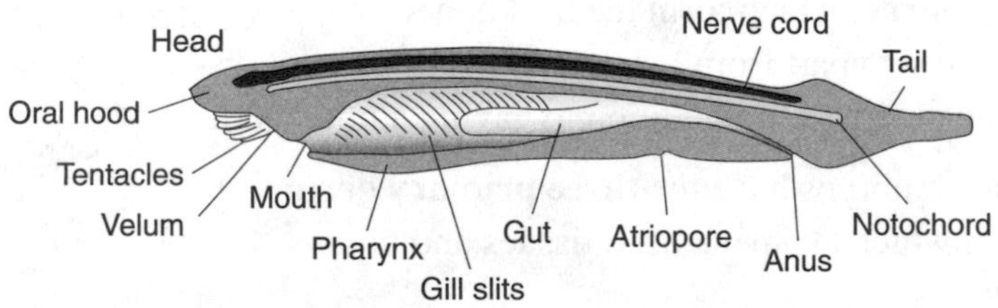

FIGURE 12.5 *Amphioxus*

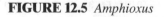

Subphylum Vertebrata–the Vertebrates

CHARACTERISTICS OF THE VERTEBRATES

Animals that have a true backbone composed of segmented parts called **vertebrae** belong to the chordate subphylum Vertebrata. The vertebrae may be made of cartilage or bone: if made of the latter, cartilage cushions prevent the bones from rubbing together. The backbone is built around the notochord and usually obliterates it. Vertebrates vary in size from large to small, but all have a living endoskeleton usually made of bone. A limited number of water-dwelling species exhibit an endoskeleton made of cartilage. All vertebrate species have marked development of the head where a brain is enclosed in a **cranium**.

Blood is pumped through a **closed circulatory system** by means of a ventral heart, having at least two chambers: an *atrium* and a *ventricle*. The *hepatic portal system* carries blood laden with food from the intestines to the liver before it reaches the body cells. Vertebrate red blood cells contain the iron-bearing pigment hemoglobin which is specialized to carry oxygen. Such a system of closed blood vessels prevents blood from entering the body cavity.

Most vertebrates (except humans) have a post-anal tail that is a continuation of the vertebral column. Although there are never more than two sets of paired appendages, some adult vertebrates show only one such set or none at all, the appendages having been lost over evolutionary time. Evidence of lost appendages may be seen in embryonic forms or may be demonstrated by *vestigial* structures. The coccyx bone in humans is a remnant (vestigial structure) of a post-anal tail. Other characteristics of vertebrates include a mouth that is closed by a movable lower jaw and a thyroid gland derived from the ventral wall of the pharynx. In the invertebrate chordates the endostyle is an evolutionary signpost pointing to the development of the thyroid gland.

The structural differences between the invertebrates and the vertebrates are summarized in Table 12.2.

TABLE 12.2
DIFFERENCES IN STRUCTURE BETWEEN THE INVERTEBRATES AND THE VERTEBRATES

	Invertebrate	Vertebrate
skeleton	nonliving exoskeleton	living endoskeleton
nerve cord	ventral, double and solid; formed by delamination* from ectoderm.	dorsal, single and hollow; formed by invagination† of ectoderm.
heart	dorsal	ventral
hemoglobin	in plasma, when present	in blood cells
circulatory system	open with hemocoel and sinuses	closed, contained in vessels

* *Delamination* means "splitting off from ectoderm."

† *Invagination* refers to an inpocketing.

CLASS AGNATHA–CYCLOSTOMES

The Agnatha are the most primitive of the vertebrates. These are the jawless fishes—lampreys and hagfish—characterized by a round mouth that has earned them the general name **cyclostome**. The agnathans have only a single nasal opening at the tip of the snout and no paired fins or paired limbs of any kind. The notochord is present throughout life and the skeletal structures are made of cartilage. The skin is smooth and slimy, lacking scales. The eyes are rudimentary.

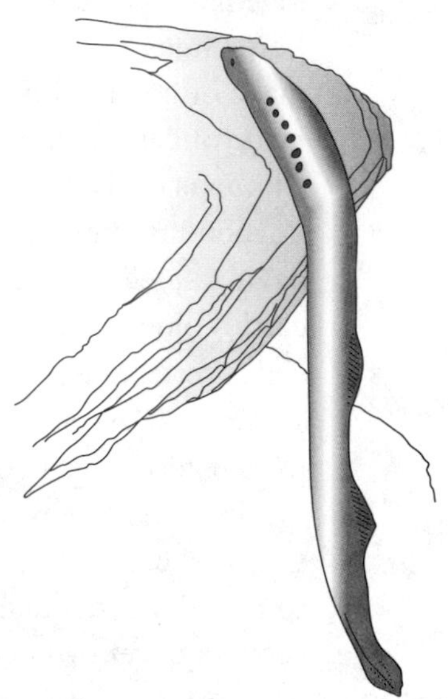

FIGURE 12.6 Lamprey

The lampreys (Figure 12.6) live in fresh water, where they are parasite-predators on bony fish. The mouth of the lamprey is a sucker disk used by the fish to attach itself to rocks or other organisms. After attaching itself to a fish, the lamprey uses a rasping tongue-like structure to break the skin and suck the blood of the host. The lampreys produce an *amnoete larva*, a filter-feeder living buried in river mud, which lasts for about seven years before changing to adult form. The adult life span is short, the adults dying after going upstream to spawn.

Hagfish live in temperate and tropical marine waters where they are scavengers, feeding on dead, disabled, and diseased fish.

CLASS CHONDRICHTHYES–CARTILAGINOUS FISH

The sharks (including the dogfish), rays, and skates are examples of the Chondrichthyes. Most species in this class live in the ocean. The Chondrichthyes have paired fins, a lower jaw, and gill arches. The body of the shark is covered with **placoid scales** which arise from the ectoderm, also forming the teeth in the jaws and on the roof of the mouth. A distinctive feature of this group of fish is that the skeleton is made of cartilage, not bone as in the Osteichthyes, or bony fish. The sharks and their relatives differ from the bony fish in other ways as well: they have no swim bladder and no true scales, and the gill slits are uncovered.

Sharks are predators and feed upon bony fish. Most of the digestion takes place in the stomach while absorption of digested food occurs in the intestine. A *spiral valve* that extends the length of the short, fat intestine increases surface area for absorption. The mouth is located ventrally (Figure 12.7).

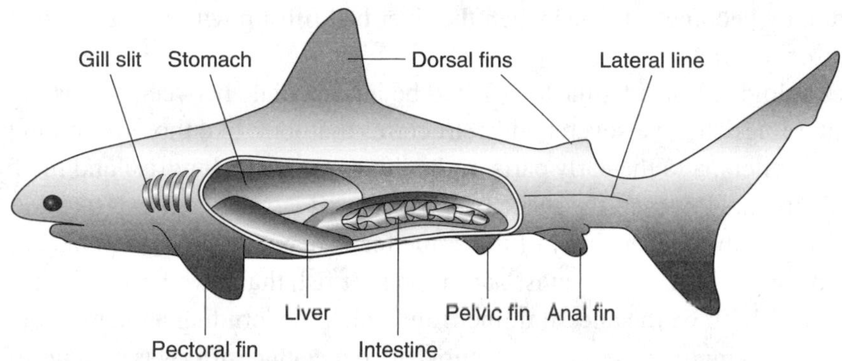

FIGURE 12.7 Structure of the shark

Of particular interest is the sensory system of the shark. The eye of the shark, except for shape, size and absence of eyelids, is very much like the human eye. The paired nostrils are pits on the ventral surface of the head that open externally only and do not empty into the throat. The nostrils are lined with an olfactory membrane which is connected by nerve fibers to the olfactory lobes in the brain. Along the sides of the body are pressure receptors known as **lateral line systems**. In the head pits of the lateral line, organs have become modified into long canals filled with mucus. These canals, known as **ampullae of Lorenzini**, are sensitive electrochemical receptors.

In all cartilaginous fish, fertilization is internal with the male using **claspers**, modified pelvic fins, to place the sperm in the female's body. In some species the embryos are nourished **ovoviviparously**, obtaining food from the egg. In other species, the embryo is maintained inside the mother's body and nourishment is passed from the blood vessels of the mother into the blood system of the developing embryo; this method of embryo nutrition is known as **viviparous**.

SUPERCLASS PISCES

Superclass Pisces includes all of the bony fish. Like the cartilaginous fish, the Pisces are water-dwelling vertebrates that breathe by means of gills and have paired eyes and a two-chambered heart, with blood flowing from the heart through the gills and then to the other parts of the body. The Pisces differ from cartilaginous fish in having *dermal* scales and bone, not cartilage, in the skeleton.

MAJOR GROUPS

The superclass Pisces is divided into four subgroups: classes Dipnoi, Crossopterygii, Ganoidei, and Osteichthyes.

The Dipnoi are lungfish that live in seasonally dry estuaries in Africa, Australia, and South America. These fish breathe by means of lungs and by gills. The lung allows these animals to obtain oxygen from the air and permits them to live in water that is too foul for gill breathing. During the dry season, they aestivate (hibernate) in dry mud; they become activated when the river bed fills up with water during the rainy season.

Some biologists consider the lungfish to be intermediate between the fish and the amphibia. Lungfish have several body structures that look like those found in primitive land dwellers, and the early parts of the life cycle of the lungfish and the frog are almost identical. However, the shape of the teeth and jaws are not froglike and indicate that direct line descendency of the amphibians may not have taken place.

The Crossopterygii, for the most part, represent fish that have become extinct. This group is thought to be the ancestral line from which modern fish and amphibia descended. A living member of this group, called a **coelocanth**, was caught off the coast of South Africa. It was a large, lobe-finned fish with bluish scales.

Members of the Ganoidei are the most primitive of the bony fish. Most species in this class live in fresh water and their distinctive feature is the heavy armor of *ganoid* scales which are flat and form heavy plates. Once the armored fish were the dominant form of fish, but now only a few scattered species remain. Living examples of the Ganoidei are sturgeon, gar, pike and freshwater dogfish. Some species have partly cartilaginous skeletons and a degenerating spiral valve indicating possible relationship to the sharks.

The Osteichthyes, or true bony fish, are the most highly organized of all the fish. They belong to the order Teleostei and are referred to as *teleosts*. They are a very successful and widely distributed group, differing widely in aquatic habitat, size, feeding patterns, and shape. Figure 12.8 illustrates the physical diversity among the teleosts. However, characteristics are typical of all species of bony fish.

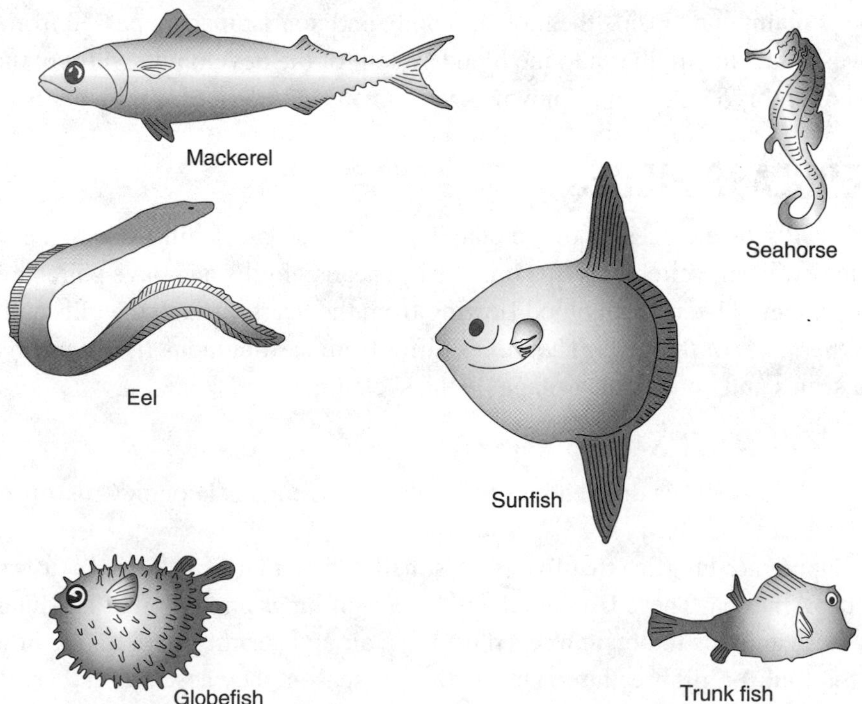

FIGURE 12.8 A diversity of bony fishes

DISTINCTIVE FEATURES

The distinctive features of the teleosts that make them different from the cartilaginous fish are primarily the skeleton, the scales and the gills. The internal skeleton is made of bone. The scales are either **cycloid** (smooth) or **ctenoid** (rough) and are constructed as thin bony plates rather than like the thick plates of the ganoid type. The gills are reduced in number to four pairs, instead of the seven pairs in sharks, and are covered with a flap of bone called the **operculum**. The fins are paired. Figure 12.9 shows the locations of the dorsal, anal, pelvic, and pectoral fins. The *swim bladder* is an outgrowth of the pharynx, an oxygen-filled structure that enables the fish to float.

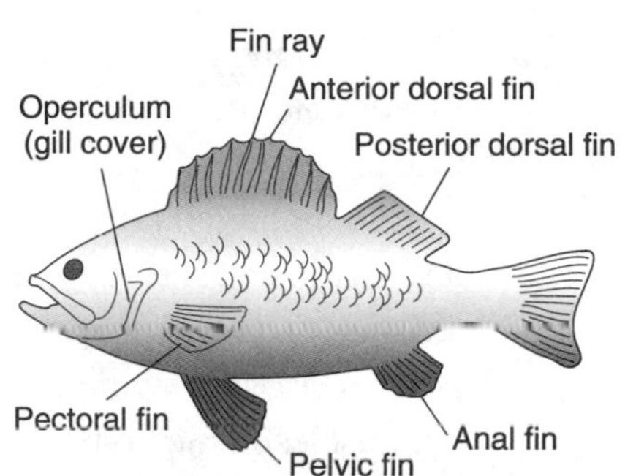

FIGURE 12.9 The bony fish

Some sense organs in bony fish are well developed; others are not. The lateral line receptors are sensitive to movements of current. The olfactory pits are not connected to the mouth and therefore have no respiratory function, but they do enable fish to respond to chemical stimuli. Eyes of fish vary in size according to species. The taste buds located in the mucosa of the mouth are poorly developed. Fish probably cannot hear in air. There are three semicircular canals that control equilibrium, but auditory vibrations reach the ear through the skull bones.

REPRODUCTIVE BEHAVIOR

Patterns of reproduction vary among the bony fish, but in most species fertilization is external. During certain breeding seasons the female lays thousands of eggs known as **roe**. The male deposits sperm cells known as **milt** close to the eggs. The sperm, attracted by some chemical substance given off by the eggs, swim to them. One sperm enters an egg and fertilizes it. The fertilized egg (**zygote**) then goes through a series of mitotic divisions (cleavage), resulting in a new immature fish called a *fry*. The fry has attached to it a sac appropriately called the **yolk sac** because it contains yolk which nourishes the young fish until it can feed independently (Figure 12.10).

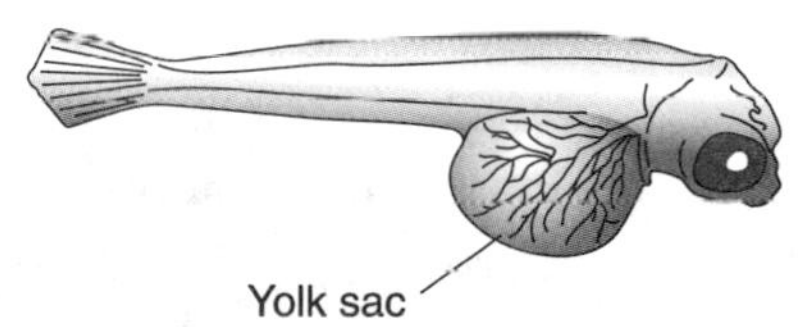

FIGURE 12.10 An immature fish

Fertilization in fish is chancy: many sperm never reach the eggs and many fertilized eggs die before development. Hence there is an overproduction of gametes to ensure the survival of the species.

In a few fish species fertilization is internal, and in a few species parental care is given to the fertilized eggs. The stickleback male, for example, takes care of the fertilized eggs in nests, and the male seahorse carries them around in a brood pouch.

CLASS AMPHIBIA–FROGS AND THEIR RELATIVES

From an evolutionary perspective, amphibia are transitional animals, living first in water and then on land during specific times in the life cycle. They are equipped for this double life, so to speak, by body structures and organs that change to meet their needs.

An amphibian must spend part of its life cycle in the water where its eggs are laid and fertilized. The eggs develop into a larval stage, or tadpole, that has fishlike characteristics. In tadpoles breathing is by means of gills, blood is pumped by a two-chambered heart, and swimming is by means of a tail and body movements made possible by muscles in the body wall. The change to adult form is known as **metamorphosis**, a process controlled by the thyroid gland. The adult amphibian loses the gills, lateral line senses, tail, unpaired fins and muscles controlling them—the fish characteristics—and develops structures adapted for life on land. An adult amphibian breathes by means of lungs and has a three-chambered heart which is more efficient at pumping blood between the lungs, the heart, and the rest of the body. In most species the adult also has limbs for movement, but no tail (Figure 12.11).

There are three general types of amphibia. The Apoda are worm-like, legless, ground-burrowing forms found in tropical and semitropical regions; *Caecilia* (Figure 12.12) is a representative genus. The Urodela are amphibians that do not lose their tadpole-like tail in metamorphosis; salamanders, newts, and mud puppies (Figure 12.13) are examples. The Anura, represented by frogs and toads, lose their tails on becoming adults. Frogs and toads are the first vertebrates to become vocal.

In frogs, fertilization is external. Sperm leave the testes through tubules called *vasa efferentia* which communicate with the kidney. The sperm cells then pass into the **Wolffian duct** which leads to the *cloaca*, a passageway that opens to the outside of the body. In the female large egg masses are released into the body cavity from two ovaries, located at the anterior end of each kidney. Beating cilia sweep the eggs into coiled tubules known as **oviducts** where they are propelled to the cloaca and then out of the body. As the eggs pass through the oviducts they are coated with a thin layer of jelly-like material. At the time when the female is depositing eggs in the shallow waters of a pond or brook, the male deposits sperm over them. The sperm swim to the eggs; then, as each sperm reaches an egg, it digests its way through the jelly and into the egg, effecting fertilization. After fertilization the jelly coating on the eggs swells due to the absorption of large amounts of water. The swelling of the black jelly causes the eggs to adhere together and protects them from predation by fish and other animals. The fertilized egg, or **zygote**, undergoes cleavage, forming a tadpole.

Amphibians demonstrate some interesting mechanisms for prolonging life. Many have powers of regeneration to the extent that entire organs can grow back after having been lost. The axolotyl and the urodele (*Triton cristatus*) are examples of animals in which regeneration occurs. Many amphibians are able to hibernate or aestivate and survive unfavorable conditions.

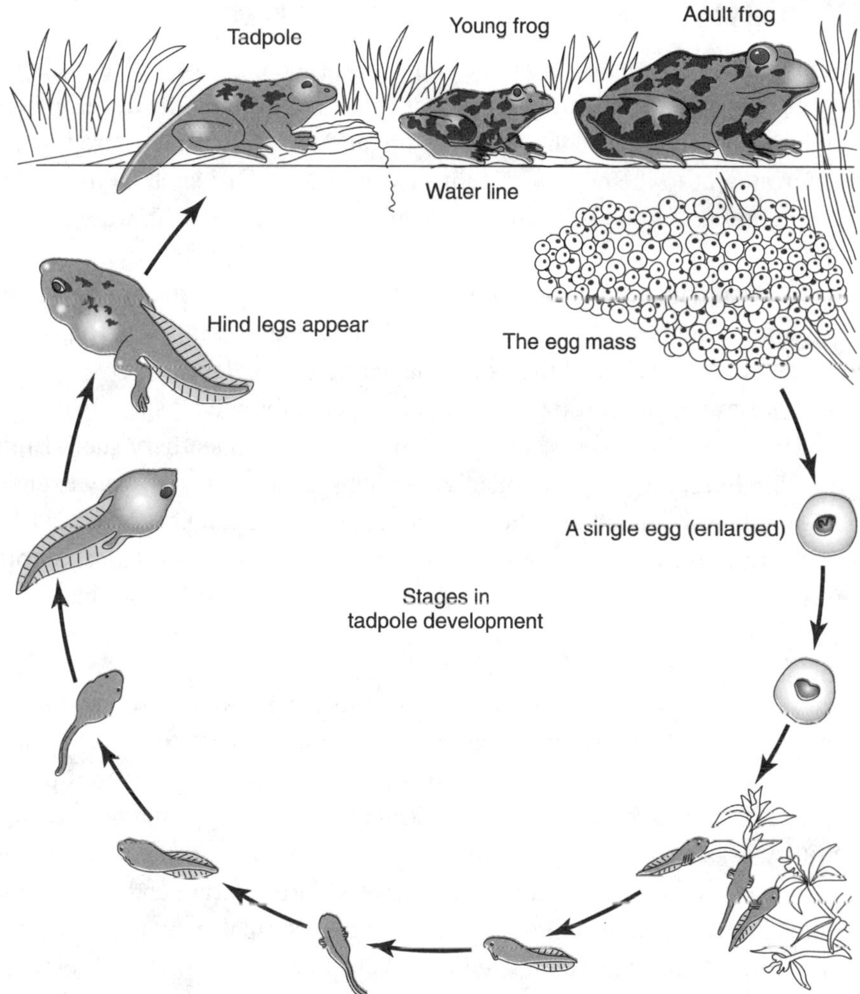

FIGURE 12.11 Metamorphosis in the frog

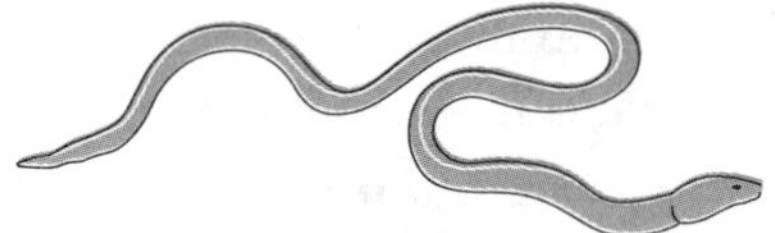

FIGURE 12.12
Caecilia, a wormlike amphibian

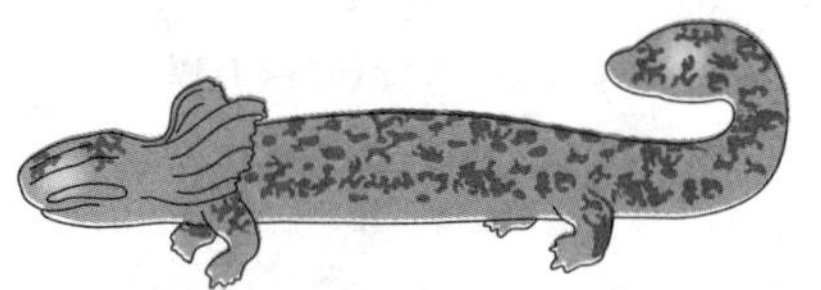

FIGURE 12.13
Necturus, the mud puppy

CLASS REPTILIA

Reptiles are the first true land vertebrates relieved of the necessity of returning to the water to reproduce. Reproducing on land was made possible by the development of special embryonic membranes that preserve on land the protection offered by the aquatic environment to the embryos of lower vertebrates. One such membrane is the **amnion**, a membranous sac surrounding the embryo and filled with water. This sac of water prevents the delicate, rapidly dividing embryonic cells from drying out and protects them from shock and mechanical injury. Animals having the amnion are called *amniotes*; those animals without an amnion are known as *anamniotes*. Reptiles, birds and mammals are amniotes and therefore adapted to life on land.

The eggs of reptiles are fertilized internally, and the female lays fertilized eggs. Reptile embryos develop encased in an egg surrounded by a leathery shell. Lining the shell is the **chorion**, an embryonic membrane that mediates the two-way exchange of gases between the outside air and the embryo. Serving as a temporary organ of respiration and excretion is a third embryonic membrane, the **allantois**. The allantois, bearing a capillary network, grows into the space between the chorion and the amnion.

GENERAL CHARACTERISTICS

Reptiles have a dry leathery skin covered with epidermal scales. A somewhat flattened skull contains a brain having a cerebrum much larger than that of the fish or amphibians. The eyes have secreting glands which keep the surface moist. Some species of reptiles have an external pore (**meatus**) positioned on the side of the head connected to the auditory ossicle or bone in the middle ear. Reptiles are air-breathers and have rather well-developed lungs. The heart is composed of two **atria** and a **ventricle**; in some species the ventricle is almost divided into two compartments, an evolutionary signpost pointing to the four-chambered heart. The body temperature of reptiles is not constant, changing with the external environment. In popular speech, such animals are called cold-blooded; in technical language, they are **poikilotherms**.

MAJOR GROUPS

Extinct Forms

Reptiles flourished during the Mesozoic Era which lasted for about 130 million years and came to an end about 65 million years ago. At that time there were more than twelve orders of reptiles. Today only four orders of reptiles remain. Among the ancient reptiles there was a great diversity of form and adaptations that permitted life in a variety of habitats: dry land, water, swamps, and air. The **stem reptiles** had many features in common with primitive amphibians and resembled modern-day lizards externally. The **therapsids**, however, were more advanced, having teeth which looked much like those of the mammals. The dinosaurs are remembered for being very large and yet some were quite small. It is believed that many of the dinosaur species were able to walk on two legs, using the enormous tail for balance. The shorter front legs were probably used when walking slowly or resting. Some of the larger dinosaurs returned to the sea. The **icthyosaurs** were best adapted for aquatic life. A few of the ancient reptiles were able to fly; the *pterosaurs* had wings but they probably did more gliding than true flying.

Living Forms

The living reptiles are classified in four orders: Rhynchocephalia, Chelonia, Squamata, and Crocodilia.

Order Rhynchocephalia—Tuatara The only living representative of this order is the New Zealand "tuatara" of the genus *Sphenodon*. This lizard is about 1.5 meters long and has a median eyestalk at the top of the head. *Sphenodon* has a number of primitive reptilian characteristics, including no tear glands, teeth in the roof of the mouth, a lung that resembles a cluster of toad's lungs, abdominal ribs, unfused frontal skull bones, and a vertebral column with a fish-like structure.

Order Chelonia—Turtles and Tortoises The chelonians are the turtles and tortoises. The skeleton is modified to form a box-like covering, the upper curved portion of which is called the **carapace**, the lower part, the **plastron**. The head and the tail are the only movable parts of the animal. The jaws are horny and toothless. Chelonians live on land, in fresh water, and in the sea. More turtles live on the American continent than anywhere else (Figure 12.14).

FIGURE 12.14 Turtle

Order Squamata—Lizards and Snakes Lizards and snakes are squamates. Their bodies are covered with a great number of small, flexible scales that, unlike the scales of bony fish, cannot be removed easily. Lizards have movable eyelids, visible earpits, and usually legs. Snakes, on the other hand, are legless and do not have eyelids or earpits.

Lizards are typically land dwellers requiring the warmth of the sunshine. The iguana is a large tree-living lizard native to Mexico. The chameleon, native to Africa, has a flexible grasping tail and grasping feet; its coat is capable of changing colors to match the background. The gila monster *Heloderna* is the only lizard whose bite is poisonous.

Although snakes are cold to the touch, they are not slimy. The body is heavily muscled and strongly ribbed. Snakes use the ribs in walking. Of the 110 species of snakes in the United States only 20 species are poisonous. Among the poisonous snakes are the copperhead, rattlesnake, water moccasin, and the coral snake. The harmless varieties include the puff adder, the garter snake, the black snake, and the milk snake.

Order Crocodilia—Crocodiles and Alligators This order includes the crocodiles and the alligators, large thick-bodied reptiles. These are considered to be the most advanced of the reptiles. The heart has a ventricle that is almost completely divided into two compartments. The lungs are very well developed and the brain has a rather large cerebrum. They live in shallow salt or fresh water where they are better able to locomote than on land (Figure 12.15).

FIGURE 12.15 Crocodile

CLASS AVES–BIRDS

CHARACTERISTICS

Birds are terrestrial vertebrates with feathers. Feathers are the distinctive feature of birds: all birds have them and no other animals are so covered. The forelimb is modified into wings for flight, leaving the hindlimbs for walking (**bipedal** locomotion). Birds are built for flight; special adaptations in body structure effect lightness in weight, efficiency and strength. Not only are the feathers light in weight and easily moved and lifted by wind, but they also create warmth next to the body. Body heat warms the air that is in contact with the bird's body. Warm air becomes lighter and rises. Other adaptations for flight are the compact, but hollow, bones, numerous air sacs occupying all available body spaces, reduced rectum, absence of teeth, and tail feathers that replace a bony tail (Figure 12.16).

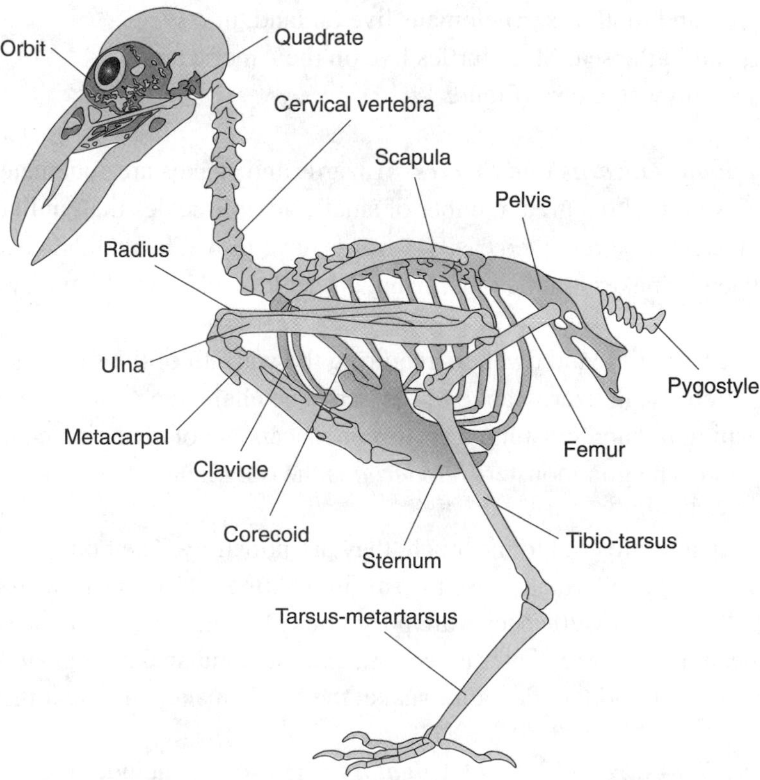

FIGURE 12.16 Bone structure of the bird

In most bird species, the wings are organs of flight. However, in some species, the wings are modified for other purposes. For example, the wings of penguins serve as flippers for swimming.

According to fossil evidence, the evolutionary link between the reptiles and the birds is *Archaeopteryx*. Biologists consider *Archaeopteryx* to have been a bird because it had feathers, but it also had some very distinctive reptilian features, such as teeth and a long, bony tail.

Modern birds have lost reptilian characteristics. The feet of birds of various species are modified for all kinds of uses: running, scratching, grasping, swimming and wading. The horny bill is a characteristic of all birds. It is toothless and modified in various shapes to carry out special functions. The colors of bird feathers are due to pigments or irridescence and indicate **sexual dimorphism**, a state in which the male is often brightly and elaborately colored while the female is drab.

BODY STRUCTURE

The brain of the bird shows greater development than the brain of the reptile, and the heart has four chambers—two atria and two ventricles, permitting complete separation of oxygenated and deoxygenated blood. Unlike the reptiles, birds maintain constant body temperature: they are warm-blooded organisms and are also known as endotherms or homeotherms. The digestive system consists of a crop, a stomach, a gizzard, an intestine, and a much reduced rectum. Birds do not retain their feces, a condition which is an adaptation for flight. There is a well developed kidney that empties by way of the ureter. Birds have remarkable eyesight, a feature that makes hawks and eagles such successful predators. On the other hand, the sense of smell is very poorly developed in birds. The **nictitating membrane** (third eyelid) is a translucent membrane that covers the eye crosswise serving as protection during flight.

BEHAVIOR

Birds show another advancement over reptiles in behavior. Song is used to establish territorial rights, an important aspect of bird social behavior. The song of birds is produced by air passing over the **syrinx**, a secondary larynx located at the lower end of the windpipe (trachea) at its junction with the bronchi. The pattern of reproduction in birds also differs from that of reptiles and amphibians. Birds practice monogamy, the mating of one male with one female either for life or for the duration of the breeding season. Before mating birds go through courtship; a special type of behavior is displayed by males trying to gain the attention of reproductive females. In birds, **fertilization** is **internal**; the **development** of the young, **external**. In most species the female lays the eggs in a characteristic-type nest and both the male and the female take turns *setting*, a process of incubation by which the eggs are kept warm, and later caring for the young.

REPRODUCTION

The testes of birds are positioned in the back just above the kidneys. Mature sperm leave the testes through tubules called the *ductus deferens* which lead into the cloaca. The gonads of females are the **ovaries**, glands where eggs are produced. In female birds, there is one functioning ovary; the other having degenerated early in the bird's life. A bird's egg cell consists of a nucleus and a little protoplasm. The cell is surrounded by yolk. A mature cell and its yolk are drawn into the upper end of the oviduct, which is funnel-shaped and contains waving cilia. As the cell with its yolk travels through the oviduct, it is surrounded with layers of the protein *albumen*. The albumen is the white of the egg. The albumen is surrounded by a thin membrane which then is covered by a calcareous shell secreted by lime-producing glands that line the lower end of the oviduct (Figure 12.17).

Fertilization is accomplished during mating at which time the male and female place their cloacas close together. Sperm swim from the cloaca of the male into the female cloaca and up into the oviduct. Fertilization takes place high up in the oviduct before the albumen and the other surrounding membranes are secreted by the oviduct cells. Most birds lay a *clutch* of fewer than six eggs. However, ducks may lay as many as 15 eggs at one time (Figure 12.18).

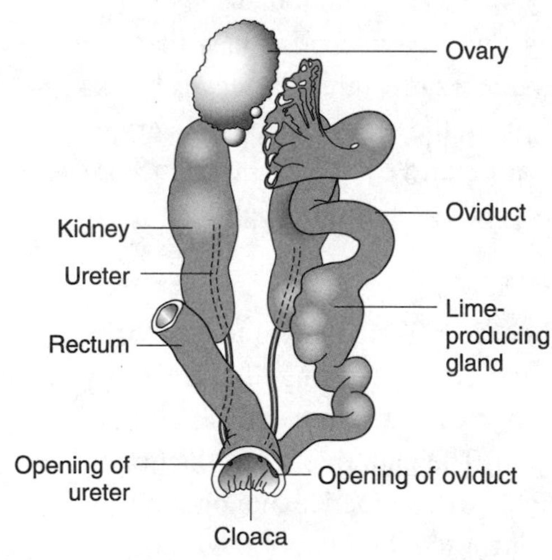

FIGURE 12.17
Reproductive system of the female bird

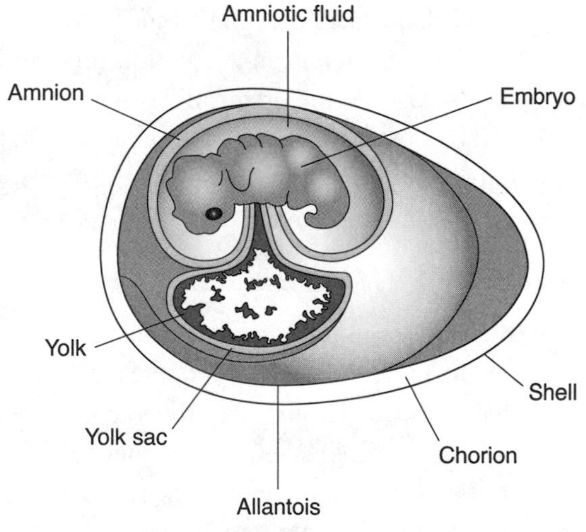

FIGURE 12.18
Development of the bird embryo

CLASS MAMMALIA

The first true mammals appeared on Earth about 210 million years ago. The early mammals were small and insignificant, but had developed adaptations enabling them to survive at the same time that the giant reptiles roamed Earth. It is believed that many of the early mammals were no larger than field mice, feeding on insects and scraps of vegetation. The large eye sockets of the fossil forms indicate that these primitive mammals were nocturnal and probably lived in trees. The extinction of the dinosaurs, 65 million years ago, appears to have given early mammals an opportunity to diversify and thrive.

Scientists present various reasons for the success of the mammals. First among these is the fact that they were fast movers and their food requirements and behavioral patterns kept them out of the way of carnivorous (flesh-eating) reptiles. Second, the early mammals have been accused of contributing to the demise of the dinosaurs by eating the unprotected eggs of these reptiles. Third, the change in form of the skeleton which allowed the positioning of relatively thin legs underneath the animal led to a narrow walking track and contributed to faster movement. (Remember how the thick legs of a crocodile are spread out from the body on a wide track.) Fourth, the refinement of the lower jaw—movable and composed of a single bone—coupled with the development of different kinds of teeth for biting, tearing and chewing made the mammals successful carnivores.

Modern mammals show a great deal of diversity. They vary in size from a shrew hardly more than 2.5 centimeters in length to the enormous blue whale which may measure more than 30 meters in length. Mammals are widely distributed over Earth and are adapted to live in diverse habitats on land (wolves), in water (sea lions), in underground burrows (gophers), in desert burrows (kangaroo rat), in open seas (whales), in the air (bats), and in forests (baboons).

GENERAL CHARACTERISTICS OF MAMMALS

The characteristics that set mammals apart from other animals and make them adaptable to a wide range of habitats are as follows:

1. Mammals have **mammary glands** (whence the name *mammal*) that supply the young with milk directly after birth. Newborn mammals are usually quite helpless and depend upon the mother for nourishment.
2. At some time during the life cycle, all mammals have **hair**. Hair is as typical of mammals as feathers are of birds and scales of bony fish and reptiles.
3. Mammals are **warm-blooded**. Constant body temperature is due, in part, to the four-chambered heart, a device which prevents the mixing of oxygenated and deoxygenated blood. The four-chambered heart first appeared in birds. However, in mammals another characteristic of blood contributes mightily to stable body temperature. The red blood cells on maturity lose their nuclei and mitochondria, restricting them to function as oxygen carriers, not as users of oxygen.

4. Most species of mammals have **sweat glands**, which provide a secondary means of excreting water and salts.
5 Mammalian teeth have evolved into three different types: **incisors** for tearing; **canines** for biting; **molars** and **premolars** for grinding.
6. All but a few species have seven vertebrae in the neck. These neck bones are known as **cervical vertebrae**.
7. A **muscular diaphragm** separates the thoracic cavity (containing the lungs and the heart) from the abdominal cavity (housing part of the digestive system, the reproductive organs and the excretory system).

MAJOR GROUPS

The class Mammalia is divided into three major groups: Protheria, Metatheria, and Eutheria.

Subclass Prototheria—Monotremes

The subclass Prototheria consists of a single order, Monotremata. The monotremes are primitive egg-laying mammals. The eggs, large and full of yolk, house the developing monotreme embryos. Examples of the monotremes are the duck-billed platypus (*Ornithorhynchus*) indigenous to Australia and Tasmania (Figure 12.19); the spiny anteater (*Echidna*), also an inhabitant of Australia; and a long-snouted anteater (*Proechidna*) indigenous to New Guinea. Modified sweat glands of the anteater secrete a milk substitute that the young lick up from tufts of hair on the mother's belly.

FIGURE 12.19
Duck-billed platypus

Subclass Metatheria—Marsupials

The subclass Metatheria consists of a single order, *Marsupialia*. The marsupials are primitive mammals that do not have a placenta. The young are about 5 centimeters long at birth and are in an extremely immature condition. At birth they crawl into the mother's pouch or *marsupium*. The rounded mouth is attached to a nipple and the mother expresses milk down the throat of the helpless fetus. As development occurs, the young marsupial is then able to obtain milk by sucking. There are 29 living genera of marsupials, 28 of which live in Australia. The oppossum *Didelphys* is indigenous to North, South, and Central America, and *Caenolestes* inhabits regions of Central America only. Besides the oppossum, other marsupials are the kangaroo, koala bear, Tasmanian wolf, wombat, wallaby, and native cat. It is believed that at one time a land bridge connected South America to Australia. With the disappearance of this land link, Australia became geographically isolated, permitting the existence of marsupials that do not have to compete with more advanced mammals.

Subclass Eutheria—Placental Mammals

The subclass Eutheria includes all of the modern mammals that have a placenta. You will recall that in the reptile and the bird a membranous sac called the **allantois** serves as an organ of respiration and excretion for the developing embryo. The reptile and bird embryos develop outside the mother's body and are nourished on stored yolk in the egg. The eutherians develop inside of the mother's body in a muscular sac, the **uterus**. The region where the allantois comes in contact with the uterine wall is heavily supplied with capillaries, the smallest blood vessels in the body. At this particular site the **placenta** is formed. The embryo is attached to the placenta by an *umbilical cord*, which contains a fetal artery and a fetal vein. By diffusion from the capillaries of the mother, food nutrients and oxygen travel into the capillaries in the embryonic placenta. The fetal artery collects blood from the capillaries of the placenta and conducts it to the capillaries of the embryo. The fetal vein collects the blood of the embryo laden with the wastes of fetal respiration and conducts it away from the embryo. It is important to emphasize that the blood of the mother and the blood of the embryo do not mix. Gasses and nutrients from the maternal circulation pass to the fetal circulation by diffusion across capillary walls.

Table 12.3 provides a quick summary of some of the placental mammals.

TABLE 12.3
SOME PLACENTAL MAMMALS

Order	Characteristics	Examples
Insectivora	Small; nocturnal; burrowing or tree-living; feed on insects; sharp-snouted.	Hedgehog, moles, shrews
Dermoptera	Link between the insectivores and the bats.	Flying lemur
Chiroptera	Winged, only mammals able to fly; nocturnal; identify objects by echolocation; according to genus feed on insects, blood, fruit.	Bats
Carnivora	Predators; swift of foot, collar bones reduced, teeth specialized for meat-eating; cerebral development.	Dogs, wolves, coyotes, bears, cats, lions, cheetah, foxes, raccoons, weasels, skunks, etc.
Rodentia	Gnawing mammals; sharp incisor teeth; plant eaters; most numerous of all living mammals.	Hares, squirrels, guinea pigs, rats and mice, beavers, muskrats, porcupines, prairie dogs, woodchucks
Primates	Evolved from tree-dwelling Insectivora ancestors; most species arboreal; teeth unspecialized; marked development of eyes and brain, especially the cerebrum; quadrupeds, but upright sitting posture.	

Order	Characteristics	Examples
Lemuroidea	Tree-dwelling; primitive; small, pointed ears; long snout; big toes and thumbs set apart from other digits.	Lemurs
Tarsioidea	Binocular vision; reduced sense of smell.	Tarsiers
Anthropoidea	Enlarged, convoluted cerebrum; well-developed eyes; prehensile tails; broad, flat noses; thumbs reduced; external nostrils close together.	Monkeys, baboons, macaques, gibbons, orangutans, gorillas, chimpanzees, humans

REVIEW EXERCISES FOR CHAPTER 12

WORD-STUDY CONNECTION

albumen
allantois
amnion
ampulla
ampullae of Lorenzini
antimeres
atriopore
atrium
auricle
bipedal locomotion
carapace
chordate
chorion
claspers
cloaca
clutch
coelocanth
coelomate
cranium
ctenoid
cyclostome
deuterostome
dorsal hollow nerve cord
ductus deferens
endostyle
fry
ganoid
haltere
hepatic portal system
homiotherm
lateral line
madreporite
mammary gland
marsupial
metamorphosis
milt
monotreme
neurocoel
nictitating membrane
notochord
operculum
oviduct
oviparous
ovoviviparous
papulae
pharyngeal gill slits
pharynx
placenta
placoid scales
plastron
poikilotherm
proboscis
regeneration
roe
sea squirts
sea star
sexual dimorphism
spiral valve
swim bladder
syrinx
teleost
tube feet
umbilical cord
uterus
vasa efferentia
ventricle
vertebrae
vestigial
viviparous
water vascular systems
Wolffian duct
yolk sac
zygote

SELF-TEST CONNECTION

PART A. Completion. *Write in the word that correctly completes each statement.*

1. The _____ stage of the echinoderms resembles that of the chordates.
2. Spines and plates made of calcium form the echinoderm _____.
3. Antimeres are the radiating _____ of the sea stars
4. All hemichords live in a _____ environment.
5. The hemichord proboscis is used to _____.
6. The most primitive of the vertebrates are the _____ fish because of their feeding patterns.
7. Hagfish are best described as _____.
8. The lamprey has _____ paired fins.
9. The amnocoete lives buried in _____.
10. The type of scales that cover the shark's body is _____.
11. The pressure receptors along the sides of fish are known as _____ systems.
12. A living "fossil" fish is the _____.
13. The teleosts are the _____ fish.
14. The larval stage of the frog is the _____.
15. The first vertebrates to become vocal are the _____.
16. The adaptation that protects the embryos of terrestrial animals against drying and dessication is the _____.
17. Dinosaurs are best classified as _____.
18. The carapace and the plastron are best associated with the _____.
19. All birds have _____.
20. An example of a monotreme is the _____.
21. An immature fish is called a _____.

22. Egg masses of frogs leave the body cavity through tubes known as _____.

23. Before mating, birds go through _____ behavior.

24. The sperm of fish are known as _____.

25. The marsupial young are born immature because there has been no development of the _____.

PART B. Multiple Choice. *Circle the letter of the item that correctly completes each statement.*

1. A distinctive feature of the phylum Echinodermata is the
   (a) excretory system
   (b) spiny skin
   (c) antimeres
   (d) notochord

2. Sea stars have a functional
   (a) excretory system
   (b) respiratory system
   (c) water vascular system
   (d) head

3. The nonfunctioning organs of the sea star are the
   (a) stomach and anus
   (b) mouth and intestine
   (c) tube feet and stomach
   (d) anus and intestine

4. Water enters the sea star by way of the
   (a) madreporite
   (b) stone canals
   (c) ring canals
   (d) cardiac stomach

5. In the sea star a light-sensitive eyespot is located
   (a) near the madreporite
   (b) adjacent to the stomach
   (c) next to the ampulla
   (d) at the tip of each arm

6. The function of the spiral valve is to
   (a) increase surface area
   (b) digest food
   (c) circulate blood
   (d) store wastes

7. The process in which embryos obtain nourishment from the yolk in the egg is described as
   (a) viviparous
   (b) ovoviviparous
   (c) amnocoete
   (d) osmosis

8. The dogfish is a (an)
   (a) amphibian
   (b) lamprey
   (c) bony fish
   (d) shark

9. Dipnoi is the class of the
   (a) sharks
   (b) skates
   (c) lungfish
   (d) cyclostomes

10. Teleosts are the
    (a) sharks
    (b) lung fish
    (c) lamprey
    (d) bony fish

11. The gill covering of the bony fish is the
    (a) spiral valve
    (b) operculum
    (c) cranium
    (d) chorion

12. It is true that fish
    (a) have well-developed sense organs
    (b) usually are farsighted
    (c) have three semicircular canals
    (d) have mouth-connected olfactory pits

13. The urodeles
    (a) give birth to live young
    (b) have tails
    (c) are usually legless
    (d) represent the reptiles

14. The fetal membrane that functions as an organ of respiration and excretion is the
    (a) amnion
    (b) allantois
    (c) chorion
    (d) shell

15. When an organ or structure "bears a capillary network," it
    (a) has hair
    (b) is divided
    (c) becomes septate
    (d) has blood vessels

16. The reptiles are true land animals because they
    (a) breathe free air
    (b) have strong walking legs
    (c) reproduce on land
    (d) feed on vegetation

17. Bipedal locomotion is characteristic of
    (a) toads
    (b) turtles
    (c) crocodiles
    (d) birds

18. Hollow bones, numerous air sacs and loss of teeth are adaptations for
    (a) floating
    (b) swimming
    (c) diving
    (d) flying

19. The four-chambered heart first appeared in
    (a) mammals
    (b) birds
    (c) reptiles
    (d) amphibians

20. Metamorphosis is best associated with
    (a) fish
    (b) rabbits
    (c) frogs
    (d) strawberries

21. Roe and milt are best associated with
    (a) fish
    (b) frogs
    (c) sea stars
    (d) jellyfish

22. The number of sperm cells necessary to fertilize an egg is
    (a) one
    (b) two
    (c) three
    (d) four

23. The number of ovaries in a female bird is
    (a) one
    (b) two
    (c) three
    (d) four

24. Egg-laying mammals are known as
    (a) marsupials
    (b) tadpoles
    (c) monotremes
    (d) placentals

25. The production of song in birds is associated with the
    (a) nictitating membrane
    (b) syrinx
    (c) gizzard
    (d) bill

PART C. Modified True-False. *If a statement is true, write "true" for your answer. If a statement is incorrect, change the* <u>*underlined*</u> *expression to one that will make the statement true.*

1. An important chordate characteristic of the hemichords is the presence of <u>a hollow notochord</u>.

2. The <u>trunk</u> of the hemichord sweeps water into the mouth.

3. The function of the gill slits is to remove excess food from the body.

4. In sea stars, the sexes are separate.

5. Acorn worms belong to the echinoderms.

6. All vertebrates have an endoskeleton made of bone.

7. Vertebrates never have more than 4 sets of paired appendages.

8. A cyclostome refers to a type of organ.

9. Lamprey eels are scavengers.

10. The shark's skeleton is made of bone.

11. The words *placoid*, *dermal*, and *ganoid* describe types of fins.

12. The most highly organized of all the fish are the bony fish.

13. Most of the teleosts are ovoviviparous.

14. Metamorphosis of the tadpole is controlled by the gastric gland.

15. *Sphenodon* is the most advanced of the reptiles.

16. In mammals the allantois evolved into the chorion.

17. Fertilization in most fish takes place internally.

18. A cloaca is a (an) tube.

19. Egg white is the protein gelatin.

20. A group of eggs laid at one time by a bird is a catch.

21. A tadpole has a three-chambered heart.

22. Frog eggs are coated with a cellulose-like substance.

23. In mammals the abdominal cavity houses the lungs and heart.

24. Mammals are warm-blooded animals.

25. Only reptiles and mammals have a four-chambered heart.

CONNECTING TO CONCEPTS

1. What distinguishing characteristics set the chordates apart from lower animals?
2. Why is "sea star" a better name than "starfish"?
3. Why is a closed circulatory system more efficient than an open circulatory system?
4. How is the body of a bird structured for flight?
5. Why is the blue whale correctly classified as a mammal, not as a fish?

ANSWERS TO SELF-TEST CONNECTION

PART A

1. larval
2. exoskeleton
3. arms
4. marine
5. burrow
6. jawless
7. scavengers
8. no
9. mud
10. placoid
11. lateral line
12. coelacanth
13. bony
14. tadpole
15. frogs and toads
16. amnion
17. reptiles
18 turtle
19. feathers
20. duck-billed or platypus
21. fry
22. oviducts
23. courtship
24. milt
25. placenta

PART B

1. **(b)**
2. **(c)**
3. **(d)**
4. **(a)**
5. **(d)**
6. **(a)**
7. **(b)**
8. **(d)**
9. **(c)**
10. **(d)**
11. **(b)**
12. **(c)**
13. **(b)**
14. **(b)**
15. **(d)**
16. **(c)**
17. **(d)**
18. **(d)**
19. **(b)**
20. **(c)**
21. **(a)**
22. **(a)**
23. **(a)**
24. **(c)**
25. **(b)**

PART C

1. gill slits
2. cilia
3. water
4. true
5. hemichords
6. Most
7. two
8. fish
9. predators
10. cartilage
11. scales
12. true
13. oviparous
14. thyroid
15. primitive
16. placenta
17. externally
18. opening
19. albumen
20. clutch
21. two
22. jelly
23. thoracic
24. true
25. birds

CONNECTING TO LIFE/JOB SKILLS

Biology encompasses a wide range of specialized fields of study. Table 1.3 in Chapter 1 presents just a few areas of biological study; there are many more specialties. Biologists select a particular area for specialization. For example, **entomologists** specialize in the study of insects. Most entomologists further narrow their field of concentration to particular insect phyla. As another example, some **marine biologists** concentrate their study on sharks. Using your school or local library, you may wish to investigate the many career areas in biology. You will also find much useful information via the Internet. Then, if you are interested, you will need to consider earning a bachelor's degree in biology.

Chronology of Famous Names in Biology

1750 **Rene Antoino deReaumur** (France)—wrote a monumental work in six volumes on the anatomical structure of insects.

1757 **Pierre Lyonet** (Netherlands)—presented a brilliant piece of research on the life cycle of the goat-moth caterpillar.

1760 **Petrus Camper** (Netherlands)—published anatomical studies on the elephant, rhinoceros, and the reindeer. Also published an excellent paper on the anatomy of the orangutan.

1763 **John Hunter** (England)—introduced "modern" methods in the study of comparative anatomy of vertebrates.

1804 **Alexander Brongniart** (France)—was the first to classify the amphibia separately from the reptiles.

1830 **Johannes Peter Muller** (Germany)—devoted himself to marine research and made valuable contributions to the study of the evolution of marine forms.

1837 **Karl Ernst von Baer** (Germany)—discovered that animal tissues arise from three primary germ layers.

1876 **Max Furbringer** (Germany)—produced valuable information on the comparative anatomy of the breast, wing, and shoulder of birds.

1957 **James Gray** (England)—carried out research which elucidated the mechanisms by which small fish gain swimming speed.

1965 **Archie Carr** (England)—studied the navigational habits of the green turtle.

1967 **Neal Griffith Smith** (England)—discovered that gulls recognize species mates by visual signals.

1968 **Kjell Johansen** (Sweden)—presented a remarkable study on the evolution of air-breathing fishes.

1969 **Crawford H. Greenewalt** (United States)—elucidated the mechanical means by which birds produce song.

1971 **Knut Schmidt-Nielsen** (United States)—discovered the pathway that air takes through the lungs of birds.

1977 **T. J. Dauson** (England)—studied the adaptive strategies of kangaroos that enable them to maintain their reproductive potential.

1980 **Benacerraf Baruj** (United States), **George D. Snell** (United States), and **Jean Dausset** (France)—won the Nobel Prize for discoveries that explain how cell structure influences organ transplants and diseases.

1982 **Roger W. Sperry** (United States), **David Hubel** (United States), and **Torsten N. Weisel** (Sweden)—won the Nobel Prize for studies elucidating the organization and function of the vertebrate brain.

1983 **Edmond H. Fischer** and **Edwin Krebs** (United States)—won the Nobel Prize for discovery of a regulatory mechanism affecting all cells.

13

Homo sapiens: A Special Vertebrate

WHAT YOU WILL LEARN

In this chapter you will review the biology of human beings: the evolutionary links and body systems.

SECTIONS IN THIS CHAPTER

- Characteristics of Primates
- The Skeletal System
- Thc Muscular System
- The Nervous System
- The Endocrine System
- The Respiratory System
- The Circulatory System
- The Lymphatic System
- The Digestive System
- The Excretory System
- Sense Organs
- Reproduction
- Human Beings and Race

- Review Exercises for Chapter 13
- Connecting to Life/Job Skills
- Chronology of Famous Names in Biology

Characteristics of Primates

Humans are **primates**. They share this mammalian order with gorillas, chimpanzees, monkeys, lemurs, tarsiers, slow lorises, and the tailed bush babies. What are the special features that distinguish a primate from other mammals?

EVOLUTIONARY CHARACTERISTICS

Scientists believe that at one time all primates lived in trees. The successful species were those that developed adaptations for arboreal life. Over evolutionary time, species such as chimpanzees, baboons, and humans have left the trees and have adapted to life on land. However, certain characteristics persist in the ground-living species that show close relationship to their tree-living "cousins.

Figure 13.1 illustrates the comparative shapes of hands of some primate species. These hands were adapted for grasping objects such as tree limbs, enabling the animals to locomote by swinging from tree limb to tree limb. Even the feet and tails of some primate species are **prehensile**, adapted for grasping. A second primate characteristic is the well-developed sense of sight. The eyes of most other mammals are located at the sides of the head. The eyes of a primate are directed forward, a structural arrangement that permits **stereoscopic** *vision*. This means that primates can see in three dimensions (length, width, and breadth), a characteristic that enables them to see ahead with clarity the branches that they grasp. Another primate adaptation is the development of a larger and more **convoluted** brain. Convolutions are folds in the brain that increase surface area and allow for a greater number of nerve cells. Mention should also be made of two other primate characteristics. Primates have teeth that are less specialized for tearing and more useful for the grinding and chewing of a varied diet. Primates also have a small number of offspring and provide extended parental care for the young.

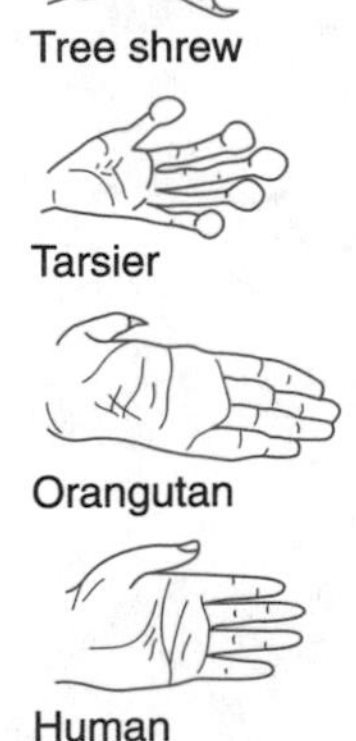

FIGURE 13.1 A primate comparison of hands

A SPECIAL PRIMATE–THE HUMAN

Though humans are primates, they are set apart from other primates by some unusual and important adaptations. **Bipedalism**, the ability to walk on two legs (instead of four), has freed the forelimbs for doing work. The human can walk great distances and carry things from region to region—things that aid in hunting, gathering, or building.

Humans can live in the most forbidding of environments and are the most ecologically versatile of the primates. The human adjustment to various climates and land surfaces is probably due to the greater brain development. In all probability, brain development is the single greatest factor contributing to human ability to form words and to speak. Speech is tied to a myriad of intricate brain functions: learning words, associating ideas, remembering. The formulation of the spoken word leads to the creation of the written word. When ideas are written, they are not lost in time. The human being is the only animal that can act in terms of history. Human civilizations function in terms of learned behavior called **culture**.

BASIC HUMAN STRUCTURE

The human body is divided into a **head**, **neck**, **trunk**, and **two pairs of appendages**—namely, the **arms** and **legs**. Hair is present on the head, under the arms, sparsely on the arms and legs, and around the pubic area; in males, there is a greater distribution of body hair, often on the chest and heavily on the back of the hands, on the arms and the legs and on the face. The face is directed forward in a vertical position; the main axis of the body is set vertically, also.

In humans, as in other mammals (birds, too), the coelomic cavity is divided into three different areas. The **pericardial** cavity encloses the heart; **two pleural cavities** contain the lungs; and the **peritoneal** cavity holds the major part of the digestive system, the reproductive system, and the urinary system.

The Skeletal System

The human skeleton, like that of all vertebrates, is a living endoskeleton that grows with the body. At birth, the human baby has a body that is made up of 270 bones. Due to the fusion of separate bones, the mature skeleton is composed of 206 bones. Figure 13.2 shows some of the bones that make up the skeleton. Table 13.1 provides a more complete summary.

The human skeleton is a magnificent feat of engineering. The primary purpose of the skeleton is to carry the weight of the body and to support and protect the internal organs. The skeleton must be strong and able to absorb reasonable amounts of shock without fracturing. At the same time, the body framework must be flexible and light enough in weight to permit movement. Skeletal bones move in response to muscles that work like levers, allowing a variety of movements such as walking, running, hopping, sitting, bending, lifting, and stooping.

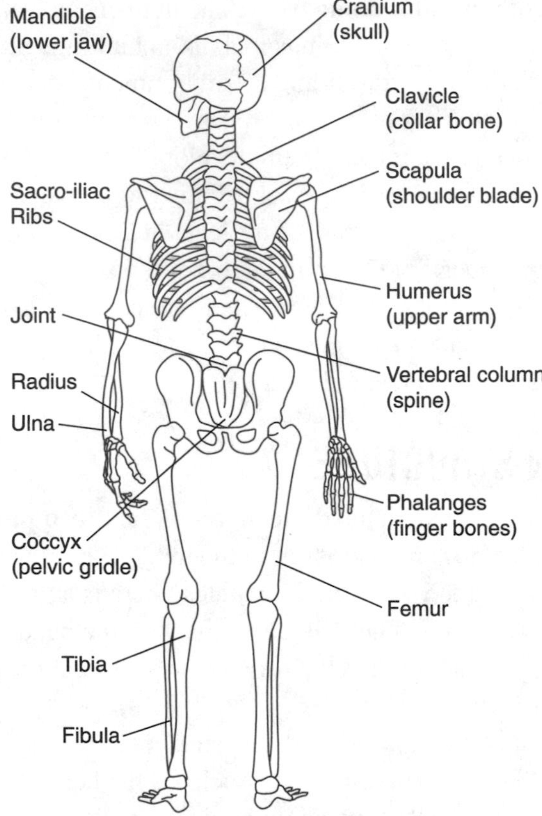

FIGURE 13.2 Bones of the human body

TABLE 13.1
THE HUMAN SKELETON

Skeleton	Region	Names of Bones
Axial Skeleton	Skull — Cranium	Frontal, parietal (2), temporal (2), occipital, sphenoid, ethmoid
	Skull — Face	Nasal, vomer, inferior turbinals, lacrimals, malars, palatines, axillae, mandible, hyoid
	Vertebral Column	Vertebrae: cervical, thoracic, lumbar, sacral, coccyx
	Thorax	Ribs, sternum
Appendicular Skeleton	Pectoral Girdle	Clavicle, scapula
	Pectoral Appendages	Humerus, radius, ulna, carpals, metacarpals, phalanges
	Pelvic Girdle	Ilium, ischium, pubis (compose the os innominatum)
	Pelvic Appendages	Femur, patella, tibia, fibula, tarsals, metatarsals, phalanges

PARTS OF THE SKELETON

The human skeleton is divided into two major parts: the **axial skeleton** and the **appendicular skeleton**.

AXIAL SKELETON

The skull, the thorax (rib cage) and the vetebral, or spinal, column are the three regions of the axial skeleton.

Skull

All the bones of the head compose the **skull**. The two regions of the skull—the **cranium** and the face—are made up of 22 flat and irregularly shaped bones. Eight bones form the cranium, which functions in the protection of the brain. The facial region, designed to protect the eyes, nose, mouth, and ears, is composed of 14 bones. The **sinuses** are air spaces in the facial bones which aid in reducing the weight of the skull. The bones of the middle ear that function in transmitting sound to the inner ear are the smallest bones in the body—namely, the *hammer*, *anvil*, and *stirrups*.

Vertebral Column

The vertebral column is composed of 26 bones known as **vertebrae**. At birth the vertebral column consists of 33 bones: seven cervical (neck) vertebrae, twelve thoracic vertebrae, five lumbar vertebrae, five sacral vertebrae, and four coccygeal (tail) vertebrae, but the five sacral bones fuse into one large triangular bone—the **sacrum**—at the back of the pelvis and the four coccygeal bones fuse into a single **coccyx**.

The vertebral column is the backbone made flexible by the cartilage and ligaments that join the individual vertebrae. Such a flexible backbone permits movement of the head and bending of the trunk. Of major importance is the backbone's function in the protection of the spinal cord, which extends downward from the brain through the opening in each vertebra. Nerves branching from the spinal cord radiate to all parts of the body through openings in the sides of the vertebrae. Figure 13.3 shows the discs of cartilage that separate the individual vertebrae. These discs prevent friction due to the rubbing of the bones and serve as shock absorbers.

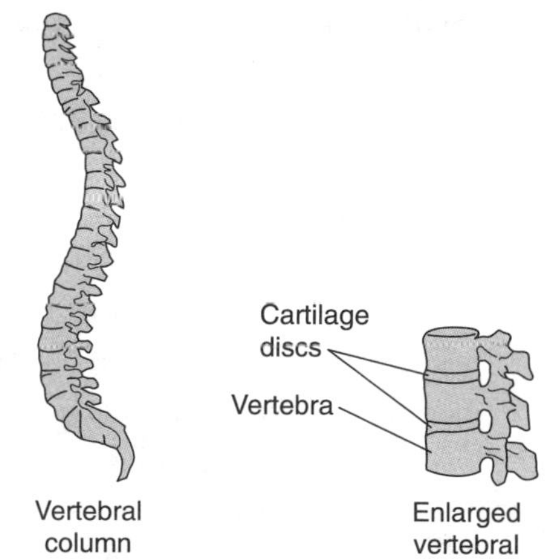

FIGURE 13.3 The human vertebral column

Thorax

Just below the neck are 12 pairs of ribs that are attached to the vertebral column. The general shape of the **thoracic basket**, or rib cage, is shown in Figure 13.4. The first ten pairs of ribs are attached to the breastbone, known also as the **sternum**, by cartilage strips, forming a structure that is smaller on top. The loose connections of the ribs to the vertebral column and the flexible cartilage connections at the sternum allow the ribs to move when the lungs are inflated. The 11th and 12th pairs of ribs are often referred to as "floating ribs" because they are attached to the vertebral column but not to the breastbone.

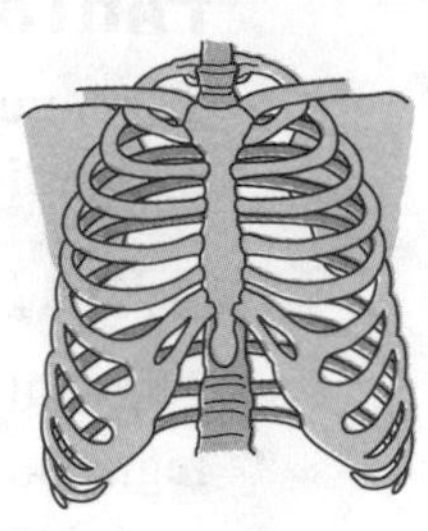

FIGURE 13.4 Human rib cage

THE APPENDICULAR SKELETON

The arms and hands, the legs and feet, and the bones of the shoulder and the pelvis make up the *appendicular skeleton*. You probably realize that "appendicular" is the adjective of the word *appendage*. An **appendage** is an attachment to a main body or structure. Legs and arms are attachments to the axial skeleton. The sites where arms and legs are attached to the axial skeleton are bones referred to as **girdles**. The *pectoral* (shoulder) girdle where the arms are attached is composed of the *scapula* (shoulder blade), a large triangular bone, and the *clavicle* (collarbone), a smaller curved bone. The legs are attached to the *pelvic girdle*, which is formed by the fusion of three bones: the ilium, the ischium, and the pubis on each side of the midline of the body (Figure 13.5).

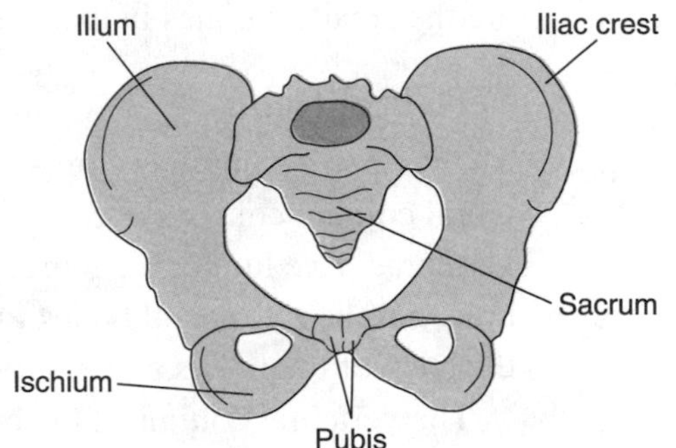

FIGURE 13.5 The pelvic girdle

The bones of the legs and arms are appropriately called the **long bones**. The **femur**, the long bone between the hip and the knee is the longest and strongest bone in the body, supporting the weight of the body. The long bones in the lower leg are the thinner **fibula** and the thicker **tibia**.

Between the shoulder blade and the elbow is the bone of the upper arm, the **humerus**, a thinner version of the femur. The two long bones of the lower arm are the **radius** and the **ulna**. Finger and toe bones are known as **phalanges**; the bones of the foot, as **metatarsals**; the bones of the hand, as **metacarpals**.

In general, long bones are shaped like tubes with rounded processes at the ends which are designed to fit into other bones to form *joints*. The ends of the long bones are filled with *spongy bone*, a structural device which makes the bones light in weight but strong. The open spaces in spongy bone are filled with red marrow; while yellow marrow fills the shaft of the long bone.

Bone marrow is an important substance. **Red marrow** in the spongy areas of the long bones and in the ribs and vertebrae is the site where red blood cells are produced at the rate of millions per minute. The yellow marrow in the shaft of the long bones contains mostly fat. However, when the blood-making capability of the red marrow is low, yellow marrow is somehow converted into red marrow.

COMPOSITION OF BONE

About 20 percent of living bone is water; the remaining 80 percent consists of mineral matter and protein. Mineral matter deposited in bone usually forms the compound calcium phosphate ($Ca_3(PO_4)_2$). Magnesium and other elements may also contribute to the mineral composition. The protein portion of the matrix is made of **collagen fibers** that are found in tendons, skin and connective tissue. The protein fibers and the mineral matter form the nonliving matrix of bone. However, bone also contains a variety of living cells and blood vessels that provide the pathways for nourishment to the cells and permit the removal of respiratory wastes from the cells.

Living bone cells are nestled in small spaces in the mineral matrix of bone. These cells are of three types, each adapted to carry out a specific function concerned with the building and maintaining of bone. The production of new bone material and the repair of broken bones are the work of **osteoblasts**, which are responsible for secreting the mineral and protein compounds that form the matrix. A second type of bone cell is the **osteoclast**, a bone breaker, able to dissolve bits of bone that are in the way of the efficient design of the skeleton. The destructive work of the osteoclasts is often followed by constructive work of the osteoblasts in the rebuilding of bone. The third type of bone cell, the **osteocyte**, functions as the caretaker of the bone tissue nearby.

The surface of nearly all parts of bone is covered by a tough membrane, the **periosteum**. The periosteum is perforated by microscopic blood vessels that supply the bone cells with nourishment. Although bone matrix appears to be solid, it is pierced by a network of **Haversian canals** through which blood vessels pass. Nerve fibers also extend into the bone interior through Haversian canals. The larger blood vessels pass directly into spongy bone and into the yellow marrow areas of the long bones.

The Muscular System

Muscles represent 40 percent of the total weight of the human body. Muscle tissue is characterized by contractility and electrical excitability, two distinctive properties that enable it to effect movement of the body and its parts. Two common disorders of locomotion and/or other movements are described in Table 13.2.

TABLE 13.2
TWO COMMON DISORDERS OF LOCOMOTION AND/OR OTHER MOVEMENT

Disorder	Description
Arthritis	Inflammation of the joints and their supporting structures
Tendonitis	Inflammation of the tendon at the point of attachment to the bone

There are three types of muscle tissue: smooth, striated, and cardiac. The movements of smooth and striated muscle tissue are controlled by contractile proteins and innervation from the nervous system; these muscles respond to the electrical stimulation of a nerve impulse. Cardiac muscle, on the other hand, functions to a great degree because of its own inherent ability to generate and conduct electrical impulses.

TYPES OF MUSCLE

SMOOTH MUSCLE

Smooth muscle is present in the walls of the internal organs, including the digestive tract, reproductive organs, bladder, arteries, and veins. Because smooth muscle is contained in organs that do not respond to the will of a person, these muscles are called **involuntary muscles**. The most common function of smooth muscle is to squeeze, exerting pressure on the space inside the tube or organ it surrounds. Food is moved down the esophagus by the squeezing action of smooth muscle. The action of smooth muscles causes urine to be expelled from the bladder, semen to be discharged from the seminal vesicles and blood to be pumped through arteries. The opening and closing of the iris of the eye in response to light is accomplished by the action of smooth muscles.

Smooth muscle is made up of cells packed with contractile proteins, the cells forming sheets of tissue. Smooth muscle tissue is *innervated* by nerve cells and fibers from the sympathetic nervous system, that part of the nervous system that controls the activities of the internal organs. The contraction of smooth muscle is in response to stimulation by nerve cells, neurohumors or hormones.

STRIATED MUSCLE

Striated muscle is variously referred to as striped muscle, **voluntary muscle** or **skeletal muscle**—terms describing its structure and function. Located in the legs, arms, back and torso, striated muscles attach to and move the skeleton; since they are moved by the will of the person, they are often termed *voluntary muscles*. A striated or skeletal muscle is made up of a great number of *muscle fibers*, each of which extends the entire length of the muscle. There are probably around six billion fibers in more than 600 muscles scattered throughout the body.

If you look at a bit of striated muscle through the microscope, you will note that the muscle fibers contain many nuclei that seem not to be separated from each other by a plasma membrane. Such an arrangement in which plasma membranes are missing is called a **syncytium**. Each muscle fiber is innervated by at least one motor neuron. A **neuromuscular junction** is the space (synapse) between a nerve cell and a muscle.

CARDIAC MUSCLE

Cardiac muscle is present only in the heart, where the cells form long rows of fibers. Unlike other muscle tissue, cardiac muscle contracts independently of nerve supply since reflex activity and electrical stimuli are contained within the cardiac muscle cells themselves. **Purkinje fibers**, part of the mechanism that controls heartbeat, are so specialized for conducting electrical impulses that they do not have contractile proteins. Each heartbeat is started by self-activating electrical activity of the heart's **pacemaker**, known as the **sinoatrial node** (S-A node), positioned in the wall of the right atrium. From the S-A node, the impulse spreads throughout the atrium to the **atrioventricular node** (A-V node), a specialized bundle of cardiac muscle located on the atrium near the ventricles. The impulse spreads from the A-V node to all parts of the ventricles, causing simultaneous contractions in the ventricles.

Cardiac muscle is innervated by the tenth cranial nerve, the **vagus** nerve. The vagus is a mixed nerve containing some nerves that speed up heartbeat and others that retard it. However, cardiac muscle contracts independently of nerve supply, and the effect of nerves on heartbeat is not completely understood.

HOW MUSCLES CONTRACT

The fine structure of a muscle fiber controls muscle contraction. A muscle fiber is a single skeletal muscle cell and is made up of a bundle of finer fibers called **myofibrils**. A single myofibril is composed of smaller units named **sarcomeres**, which are the units of contraction in muscle tissue. **Sarcomeres** are arranged in single file along the length of the myofibril. Within the sarcomere are alternating rows of thin and thick filaments. The thin filaments are attached to two vertical bands of thick protein called the **Z lines**. Figure 13.6 shows the fine structure of a muscle myofibril. A Z line marks the boundary between sarcomeres.

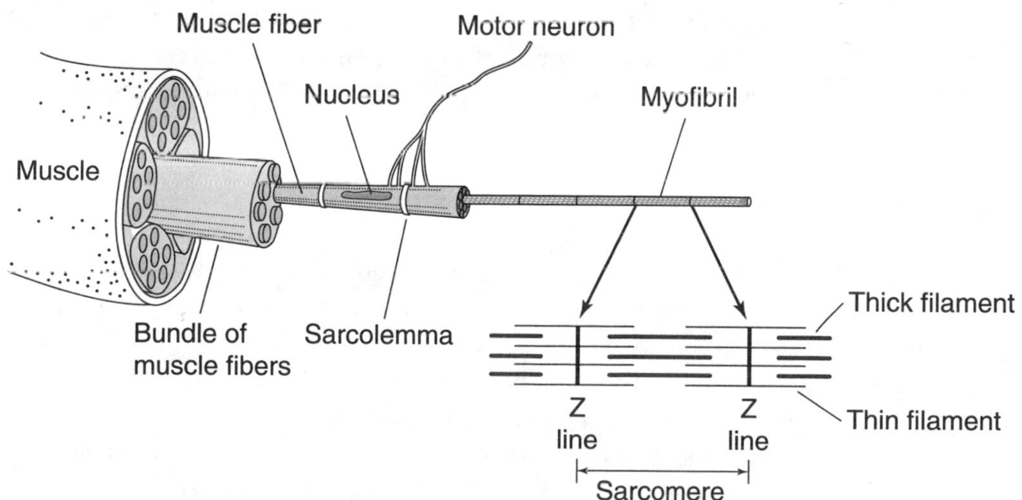

FIGURE 13.6 The fine structure of skeletal muscle

Muscles contract due to a sliding filament mechanism. When the thick and thin filaments slide past each other, the Z lines of the sarcomeres are pulled closer together.

The sliding filament theory explains the physical changes that muscles undergo as they contract. When the sarcomeres contract, the myofibrils also contract, causing the contraction of the muscle fibers. Figure 13.7 illustrates the sliding filament mechanism of muscle contraction.

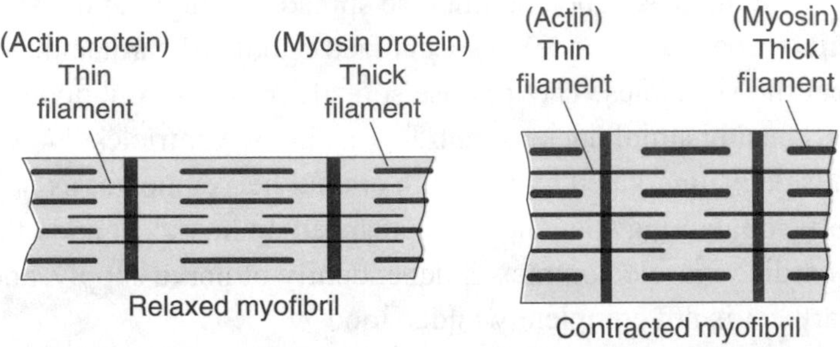

FIGURE 13.7 Sliding filament mechanism of muscle contraction

The Nervous System

The nervous system in humans is made up of two major parts: the **central nervous** system and the **peripheral nervous system**. Nervous tissue is specialized to receive stimuli from the outside environment and to conduct impulses to other body tissues. The development of the nervous system and particularly of the brain is what makes humans significantly different from other animals.

The basic unit of function of the nervous system is the **neuron**, or nerve cell. An understanding of the structure and function of this cell and of the way in which it transmits nerve impulses is important before dealing with the parts of the nervous system.

NERVE CELLS

The parts of the nerve cell are the **cyton**, or cell body, the **dendrites** and the **axon**. The dendrites receive signals from sense organs or from other nerve cells and transmit them to the cyton. The cell body passes signals to the axon, which then conducts the signals away from the dendrites and cell body. A nerve cell has only one axon but many dendrites. The axon terminating in **end brushes** (known also as terminal branches) is popularly called a **nerve fiber**. Many of the axons in the vertebrate body are covered by a fatty **myelin sheath** made of **Schwann cells**. The space between two Schwann cells is known as a **node of Ranvier**. Axons with the myelin sheath transmit impulses more quickly than those without the fatty coverings.

The nervous system has three types of neurons. **Sensory** or **afferent** neurons receive impulses from the sense organs and transmit them to the brain or spinal cord.

Associative or *interneurons* are located within the brain or spinal cord. These transmit signals from sensory neurons and pass them along to motor neurons. **Motor** or **efferent** (Figure 13.8) neurons conduct signals away from the brain or spinal cord to muscles or glands, so-called **effector** organs.

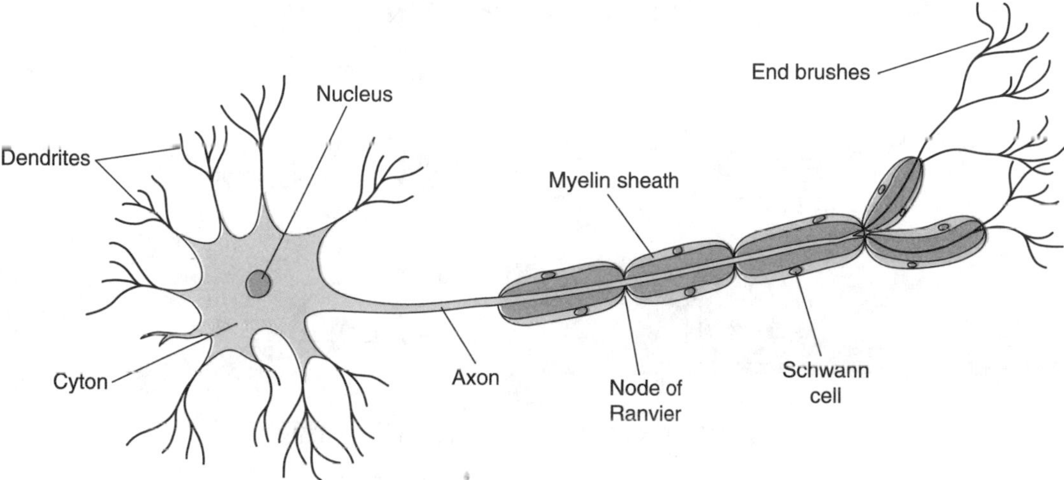

FIGURE 13.8 Motor neuron

THE NATURE OF THE NERVE IMPULSE

The primary function of the nervous system is to permit communication between the external and internal environments. Communication in the nervous system is made possible by signals or impulses carried in a one-way direction along nerve cells. These impulses are electrical and chemical in nature.

When a neuron is not carrying an impulse, it is said to be at **resting potential**. When a nerve cell is stimulated to carry an impulse, its electrical charge changes and it is said to have an **action potential**. Action potentials (nerve impulses) from any one nerve cell are always the same. All impulses are of the same size, there being no graded responses. This circumstance is known as the "**all or none response**," meaning that a nerve cell will transmit an impulse totally or not at all (Figure 13.9).

The electrical changes that occur in a nerve cell are due to differences in the distribution of certain ions, or charged particles, on either side of the nerve membrane. Three factors are responsible for the distribution of ions. The first factor is the nature of the cell membrane itself. Remember that it is highly selective. The cell membrane is impermeable to the sodium ion (NA^+); at the same time, the cell membrane is highly permeable to the potassium ion (K^+). A second factor involves diffusion, the process by which molecules move from an area of greater concentration to one of lesser concentration. A third factor involves the attraction of ions of opposite charge and the repulsion of ions of like charge.

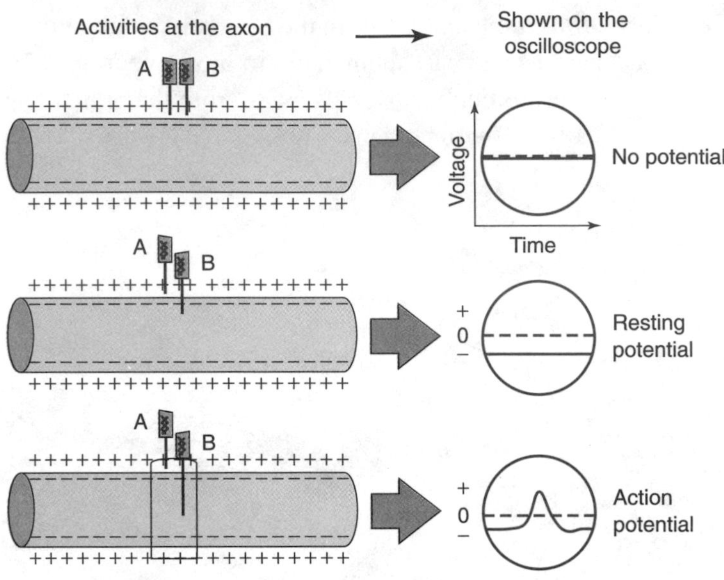

FIGURE 13.9 Resting and action potentials

Let us begin with the situation in which there is a large concentration of K^+ inside the cell and a large concentration of Na^+ outside the nerve cell membrane. By diffusion, K^+ ions will cross the plasma membrane to the outside of the cell. Na^+ ions cannot follow, however, because the cell membrane is impermeable to them. K^+ ions will continue to move to the outside until an equilibrium is established. As a result, the outside of the nerve cell membrane becomes more positive than the inside of the membrane. The cell now can be described as being at its **resting potential**. The cell is also **polarized**.

As an impulse travels through the nerve cell, several events take place. As the impulse touches a given point along the length of the plasma membrane, that site becomes permeable to Na^+ ions and Na^+ crosses the membrane and enters the cell. At the same time, the cell membrane becomes even more permeable to K^+ which then leaks out of the cell at a greater rate. These events lead to the **depolarization** of the cell in which there is a wave-like reversal of electrical charge along the length of the cell membrane. After the impulse is transmitted, the cell uses energy in mechanisms known as the **sodium-potassium pump** and **carrier-facilitated transport** to bring the cell back to its resting potential.

Axons with myelinated sheaths can conduct impulses at the rate of 200 meters per second. Naked axons may conduct impulses at the rate of a few millimeters per second.

Impulses travel from one neuron to another crossing a specialized gap called the *synapse*. The synapse is a space between nerve cells that measures about 20 nanometers in width—just enough distance to prevent the touching of nerve cells. The terminal branches of the axons have synaptic knobs at their ends. As impulses travel along an axon, the synaptic knobs release chemicals called *neurotransmitters* or *neurohu-*

mors. The neurotransmitters carry the impulse across the synapse onto the dendrites of the receiving nerve cell. The two main neurohumors are **acetylcholine** and **norepinephrine**, each with inhibitory or excitatory capabilities. After an impulse has crossed a synapse, the neurotransmitter is destroyed by an enzyme such as **cholinesterase**. (Many drugs of abuse, including LSD and cocaine, interfere with nerve impulse transmission.)

The endbrushes of the motor neurons are buried in a muscle or a gland. The junction between the nerve fibers and the muscles is known as the **neuromuscular junction**. The neurotransmitter **acetylcholine** carries the impulse from the nerve fiber to the muscle or gland. Nerve transmission across the neuromuscular junction is always excitatory, never inhibitory.

THE CENTRAL NERVOUS SYSTEM

The brain and the spinal cord compose the central nervous system. In the vertebrate body, the organs of the central nervous system are well protected by being wrapped in connective tissue and enclosed in bone. The brain, covered by the membranous **meninges**, rests in the skull cavity where it is enclosed by the cranium. The spinal cord, also covered by connective tissue, is circled by the vertebral column.

BRAIN

The human brain (Figure 13.10) contains approximately 100 billion neurons and has an average weight of 3 pounds. The brain is the master control center of the body, processing information that is gathered by the sense organs. The brain receives, processes, stores, and retrieves information. It coordinates responses by stimulating and inhibiting the activities of different body systems and organs. The brain is divided into several parts, each with special functions. Among the most important parts are the **forebrain**, the **midbrain**, and the **hindbrain**.

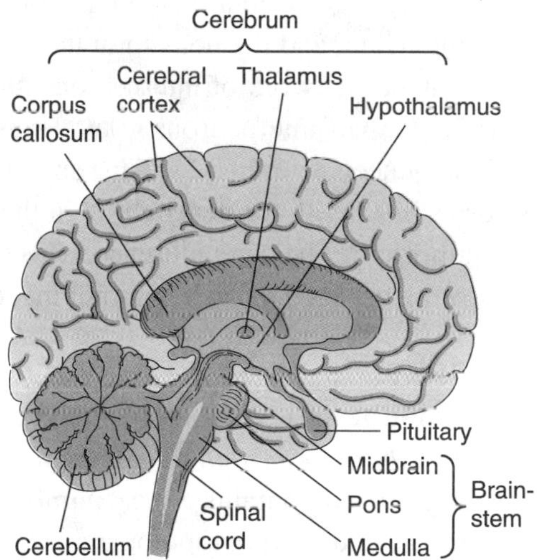

FIGURE 13.10 The human brain

Forebrain

In mammals the largest and most highly developed part of the brain is the **forebrain**. It includes the **cerebrum**, the seat of intelligence. It controls the voluntary, conscious activities of the human body. All voluntary muscle movements and all speech, thinking, and memory are controlled by the cerebrum.

The cerebrum folds back on itself in many places forming wrinkles known as **convolutions**. These convolutions increase the surface area of the brain. The cerebrum is divided into two equal halves—the right and left cerebral hemispheres. Each of these hemispheres is subdivided into distinct regions of nervous control by fissures and convolutions. These regions control sensory areas for different parts of the body and are not haphazardly arranged.

The outer layer of the cerebrum is called the *cerebral cortex* and is composed of *gray matter*—unsheathed nerve cell bodies. Under the cortex is the *white matter* made up of sheathed axons. These fibers connect the various parts of the cerebrum and the cerebrum with other parts of the brain.

In addition to the cerebrum, the forebrain includes other regions. A pair of **olfactory lobes** process the sense of smell. The **thalamus** serves as a switchboard for sensory information. Incoming signals from the sense organs are sent to clusters of neurons (called *nuclei*) in the thalamus and then directed to the cerebrum. These nuclei also integrate some outgoing motor signals. The **hypothalamus** monitors signals from internal organs and controls body temperature, osmoregulatory activities, the onset of maturity, thirst, hunger, and the sex drive. The hypothalamus is also the region where the nervous and hormonal systems interact.

Midbrain

The **midbrain** is covered with gray matter that accepts visual and sensory signals. These signals are sent on to higher brain centers for processing. The midbrain, medulla oblongata, and the pons form the **brainstem**.

A major network of interneurons runs through the entire length of the brainstem. This group of interneurons is known as the **reticular formation**. The reticular formation connects with nerves in the spinal cord that control muscle contraction, and also forms links with the **cerebellum** in the hindbrain to control balance, muscle tone, and equilibrium. The cerebellum controls the precision and coordination of voluntary movements such as walking, running, dancing, skating, writing, and keyboarding.

Hindbrain

The **hindbrain** is composed of the **medulla oblongata**, the **cerebellum**, and the **pons**. The *medulla*, or *medulla oblongata*, lies below the cerebrum and connects with the spinal cord. It contains a great number of **ganglia** (cytons) that receive sensory impulses and send out motor signals. Through the medulla pass many of the sensory and all of the motor nerves on their way to or from the higher centers in the brain. The medulla controls automatic, involuntary activities such as the contraction of smooth muscles, reflex movements, dilation and constriction of blood vessels, swallowing, and breathing.

The *cerebellum* controls body coordination, balance, and equilibrium. It receives sensory signals about body position and balance and integrates these signals with those coming from the sense organs: eyes, ears, skin, and muscle spindles.

The *pons* serves as a bridge. Bands of nerve extending from the pons fibers link the cerebrum, cerebellum, and brainstem.

SPINAL CORD

The **spinal cord** is an elongated tubular structure containing masses of nerve cells and fibers and lying within the vertebral column. It is composed of a central H-shaped core of gray matter surrounded by white matter. The spinal cord conducts impulses to and from the brain; the impulses enter and leave the spinal cord through spinal nerves which extend from the spinal cord to the other organs of the body. The spinal cord is also the center for simple reflex activity.

In a simple **reflex**, often known as a **reflex arc**, only sensory nerves, the spinal cord, and motor nerves are involved. It allows instantaneous response without involving transmission to and from the brain. An example of a reflex is pulling your hand from a hot stove. When you touch a hot stove, sensory nerves pick up the stimulus from receptors in the skin and transmit it to the spinal cord, which signals motor nerves to signal muscles for you to pull your hand away—all instantaneously.

THE PERIPHERAL NERVOUS SYSTEM

The peripheral nervous system connects the central nervous system—the brain and spinal cord—with the other organs of the body. It has two parts—somatic and autonomic.

SOMATIC NERVOUS SYSTEM

The somatic nervous system is composed of cranial nerves and spinal nerves. The fibers of both sensory and motor neurons are bundled together to form the cranial nerves. Twelve cranial nerves extend between the brain and the sense organs (eyes, ears, nose, etc.), heart, and other internal organs. Thirty-one pairs of mixed sensory and motor nerves extend from the spinal cord to the muscles and organs of the body. Each of the spinal nerves separates into sensory fibers and motor fibers as these nerves join to form the spinal cord. The sensory fibers lead into the dorsal side of the spinal cord. Some of the sensory fibers synapse with the associative (interneuron) neurons; others lead to the brain. The cell bodies of the sensory nerves are in the dorsal root ganglia outside the spinal cord. The motor nerves lead out from the spinal cord on the ventral side. Their cell bodies are in the spinal cord, where they synapse with the associative neurons in the spinal cord.

AUTONOMIC NERVOUS SYSTEM

A network of nerves known as the autonomic nervous system controls the body's involuntary activities and the smooth muscles of the internal organs, glands, and heart muscle. It is composed of motor (efferent) neurons leaving the brain and spinal cord and also of peripheral efferent neurons (Figure 13.11).

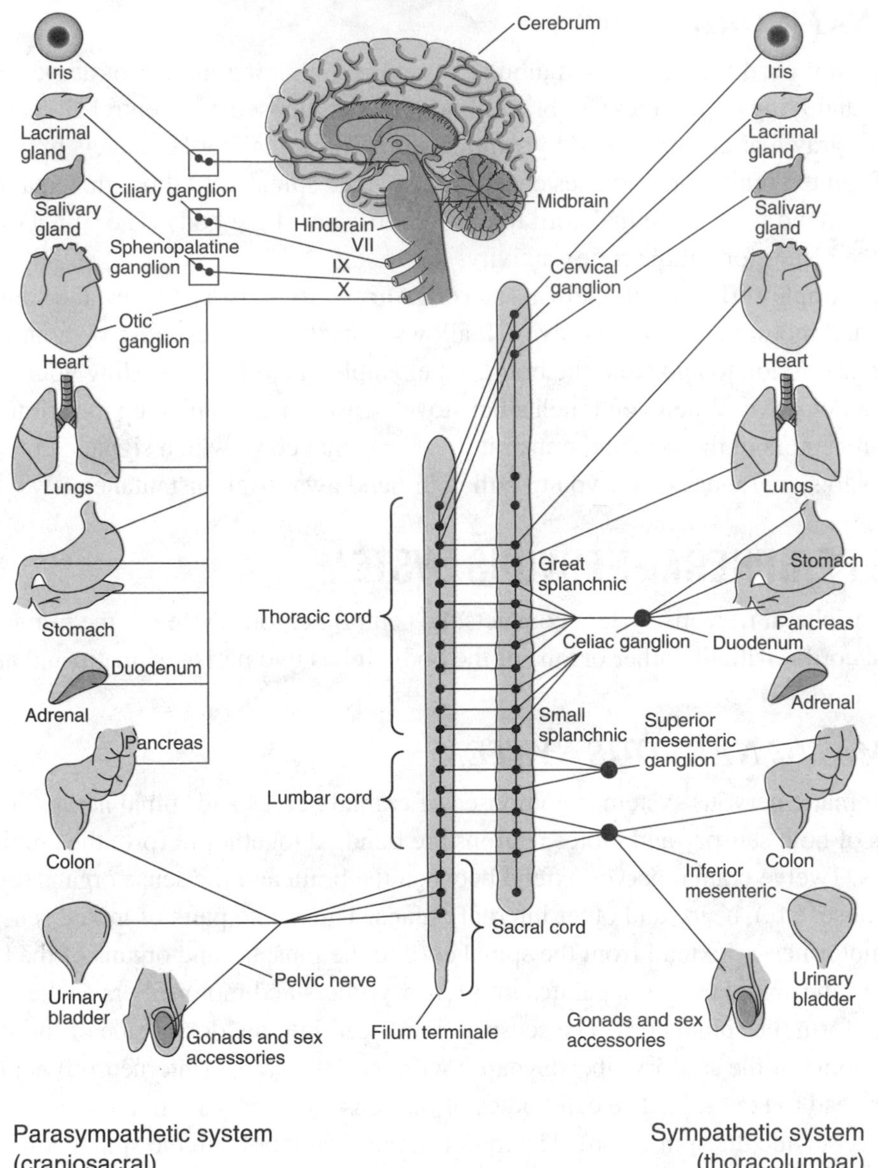

FIGURE 13.11 The autonomic nervous system

The autonomic nervous system is divided into the **symphathetic system** and the **parasympathetic system**. These subsystems are antagonists. When one set of nerves activates the smooth muscles of the body, the other set inhibits the action. For example: the parasympathetic nerves dilate the blood vessels and slow the heartbeat; the sympathetic nerves constrict the blood vessels and quicken the heartbeat.

Figure 13.12 summarizes the relationships of the parts of the nervous system.

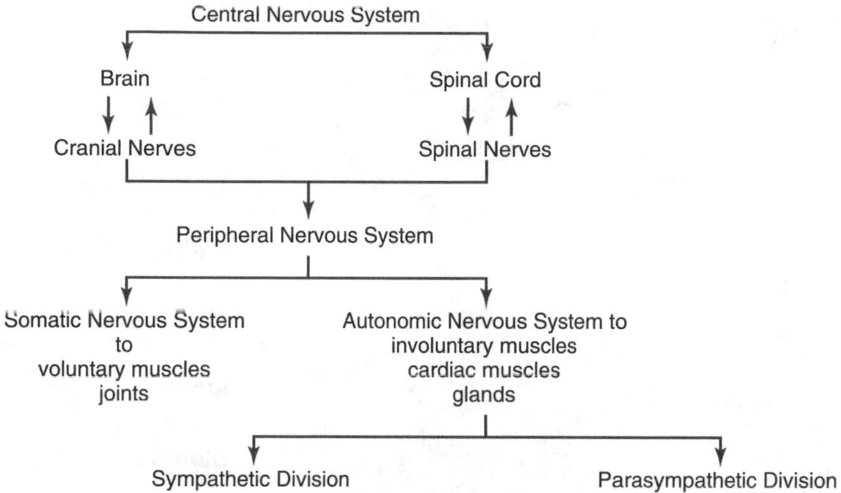

FIGURE 13.12 The nervous system the easy way

DISORDERS OF THE NERVOUS SYSTEM

Table 13.3 lists and describes four disorders of the nervous system.

TABLE 13.3
SOME DISORDERS OF THE NERVOUS SYSTEM

Disorder	Description
Cerebral palsy	A form of paralysis denoted by jerky, spastic, writhing movements resulting from damage to the portion of the brain that controls muscles. Cerebral palsy is a group of syndromes with a common result.
Meningitis	Inflammation of the meninges, the membranes that surround the brain and spinal cord
Stroke	A disorder resulting from a hemorrhage in the brain or a blood clot in a cerebral blood vessel which may cause brain damage
Polio	A viral disease that affects the central nervous system and often results in muscle damage to the legs and/or arms

The Endocrine System

Within the mammalian body there is a constellation of **ductless glands** known as the **endocrine system**. Figure 13.13 shows the locations of these glands in the human body. You will notice that these glands are not grouped together but are distributed throughout the body. Although these glands are not grouped together, they are considered to be a system because of similarities in structure and function.

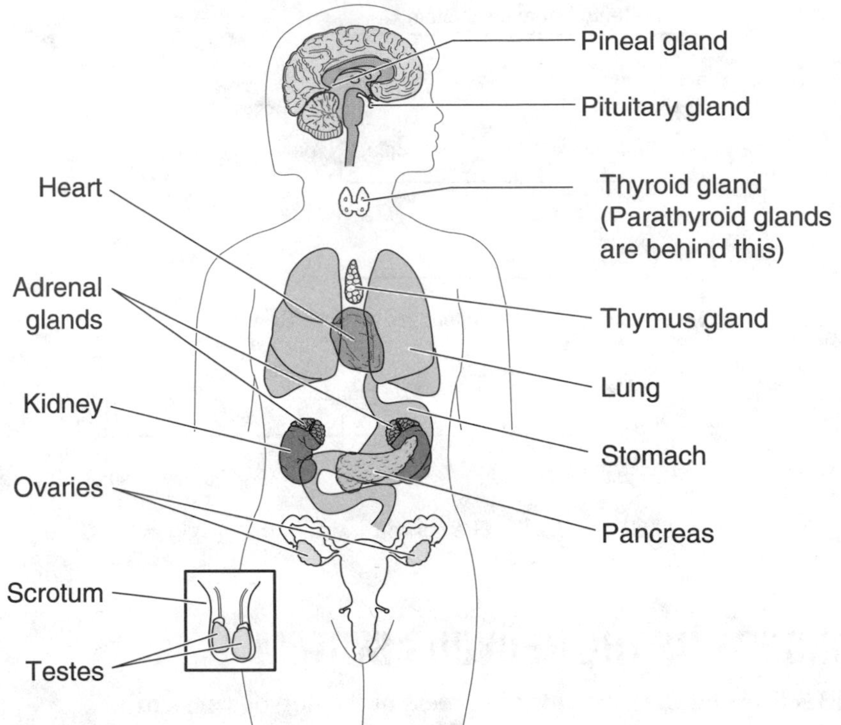

FIGURE 13.13 Diagram showing the location of the major endocrine glands: pineal, pituitary, thyroid, parathyroid, thymus, pancreas (part), lining of the small intestine, adrenal glands, and sex glands

As the name implies, ductless glands do not have ducts and therefore do not discharge their secretions directly into another organ. Most of the glands in the body, however, are duct glands, delivering their secretions directly into a contiguous or nearby organ. For example, the salivary gland delivers saliva directly into the mouth, and sweat glands conduct perspiration directly to the skin. **Endocrine glands**, also known as *glands of internal secretion*, deliver their secretions—**hormones**—into the bloodstream, which then carries them to their target organs. Hormones regulate many of the important metabolic activities of cells and organs.

The endocrine system is made up of the *pituitary gland*, the *thyroid gland*, the *parathyroid glands*, the *adrenal gland*, the *isles of Langerhans* in the pancreas, the *thymus gland*, the *pineal gland*, and the *gonads*—testes in the male and ovaries in the female. Certain secretions of the stomach and small intestine are also hormones and thus part of the endocrine system.

Through their secretions the endocrine glands regulate growth, rate of metabolism, response to stress, blood pressure, muscle contraction, digestion, immune responses, and the development and functioning of the reproductive system. Hormones exert their influence by becoming involved with the genetic machinery of cells and by affecting the metabolic activities of cells, working through the cellular respiration pathways.

Prostaglandins are a recently discovered group of hormones that are not produced in any particular gland. Prostaglandins of one kind or another are produced by most tissues in response to other hormones or to irritation of the tissues. One group of prostaglandins is responsible for the pain brought on by inflammatory responses. One specific prostaglandin causes the blood to clot; another type enhances circulation. It seems that some prostaglandins can cause adverse effects in the body, while others prevent them.

The **pituitary gland**, sometimes called the **hypophysis**, hangs from the base of the brain and is thought to exert control over much of the functioning of the other endocrine glands. It does this through its **trophic hormones**—hormones that stimulate the activity of other glands. Trophic hormones from the pituitary are known to stimulate secretions of the thyroid gland and of the adrenal gland and to regulate functioning of the sex organs. The pituitary (and, in turn, the rest of the endocrine system) is itself thought to be controlled by the hypothalamus region of the brain. The hypothalamus releases *neurosecretions*, known as *releasing factors*, which are transmitted to the pituitary where they, in turn, regulate the release of trophic hormones (Figure 13.14).

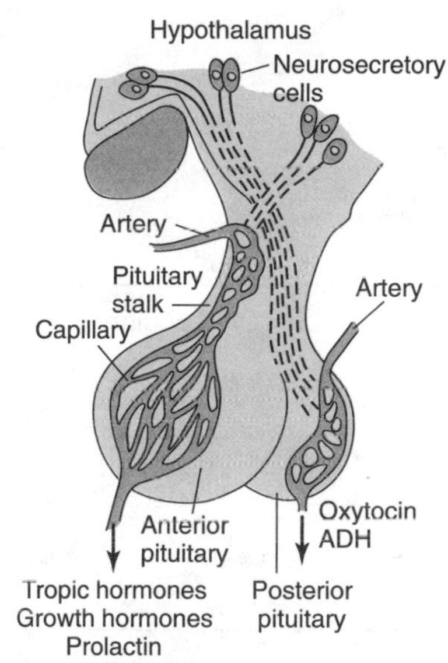

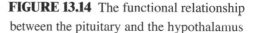

FIGURE 13.14 The functional relationship between the pituitary and the hypothalamus

The level of hormones in the blood and the release of hormones from endocrine glands are subject to a feedback control mechanism, with the blood "feeding" back to the brain information on how much hormone is circulating in the bloodstream. If, for example, the blood level of thyroxin, the thyroid hormone, falls too low, the hypothalamus secretes thyroid-releasing factor, which stimulates the pituitary to release thyroid-stimulating hormone (TSH); in turn, the thyroid is stimulated to secrete more thyroxin. As the blood levels of thyroxin rise, the release of TSH and thyroxin is slowed.

Table 13.4 summarizes the endocrine system, listing the major endocrine glands, their hormones and their functions, and the problems associated with excess or diminished secretion.

TABLE 13.4
ENDOCRINES AND THEIR HORMONES

Name of Gland	Location	Hormone	Normal Function	Problem Associated with: Excess Secretion	Problem Associated with: Diminished Secretion
Anterior pituitary	Base of brain Forward portion	Growth hormone (STH)	Affects skeletal growth, protein synthesis, blood glucose concentration	Gigantism acromegaly	Dwarfism
		Trophic hormones TSH ACTH FSH LH	Stimulate target glands thyroid adrenal cortex ovarian follicles; testes gonads	Oversecretion of glands	Undersecretion of glands
Posterior pituitary	Hind portion	Vasopressin	Controls blood pressure; reabsorption of water by kidney tubules	Increased blood pressure; glycogen converted to sugar	Decreased blood pressure; excess sugar changed to fat; kidney tubules not reabsorbing water
		Oxytocin	Causes contractions of uterus		
Thyroid	Two lobes on either side of larynx	Thyroxin (65% iodine)	Controls rate of oxidation in cells	Increased oxidation; nervous exophthalmic goiter	Lowered oxidation; in a child—cretinism; in an adult—myxedemic goiter due to lack of iodine in drinking water
Parathyroid	Four glands above thyroid	Parathyroxin	Regulates amount of calcium in blood	Trembling due to lack of muscular control	Contraction of muscles (tetany); death
Stomach	Mucous lining (mucosa)	Gastrin	Stimulates secretion of gastric juice	Promotes ulceration of stomach wall	Inhibits gastric digestion

TABLE 13.4
ENDOCRINES AND THEIR HORMONES (continued)

Name of Gland	Location	Hormone	Normal Function	Problem Associated with: Excess Secretion	Diminished Secretion
Small intestine	Mucous lining	Secretin	Activates the liver and pancreas to secrete and release their secretions	Excessive pancreatic and liver secretions	Diminished pancreatic and liver secretion
Adrenal medulla	Two glands above kidney	Adrenalin	Controls release of sugar from liver; contraction of arteries; clotting	Increases blood pressure; promotes clotting; releases glycogen; strengthens heartbeat	
Adrenal Cortex		Glucocorticoids	Affects normal functioning of gonads; helps maintain normal blood sugar levels		Addison's disease: muscular weakness, darkening of skin, low blood pressure; death
		Mineralo-corticoids	Stimulates kidney tubules to reabsorb sodium		
Pancreas Isles of Langerhans	Embedded in pancreas	Insulin	Regulates storage of glycogen in liver; accelerates oxidation of sugar in cells		Diabetes; unused sugar remains in blood and is excreted with urine
Gonads	Abdominal region	Testosterone (males) Estrogen Progesterone (females)	Regulates normal growth and development of sex glands; regulates reproduction; controls sex characteristics	Premature development of gonads; effects on secondary sex chara-chteristics	Interference with normal reproductive functions; diminished growth of sex characteristics

TABLE 13.4
ENDOCRINES AND THEIR HORMONES (continued)

Name of Gland	Location	Hormone	Normal Function	Problem Associated with: Excess Secretion	Diminished Secretion
Thymus	Chest region	Thymosin	Stimulates immunological activity of lymphoid tissue		Breakdown of immune system
Pineal	Base of brain	Melatonin	Regulates gonadotropins by anterior pituitary		

The Respiratory System

The process of **respiration** consists of external breathing and internal respiration. **Breathing** concerns the intake of air (**inhaling**) and the letting out of carbon dioxide and water vapor (**exhaling**). Internal respiration (cellular respiration) takes place in cells and is the series of biochemical events by which energy is released from food molecules.

Air is taken in through the nose. The nasal cavity is divided into two pathways by a septum made of cartilage and bone, and the bony **turbinates** increase the tissue surface along the dividing wall. The surfaces of the septum and the walls of the nasal cavities are covered with mucous membranes, some of which are lined with fine hairs. The nasal cavities have several small openings that lead to spaces in the facial bones called **sinuses**. These eight sinuses help to equalize the air pressure in the nasal cavity, reduce the weight of the skull, and contribute to the sound of the voice.

As air passes through the nasal cavity, it is warmed, humidified and filtered for dust particles. Incoming air passes through the nasal cavity into the **pharynx** (throat) through the **larynx** (voice box) and then into the **trachea** (windpipe). The trachea extends from the back of the throat down into the chest. It is held open by a series of C-shaped cartilage rings.

Just behind the middle of the breastbone, the trachea divides into two branches: the left and right **bronchi** (sing., bronchus). Each bronchus divides and subdivides into smaller tubules called **bronchioles** that ramify throughout the lungs. Each bronchiole ends in a tiny air sac called an **alveolus**. Each alveolus is surrounded by blood capillaries. Oxygen diffuses from the lungs into the bloodstream and is transported to all parts of the body by way of the red blood cells. Conversely, carbon dioxide diffuses out of the blood into the air sacs and makes the reverse trip through the respiratory tubes, finally leaving the body through the nose.

The human body has two lungs. Each of these is enclosed in a double membranous sac known as the **pleural sac**. Not only is this sac airtight, but also it contains a lubricating fluid. The pleural sac and the lubricating fluid prevent friction that might be caused by the rubbing of the lungs against the chest wall.

The respiratory centers in the medulla of the brain and in other brain regions control breathing. The size of the chest cavity is regulated by the **diaphragm**, a flat sheet of muscle that separates the chest cavity from the abdominal cavity. The diaphragm is attached to the breastbone at the front, to the spinal column at the back and to the lower ribs on the sides. When the diaphragm muscle contracts, the diaphragm is drawn downward, creating a partial vacuum in the chest cavity and thus causing air to flow through the respiratory tubes into the lungs. When the diaphragm is relaxed, the chest cavity becomes smaller, forcing the air out. The average rate of respiration in humans is about 18 breaths per minute.

Table 13.5 describes five common disorders of the respiratory system.

TABLE 13.5
SOME DISORDERS OF THE RESPIRATORY SYSTEM

Disorder	Description
Pleurisy	Inflammation of the pleural linings of the lungs, caused by an accumulation of fluid between the pleural layers
Bronchitis	Inflammation of the membranes that line the bronchial tubes
Asthma	An allergic response resulting in the constriction of the bronchial tubes
Emphysema	A condition characterized by the loss of elasticity in the muscle fibers of the air sacs in the lungs, causing their enlargement and degeneration. The result is difficulty in breathing, overwork of the heart, and very often death.
Hiccough (also spelled "hiccup")	Irritation of the nerves that control the diaphragm, resulting in an irregular intake of air which causes a peculiar noise as the glottis closes down

The Circulatory System

The human circulatory system consists of the heart and the system of blood vessels that transport blood throughout the body.

THE HEART

The human heart lies in the chest cavity behind the breastbone and slightly to the left. The heart is a bundle of cardiac muscles specialized for rhythmic contractions and relaxations known as *heartbeat*. The rate of average heartbeat is 72 times per minute.

Figure 13.15 shows the external structure of the heart. In size, the heart is about as large as a person's clenched fist. The walls are thicker on one side than on the other. The surface is covered with a number of small arteries and veins; these small arteries are the **coronary arteries**, which carry blood laden with oxygen and nutrients to the muscle fibers of the heart. A number of large arteries and veins lead into the top of the heart. These carry blood to and from the other parts of the body.

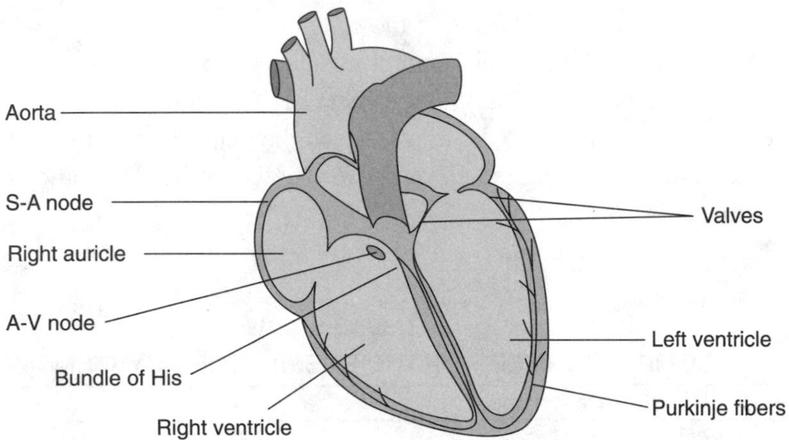

FIGURE 13.15 Structure of the heart

The inside of the heart is divided into four chambers. The two chambers at the top are the receiving chambers, or the **atria**. The lower chambers, the **ventricles** are pumping chambers. Each atrium is separated from the ventricle below by a valve. The atrium and the ventricle on the right are separated from the left atrium and ventricle by a thick wall of muscle called the **septum**.

PATH OF THE BLOOD

The heart is a double pump. Blood flows from the right atrium into the right ventricle which pumps the blood through the **pulmonary artery** to the lungs where it receives oxygen and gives up waste products. The right ventricle represents the first pump. Blood returns from the lungs to the heart by way of the **pulmonary vein** and empties into the left atrium. The left atrium then sends the oxygenated blood into the left ventricle which then pumps it out to all parts of the body. The left ventricle represents the second pump.

Blood is oxygenated in the lungs. Deoxygenated blood leaves the heart (pumped by the right ventricle) by way of the pulmonary artery. Oxygenated blood is returned to the left atrium by way of the pulmonary vein. Arteries always carry blood away from the heart; veins carry it to the heart.

Heart valves prevent the backflow of blood. Separating the right atrium from the right ventricle is the **tricuspid valve**, so named because it has three flaps of tissue or cusps. The opening and closing of these cusps is controlled by papillary muscles. A valve with two cusps, the **bicuspid valve**, separates the left atrium from the left ventricle. (The bicuspid valve is also known as the *mitral valve*). Other valves are located where the aorta and the pulmonary arteries join the ventricles. A diagram of the human circulatory system is shown in Figure 13.16.

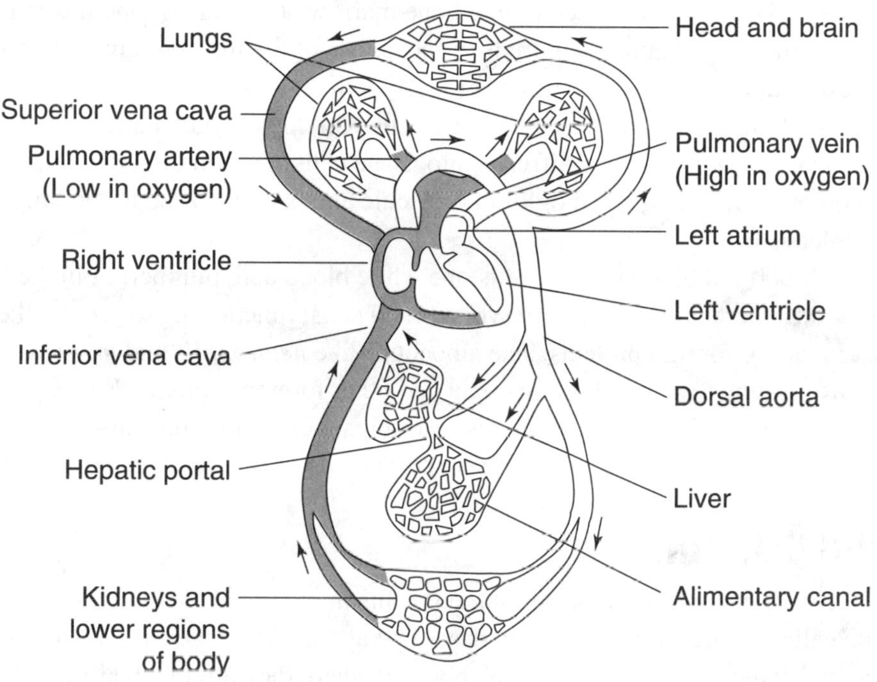

FIGURE 13.16 Path of the blood in the human circulatory system

Several effects of the heart's pumping action can provide information about the condition of the heart and circulation. When a doctor listens to the heart through a stethoscope, he or she hears a series of sounds like "lubb-dub." The "lubb" is caused by the closing of the valves between the atria and the ventricles. At the same time, the ventricles contract. The "dub" sound occurs when the semilunar valves of the aorta and the pulmonary artery close. The pause between the "dub" sound and the next "lubb" sound is the short time in which the heart rests.

The pulse is caused by the force of the blood on the arteries as the heart beats. The pulse rate, like the rate of heartbeat, is 72 beats per minute under normal conditions. Strenuous exercise increases the pulse rate.

Blood pressure measurement is a very valuable diagnostic procedure. Each time the ventricle contracts, blood is forced through an artery, increasing blood flow. The contraction phase is known as **systole**: relaxation, **diastole**. Normal systolic pressure is 120; diastolic, 80. This information is written as 120/80.

WORK OF THE BLOOD

Blood consists of a liquid medium called **plasma** and three kinds of blood cells: **red blood cells (erythrocytes)**, **white blood cells (leucocytes)**, and **platelets**.

The human body contains about 25 trillion erythrocytes, each one lasting about 120 days. New red cells are produced by the bone marrow at the rate of one million per second. Erythrocytes contain hundreds of molecules of the iron-protein compound *hemoglobin*. In the lungs, oxygen binds loosely to hemoglobin forming the compound oxyhemoglobin. As erythrocytes pass body cells with low oxygen content, oxygen is released from hemoglobin and diffuses into tissue cells. Carbon dioxide combines with another portion of the hemoglobin molecule and is transported to the lungs where it is exhaled.

For each 600 red blood cells there is one white blood cell, numbering in the billions in the blood. There are five types of white blood cells, functioning to protect the body against invading foreign proteins. The amoeboid-like *neutrophils* and *monocytes* behave as **phagocytes**, engulfing bacteria and other foreign proteins. *Eosinophils* fight bacterial and parasitic infections and also participate in allergic response mechanisms. *Lymphocytes* participate in immune responses, and *basophils* produce anticoagulants.

BLOOD CLOTTING

Platelets are the smallest of the cellular elements in the blood. Actually, platelets are not true cells and are more accurately described as *cellular fragments*. Platelets do not have nuclei. They originate in the bone marrow where they are pinched-off cytoplasmic fragments of large cells. Despite their small size, platelets play a major role in the clotting of blood.

When a capillary is cut, platelets collect at the site of the injury. There they break into smaller fragments and initiate the complicated chemical process of blood clotting, in which more than 15 factors, including thromboplastin, calcium (Ca), and fibrinogen, are involved in the formation of a clot containing blood cells in a fibrin meshwork (Figure 13.17).

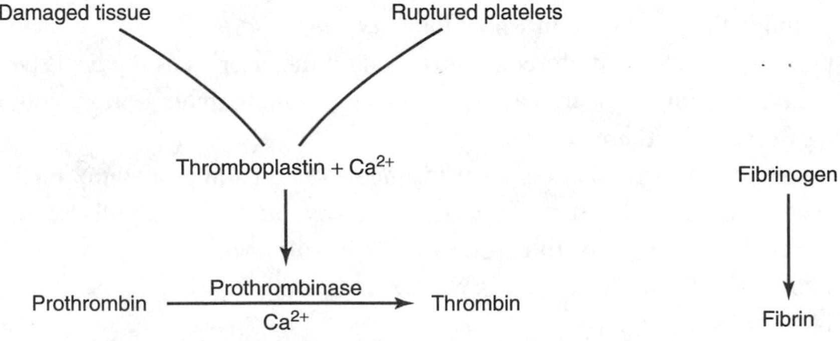

FIGURE 13.17 Some steps of the blood clotting process

BLOOD TYPES

The main types of blood are A, B, AB, and O. Transfusions of blood are possible only when the blood types of donor and recipient are compatible. The recipient of the blood transfusion cannot have antibodies to the donor blood. If the blood types are not compatible, proteins in the plasma will recognize foreign antigens on red blood cells and respond by causing the cells to *agglutinate*, or clump, a potentially fatal condition that causes blockage in small blood vessels. Table 13.6 summarizes the blood proteins involved in blood types.

TABLE 13.6
PROTEINS OF BLOOD TYPES

Blood Type*	Cell Antigen	Plasma Antibody
A	A	b
B	B	a
AB	AB	none
O	none	a and b

* Type AB—universal recipient; type O—universal donor.

DISORDERS OF THE CIRCULATORY SYSTEM

Table 13.7 describes some common disorders of the blood circulatory system.

TABLE 13.7
SOME DISORDERS OF THE BLOOD CIRCULATORY SYSTEM

Disorder	Description
Cardiovascular disorders	Disorders involving the heart and blood vessels.
Hypertension (high blood pressure)	An indication that pressure in the arteries is high. High arterial pressure may be caused by hardening of the arterial walls, stress, heredity, improper diet, cigarette smoking, and aging.
Coronary thrombosis	A form of heart attack in which a blockage in the coronary artery or one of its branches results in oxygen debt in the heart muscle.
Angina pectoris	A form of heart attack in which narrowing of the coronary arteries deprives the heart muscle of oxygen.
Blood disorders	Abnormalities in blood cells.
Anemia	Inability of the blood to carry adequate amounts of oxygen because of faulty red blood cells.
Leukemia	A disease of the bone marrow in which there is uncontrolled production of nonfunctional white blood cells.

The Lymphatic System

Homo sapiens actually has two circulatory systems. One is the blood circulatory system, the other is the lymphatic system. The body cells are bathed with tissue fluid called **lymph**. Lymph comes from the blood plasma, diffusing out of the capillaries into the tissue spaces in the body. Lymph differs from plasma in that it has 50 percent fewer proteins and does not contain red blood cells. Lymph has the important function of bringing nutrients and oxygen to cells and removing from them the waste products of respiration.

Although there is a constant flow of lymph from the blood plasma, neither the blood volume nor its protein content is diminished because, as fast as lymph is drained from the blood plasma, it is returned to the blood. Radiating throughout the body are tiny lymph capillaries, which join together to form larger lymph vessels and ducts. Tissue fluids are propelled through the body by differences in capillary pressure, muscle action, intestinal movements, and respiratory movements. These movements squeeze the lymph vessels and push the fluid along. Lymph moves in only one direction: toward the heart. Lymphatic valves prevent a backflow.

Lymph is returned to the circulating blood through a large lymphatic vessel called the **thoracic duct**, which discharges its contents into the left subclavian vein. The right lymphatic duct empties its contents into the right subclavian vein. These veins then merge into other veins that empty into the heart.

In addition to vessels, the lymphatic system has lumpy masses of cells, known as **lymph nodes**, distributed throughout the body. These lymph nodes or glands are filtering organs that clear the tissue fluids of bacteria and other foreign particles. Lymph nodes are in the head, face, neck, thoracic region, armpits, groin and pelvic and abdominal regions. These nodes help the body in defense against disease. In addition to filtering out bacteria, they produce lymphocytes and antibodies.

Edema is the swelling that results from inadequate drainage of lymph from the body tissue spaces. Edema is brought about by heart and kidney disorders, malnutrition, injury, or other causes.

The Digestive System

The human digestive system begins with the mouth and ends with the anus, and is often described as a "tube within a tube." Variously called the *gut*, *alimentary canal*, or *gastrointestinal tract*, the digestive system extends from the lower part of the head region through the entire torso (Figure 13.18).

Essentially, this system carries out five separate jobs that have to do with the processing and distribution of nutrients. First, it governs ingestion, or food intake. Second, it transports food to organs for temporary storage. Third, it controls the mechanical breakdown of food and its chemical digestion. Fourth, it is responsible for the absorption of nutrient molecules. Fifth, it provides for the temporary storage and then the elimination of waste products.

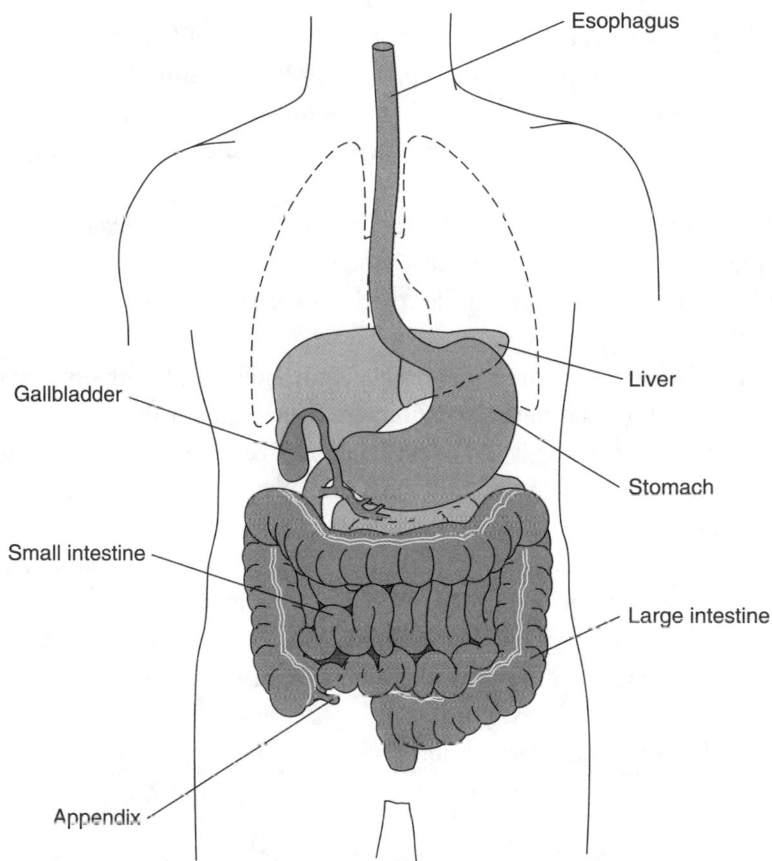

FIGURE 13.18 The human digestive system

Digestion begins in the mouth. Teeth grind the food while three pairs of **salivary glands** pour salivary juice (saliva) into the mouth. Saliva contains the enzyme *salivary amylase* (ptyalin), which begins the digestion of starch. The moistened, chewed food is swallowed and moves through the throat into the food tube, or **esophagus**. The esophagus has no digestive function but moves the food into the stomach by waves of muscle contractions called **peristalsis**.

Chemical digestion is also known as **hydrolysis**. As the name indicates, hydrolysis is the splitting of large, insoluble molecules into small molecules that are able to dissolve in water. In the digestive system, hydrolysis is regulated by digestive enzymes, as shown in the examples that follow:

$$\text{maltose} + \text{water} \xrightarrow{\text{(maltase)}} \text{glucose} + \text{glucose}$$

$$\text{proteins} + \text{water} \xrightarrow{\text{(protease)}} \text{amino acids}$$

$$\text{lipids} + \text{water} \xrightarrow{\text{(lipase)}} \text{3 fatty acids and 1 glycerol}$$

The ***stomach*** is the widest organ in the alimentary canal. It stores food while it churns and squeezes it, turning it into the consistency of a thick pea soup. In this semiliquid form, food can be worked upon by enzymes. **Gastric glands** embedded in the walls of the stomach secrete **gastric juice**, a combination of hydrochloric acid and two enzymes: rennin and pepsin. Rennin is specialized for digesting the protein in milk; pepsin, for hydrolyzing several plant and animal proteins. The semiliquid food, often referred to as a *bolus* or *chyme*, is released a little at a time into the upper part of the small intestine.

Between the stomach and the small intestine is a ring of muscle called the **pyloric sphincter** that closes the stomach off from the **duodenum**, the upper part of the small intestine. In effect, the sphincter muscle regulates the flow of chyme from the stomach into the intestine.

The major work of digestion occurs in the small intestine. Lying outside of the alimentary canal are two important glands that are necessary for the many processes of digestion. The largest of these is the **liver**. It synthesizes *bile* and stores it in a pouch known as the **gallbladder**. Through bile ducts, bile is released into the small intestine where it serves as an emulsifier of fat, enabling it to be acted upon by the fat-digesting enzyme lipase. The other accessory gland is the **pancreas**, a dual gland that synthesizes both hormones and enzymes. The pancreas releases pancreatic juice into the small intestine. This digestive juice is a combination of water and several digestive enzymes, each of which is specific for the digestion of fat, carbohydrate, or protein.

In the walls of the small intestine are *intestinal glands* which manufacture and secrete intestinal juice, a combination of enzymes that digest starches, sugars, and proteins. The outcome of all digestion is that nutrient molecules are reduced to soluble forms that enable them to cross cell membranes. Carbohydrates are digested into glucose or fructose. Proteins are broken down into amino acids. Fats are hydrolyzed into fatty acids and glycerol. These nutrients are absorbed by finger-like *villi* (sing., villus) an adaptation in the small intestine for increasing surface area (Figure 13.19).

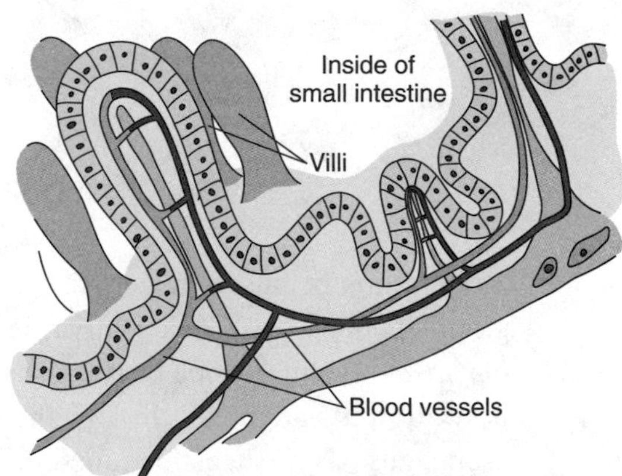

FIGURE 13.19 Villi

Digested food diffuses into the capillaries of the villi. Blood then carries the food molecules to the liver through the **portal vein**. In the liver, sugar is removed from the blood and stored as *glycogen*. Digested fat molecules are absorbed into the **lacteals** (lymph vessels) and then enter the bloodstream through the **thoracic duct**, which is in the chest cavity.

Food that is not digested passes into the large intestine, also called the **colon**. The large intestine absorbs a great deal of water and dissolved minerals. Undigested food, called *feces*, is pushed into the rectum, where it is stored temporarily until eliminated through the anus.

Some common disorders of the digestive system are described in Table 13.8.

TABLE 13.8
SOME DISORDERS OF THE DIGESTIVE SYSTEM

Disorder	Description
Ulcer	A punched-out defect in the wall of the alimentary canal, caused by the digestive action of gastric juice
Constipation	Difficult passage of stools from the large intestine, caused by excessive absorption of water
Diarrhea	Passage of frequent and watery stools, associated with decreased water absorption and possibly resulting in dehydration
Appendicitis	Acute inflammation of the appendix
Diverticulosis	Grapelike outpocketings on the colon wall, which may become infected and obstruct the bowel
Gallstones	Aggregations of hardened bile salts, cholesterol, and calcium in the gallbladder

The Excretory System

In human beings, the lungs, the skin, the liver, and the urinary system work to expel the wastes produced in metabolic activities. The lungs excrete carbon dioxide and water. The skin expels water and salts from the sweat glands and a small amount of oil from the sebaceous glands. The liver contributes to the excretory system by breaking down waste and excess proteins to produce **urea** in a process called *deamination*. The urinary system handles the major work of excretion.

The human urinary system is located dorsally in the abdomen. Figure 13.20 shows the organs that make up the urinary system and their locations. This system consists of two *kidneys*, tubes known as *ureters* extending from each kidney to a **urinary bladder**, and a single **urethra**, a tube that leads out of the bladder.

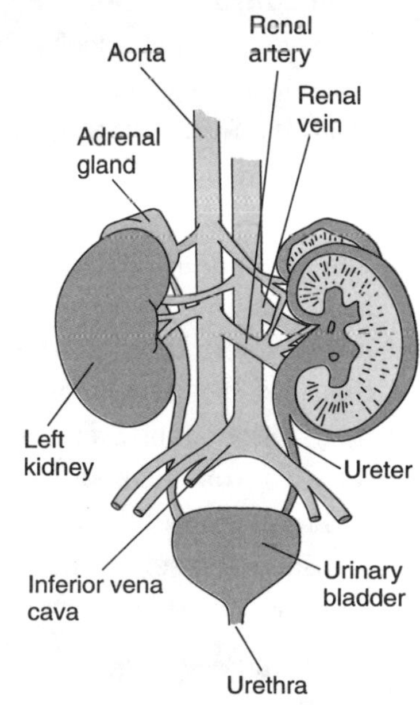

FIGURE 13.20
The urinary system

The unit of structure and function of the kidney is the **nephron**. There are about one million of these microscopic units in each kidney. They actively remove waste products from the blood and return water, glucose, sodium ions and chloride ions to the blood. Figure 13.21 shows the structure of a nephron.

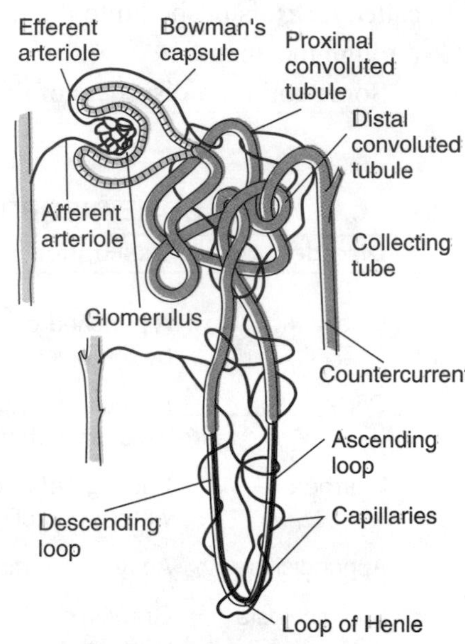

FIGURE 13.21 The kidney nephron

Actually, the nephron is made up of several structures. The first is a knot of capillaries called the **glomerulus**. The glomerulus fits into a second portion—the **Bowman's capsule**, a cup-shaped cellular structure that leads into the third part, the kidney tubule. There are four main parts of each kidney tubule: the **proximal convoluted tubule**, the **loop of Henle**, the **distal convoluted tubule**, and the **collecting duct**.

The **renal artery** transports blood into the kidney. The artery divides into smaller vessels, called **arterioles**, which divide into still smaller blood vessels, called **capillaries**. It is this network of capillaries that forms the glomerulus. The pressure of the blood in the glomerulus is quite high and forces fluid from the blood through the walls of the capillaries into the hollow cup of the Bowman's capsule. This process is known as **pressure filtration**. The fluid entering the nephron has the same composition as the blood except that it lacks blood cells, large protein molecules, and lipids, which are not able to cross the membranes of the cells that compose the capillary walls.

As the filtrate (the water filtered out of the blood) moves into the proximal convoluted tubule, much of the water is reabsorbed back into the blood. Glucose and sodium are reabsorbed by active transport. As the water that is left in the tubule moves down into the **loop of Henle**, chloride ions and sodium are reabsorbed into the bloodstream by active transport. The remaining water and urea move into the distal convoluted tubule where additional sodium ions are reabsorbed into the bloodstream by active transport. The remaining water and urea waste pass into the collecting duct and, from there, through the ureters to the bladder. This **urine** is now stored in the bladder until released from the body through the urethra.

The actions of the posterior pituitary hormone vasopressin and the steroid hormone aldosterone from the adrenal cortex regulate absorption in the kidney.

Table 13.9 describes three disorders of the urinary system.

TABLE 13.9
THREE DISORDERS OF THE URINARY SYSTEM

Disorder	Description
Kidney stones	An accumulation of mineral salts that precipitate out from the urine and form stones in the kidneys and other parts of the urinary tract
Nephritis	Inflammation of the kidney, often caused by toxins from bacteria that infect other parts of the body
Gout	A disease denoted by the accumulation of the chalky salts of uric acid in joints and at the ends of bones, and caused by malfunction of the kidney in secreting uric acid

Sense Organs

The human body has five major senses—sight, hearing, taste, smell, and touch—that provide information about the external environment and transmit the stimuli to sensory nerves and ultimately to the brain for processing.

THE EYE AND VISION

The human **eyeball** measures about 2.5 centimeters in diameter. Most of the eyeball rests in the bony eye-socket of the skull. Only about one sixth of the eye is exposed. External structures associated with the eye are eyelids and lashes and eyebrows. A delicate protective membrane, the **conjunctiva**, covers the eye. Three pairs of small muscles attach the eye to the eye-socket. Secretions from tear glands help to keep the eye moist.

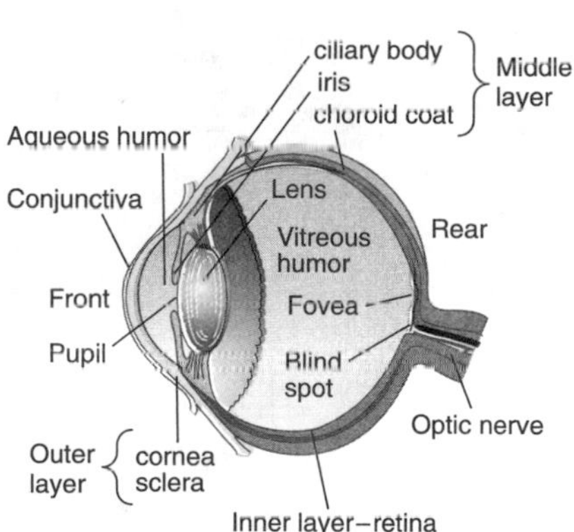

FIGURE 13.22 Structure of the human eye

Figure 13.22 shows the structure of the human eye. Notice that the eyeball is divided into two chambers that are separated by a **lens**. The front chamber contains a clear fluid called the **aqueous humor**. The back chamber contains a transparent jelly-like material called the **vitreous humor**.

The lens is transparent and is made of a great many layers of protein fibers. It measures about 8 millimeters in diameter. The function of the lens is to focus light on the **retina** at the back of the eyeball. The shape of the lens changes; it flattens when focusing on distant objects and thickens when focusing on near objects. The ability to bring objects into focus although they are located at different distances is called **accommodation**.

The colored portion of the front of the eye is the **iris**; in its center is a hole called the **pupil**. Light enters the eye through the pupil and passes through the cornea, the aqueous humor, the lens, and the vitreous humor. Light reaches the *retina*, where a barrage of signals is set up. These signals are conducted to the *optic nerve*, which carries them to the visual portions of the brain. The lens turns the image upside down and reverses it from left to right. The visual centers in the brain correct the inversions and reversals of the lens to make the image right side up.

Buried in the retina are cells called **rods** and **cones**. There are about 7 million cones and 120 million rods. The rods help the eye to accommodate in dim light and aid in night vision. The cones are responsible for color vision.

THE EAR AND HEARING

The human ear is made up of three divisions: the **outer ear**, the **middle ear** and the **inner ear**. Figure 13.23 shows the structure of the ear. The outer ear catches sound waves and transports them to the **eardrum**, a membrane that stretches across the outer canal separating it from the middle ear. Sound waves cause the eardrum to vibrate.

The middle ear is a small cavity that is filled with air. It lies inside of the skull bone between the outer ear and the inner ear. At the bottom of the middle ear is an opening that leads into a canal. This canal, called the **Eustachian tube**, is a passageway that connects the middle ear to the throat. This tube equalizes the air pressure in the ear with that in the throat. This equalization is accomplished by yawning or swallowing.

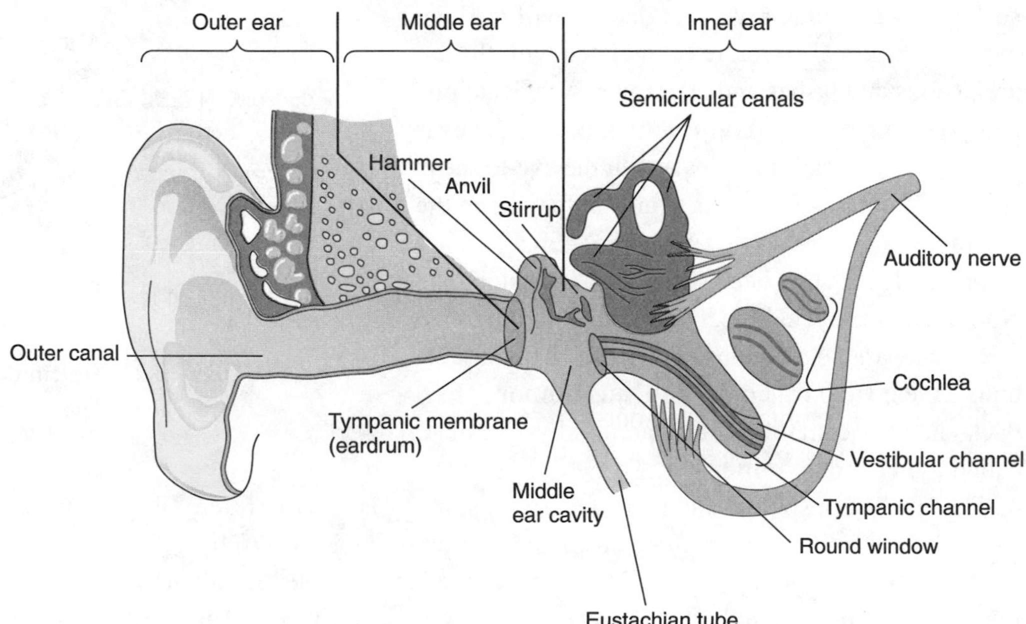

FIGURE 13.23 Structure of the human ear

The middle ear contains three bones called the **hammer**, the **anvil**, and the **stirrup**, which are the smallest bones in the body. These bones accept the vibrations from the eardrum and transmit them to the oval window, one of two small membrane-covered openings between the middle ear and the inner ear.

The inner ear, which is entirely encased in bone, has a fluid-filled structure called the **cochlea**, so named because it resembles a snail in shape. The cochlea has numerous canals that are lined with hair cells. The vibrations from the oval window are transmitted to the hair cells in the cochlea and thence on to the **auditory nerve**, which conducts the vibrations to the brain. In the brain, these signals are interpreted into sounds.

THE OTHER MAJOR SENSES

The organs of smell are located in the mucous membranes of the upper part of the nasal cavities. Special **olfactory cells** respond to odors and pass the impulse along the **olfactory nerve** to the brain. The sense of smell is far more important in lower animals than it is in humans.

The organs of taste are found chiefly on the tongue. These **taste buds**, as they are called, distinguish basic qualities such as bitter, sweet, sour, and salty.

The organs of touch are located on the skin surface and they respond to temperature, pain, and pressure.

Reproduction

In this section we will discuss the human reproductive system and, very briefly, the development of human young.

MALE REPRODUCTIVE SYSTEM

In the male reproductive system some organs are located outside of the body and others are positioned internally. Look at Figure 13.24, and note the scrotum, a sac-like organ located outside the body. The **scrotum** contains the **testes**, glands that produce sperm and the male hormone *testosterone*. Also positioned outside the body is the **penis**, the organ that delivers the sperm into the body of the female.

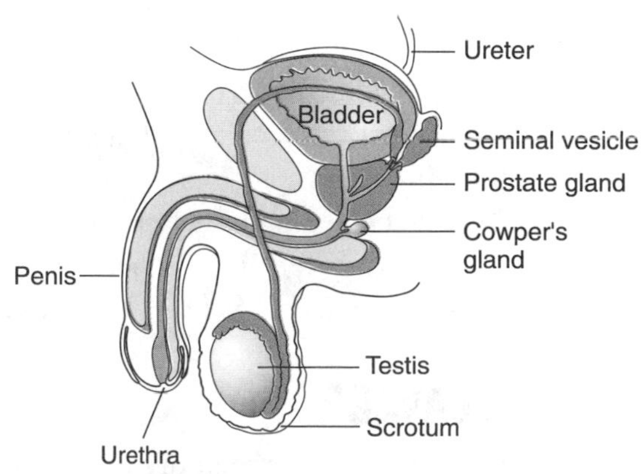

FIGURE 13.24 Male reproductive system

Each testis contains thousands of very small tubes called **seminiferous tubules**. Within these tubules, the sperm cells are manufactured. At the top of and behind each testis is a mass of coiled tubules called the **epididymis**. Each epididymis tubule functions as a storage place for sperm and also serves as a pathway which carries the sperm to a duct called the **vas deferens**. In its travels to the vas deferens, the sperm pass the **seminal vesicles** where they obtain nutrients. From the vas deferens, the sperm are conducted to the urethra, a single tube that extends from the bladder through the penis. Sperm cells leave the body through the penis. Erection of the penis and ejaculation of **semen** (a mixture of sperm and secretions from the seminal vesicle, the prostate gland, and Cowper's gland) are processes necessary for the placement of sperm in the female reproductive tract.

FEMALE REPRODUCTIVE SYSTEM

The female reproductive system serves three important functions: the production of egg cells, the disintegration of nonfertilized egg cells, and the protection of the developing embryo. The reproductive system has specialized organs to carry out these functions.

Two oval-shaped **ovaries** lie one on each side of the midline of the body in the lower region of the abdomen. On a monthly alternating basis each ovary produces a mature egg. Eggs are located in spaces in the ovary called **follicles**. As an egg matures, it bursts out of the ovarian follicle and is released into the appropriate branch of the **fallopian tube**, a tube that leads from the region of the ovary to the uterus (Figure 13.25).

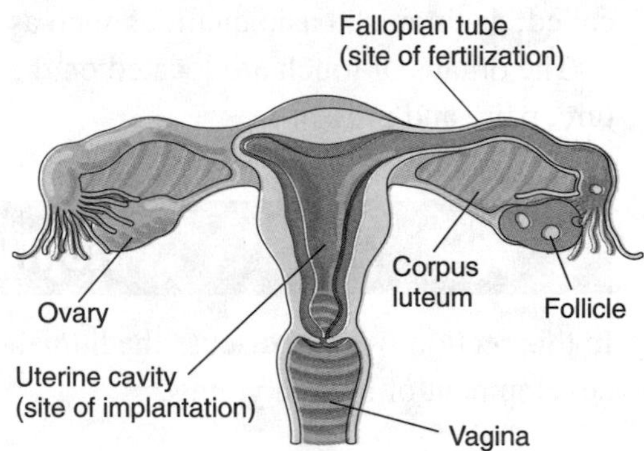

FIGURE 13.25 Female reproductive system

If the egg is not fertilized, it is discharged from the body in a process called **menstruation**. The vascularized lining of the uterus, known as the **endometrium**, disintegrates in response to decreased levels of estrogen and progesterone in the blood. Menstrual bleeding lasts four to seven days.

If the egg is fertilized, it becomes implanted in the uterus where it goes through a series of cell divisions known as **cleavage**.

CLEAVAGE

During the series of cell divisions that occur in cleavage, there is no growth in size of the zygote or any separation of the cells. Figure 13.26 shows the stages of cleavage.

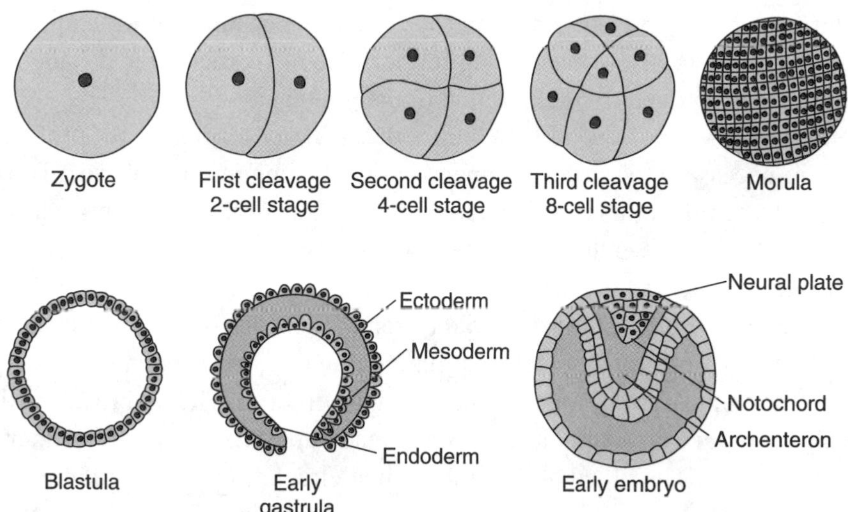

FIGURE 13.26 Stages of cleavage

The first division of cleavage results in the formation of two cells; the second, four. Succeeding divisions result in eight cells, then 16, 32, and so forth, until a solid ball of cells called a **morula** is formed. Cells in the morula migrate to the periphery, and the solid ball of cells changes to a hollow ball called a **blastula**.

Within a very short time, one side of the blastula pushes inward, forming what resembles a double-walled cup. This stage of cleavage is known as the **gastrula**. During the gastrula stage three distinct layers of cells—the **ectoderm** (the outer layer), the **mesoderm** (the middle layer), and the **endoderm** (the inner layer)—are formed. These layers, known as **primary germ layers**, develop into the tissues and organs of the body through a process known as **differentiation**. An **embryo** is now forming. Table 13.10 summarizes the development of the primary germ layers.

TABLE 13.10
DIFFERENTIATION OF THE THREE PRIMARY GERM LAYERS

Ectoderm	Endoderm	Mesoderm
skin nervous system sense organs	lining of lungs lining of digestive system pancreas liver respiratory system	muscles, skeleton, heart, blood vessels, blood, ovaries, testes, kidneys

DEVELOPMENT OF THE HUMAN EMBRYO

As a result of cleavage of the fertilized egg and its implantation in the uterus, many changes occur in the body. The follicle from which the egg cell bursts becomes filled with some yellowish glandular material and is now known as the **corpus luteum**. The corpus luteum acts as an endocrine gland, secreting **progesterone**, which prevents any other eggs in the ovary from developing further. The menstrual cycle is halted. No more eggs are discharged for the duration of the pregnancy.

The embryo produces several membranes that will not form any part of the new baby but are necessary to the development and well being of the embryo. (The term **fetus** is used when the embryo takes human form.) One of these membranes is the **amnion**, a water-filled sac that completely surrounds and protects the embryo. The water absorbs shocks and prevents friction that might damage the embryo.

The implanted embryo is attached to the uterus by means of the **umbilical cord**, a structure that contains blood vessels that function in carrying nutrients and oxygen to the embryo and transporting wastes away from the embryo. The umbilical cord connects with the **placenta**, a vascularized organ made up of tissues of the mother and tissues of the embryo. The blood of the embryo that circulates in the capillaries of the placenta is separated from the blood of the mother by layers of cells thin enough to allow diffusion between the two circulatory systems. There is no mixing of the blood of the mother with the blood of the embryo.

PARTURITION

The birth process is known as **parturition**. In humans the period of gestation (period of embryonic and fetal development) is about 9 months or 40 weeks. At the end of that time, the uterus begins to contract in a process called **labor** to expel the baby. The onset of uterine contractions is probably caused by the release of oxytocin into the bloodstream by the posterior pituitary. The human newborn passes through the neck of the uterus (cervix) head first and then through the vagina to the outside.

MULTIPLE BIRTHS

Although humans usually produce only one offspring at a time, sometimes two, three, or even more young may be born at the same time, particularly if a woman has been taking fertility drugs. Of these multiple births, twins are the most common. There are two types of twins: identical and fraternal. Identical twins result from the fertilization of one egg and have the same genetic makeup. They are of the same sex and are almost identical in appearance. They develop in a common chorionic sac and share a common placenta. However, the umbilical cords are separate.

Fraternal twins develop from two separate fertilized eggs. They do not share a common genetic makeup and are no more alike than siblings born at separate times. The sexes may be different. Each fraternal twin has its own chorionic membrane and its own placenta.

Human Beings and Race

All humans belong to the species ***Homo sapiens***. As a result, the genetic material of all people is so similar that all humans can interbreed and produce fertile offspring. The human species is really a group of interbreeding populations. Populations that have adapted to certain environments become genetically different based on the frequency with which certain genes appear. Skin color, hair texture, body build and facial bone structure are a few of the characteristics that identify human population groups known as **races**.

Although we can make broad generalizations about the identifying characteristics of racial groups, not every member of each group fits these specifications. A set of physical characteristics can be drawn up that will fit individuals of several different races. Therefore, it is difficult for biologists and anthropologists to agree on the number of human races. Modern anthropologists divide *Homo sapiens* into three major stocks: Caucasoid, Mongoloid, and African. Each of these groupings is subdivided into several human populations distinguishable by certain pronounced characteristics. The Caucasoid stock is composed of four white races: Nordic, Alpine, Mediterranean, and Hindu. The Asian stock is divided into the Malaysian, American Indian, and Mongolian populations. The African stock is separated into the Black, Melanesian, Pygmy Black, and Bushman populations. Of doubtful groupings are the Polynesian and the Australoid peoples. Many anthropologists classify these groups as Asian; others disagree. No matter the grouping, the differences among human races are very small.

REVIEW EXERCISES FOR CHAPTER 13

WORD-STUDY CONNECTION

accommodation
action potential
afferent neuron
agglutinate
alimentary canal
alveolus
antibody
appendage
atrioventricular node
atrium
axon
basophil
bipedalism
biscuspid valve
blastula
Bowman's capsule
bronchiole
bronchus
cardiac muscle
cerebellum
cerebrum
chyme
clavicle
cleavage
coccyx
cochlea
cones

convolution
coronary artery
corpus luteum
cranium
culture
cyton
dendrite
diaphragm
diastole
differentiation
duodenum
eardrum
edema
effector organ
embryo
endocrine gland
eosinophil
epididymis
erythrocyte
Eustachian tube
fallopian tube
femur
fetus
fibula
follicle
gastrula
girdle
glomerulus
gray matter
Haversian canals
hormone
humerus
hydrolysis
hypophysis
hypothalamus
interneuron
involuntary muscle
labor
lacteals
lens
leucocyte
lymph
lymphocyte
medulla
meninges
menstruation
metacarpal
metatarsal
mirral valve
monocyte
morula
myofibril
nephron
neuromuscular junction
neuron
neurotransmitter
neutrophil
node of Ranvier
olfactory nerve
optic nerve
osteoblast
osteoclast
osteocyte
oval window
ovary vagus
parturition
pericardial
periosteum
peritoneal
phalanges
placenta
platelet
pleural sac
prehensile
primate
prostaglandins
Purkinje fibers
pyloric sphincter
race
radius
reflex
releasing factor
resting potential
retina
rods
round window
sacrum
sarcomere
scapula
Schwann cell
scrotum
semen
sinoatrial node
sinus
skull
smooth muscle
sodium-potassium pump
stereoscopic vision
sternum
striated muscle
synapse
syncytium
systole
thalamus
thoracic duct
thorax
tibia
tricuspid valve
trophic hormone
ulna
umbilical cord
ureter
urethra
nerve
vas deferens
ventricle
vertebrae
vertebral column
villus
voluntary muscle
white matter

SELF-TEST CONNECTION

PART A. Completion. *Write in the word that correctly completes each statement.*

1. Humans belong to the species named _____.
2. The hands of primates are adapted for _____.
3. The ability to walk on two legs is known as _____.
4. Human patterns of learned behavior are known collectively as _____.
5. The human pericardial cavity encloses the _____.
6. One function of the facial sinuses is to reduce the _____ of the skull.
7. A common name for the thoracic basket is the _____.
8. Bones that attach arms and legs to the axial skeleton are known as _____.
9. Cells that build bone matrix are called _____.
10. Cells that dissolve the mineral matrix of bone are the _____.
11. The function of smooth muscle is to _____. organs it surrounds.
12. The type of muscle that moves the skeleton is called voluntary or _____ muscle.
13. The space between a nerve cell and a muscle is the _____ junction.
14. Two distinctive properties of muscle are electrical excitability and _____.
15. The S-A node acts as the _____ of the heart.
16. Myofibrils are correctly associated with _____ cells.
17. Sensory information is integrated in the brain region known as the _____.
18. The nature of a nerve impulse is both electrical and _____.
19. Balance and coordination are controlled by the part of the brain known as the _____.
20. The peripheral nervous system is divided into the somatic and _____ systems.
21. A distinctive structural characteristic of endocrine glands is that they do not have _____.

22. The homeostatic mechanism that regulates the pituitary gland and its target organs is known as a _____ system.

23. The endocrine glands that cap the kidneys are the _____.

24. The contraction of a ventricle in the heart is known as _____.

25. The force of the blood on the arteries is known as _____.

26. The largest lymphatic vessel in the body is the _____.

27. The unit of structure and function in the kidney is the _____.

28. The part of the eye that bends light rays is the _____.

29. The _____ in the eye are responsible for color vision.

30. Pressure in the middle ear is regulated by the _____ tube.

31. The solid ball of cells formed during cleavage is known as the _____.

32. The corpus luteum behaves as a _____ gland.

33. The birth process is known as _____.

34. Food moves through the digestive tract by waves of muscle contraction known as _____.

35. The _____ closes the stomach from the duodenum.

36. Bile is stored in the _____.

37. The large intestine is also called the _____.

38. Tubes leading from the kidney to the urinary bladder are called _____.

39. The _____ nerve carries signals from the eye to the brain.

40. Semen is a mixture of sperm and secretions from the seminal vesicle, Cowper's gland and the _____ gland.

PART B. Multiple Choice. *Circle the letter of the item that correctly completes each statement.*

1. The eutheria are
   (a) flying birds
   (b) infertile frogs
   (c) female kangaroos
   (d) placental mammals

2. A prehensile tail is adapted for
   (a) swimming
   (b) grasping
   (c) gliding
   (d) balancing

3. Convolutions are best associated with the
   (a) pharynx
   (b) lung
   (c) brain
   (d) nerve cord

4. The number of bones in the mature human skeleton is
   (a) 206
   (b) 270
   (c) 290
   (d) 305

5. The major function of the backbone is to
   (a) protect the spinal cord
   (b) bend the trunk
   (c) hold the neck and head
   (d) support the legs

6. The strongest bone in the body is the
   (a) clavicle
   (b) sternum
   (c) femur
   (d) metatarsal

7. A true statement about bone is
   (a) The matrix is composed of living material.
   (b) Yellow marrow is in the spongy areas of the long bones.
   (c) Bone has a variety of living cells.
   (d) The ends of long bones are made of solid bone.

8. Bone is covered by a membrane known as the
   (a) peritoneum
   (b) pericardium
   (c) perithecium
   (d) periosteum

9. Three names for voluntary muscle are
   (a) striped, cardiac, smooth
   (b) striated, skeletal, smooth
   (c) skeletal, cardiac, striated
   (d) striated, striped, skeletal

10. A syncytium is
   (a) a group of cells without plasma membrane boundaries
   (b) a segment of smooth muscle that rings the trachea
   (c) a group of nerve cells that stimulates muscle
   (d) a single muscle fiber when stimulated by a neuron

11. A true statement about Purkinje fibers is
   (a) They are found in smooth muscle.
   (b) They lack contractile proteins.
   (c) They form a syncytium with skeletal muscle.
   (d) They cannot conduct electrical impulses.

12. If nerves to the heart are severed, the heart will
   (a) cease functioning
   (b) beat faster
   (c) beat regularly
   (d) beat erratically

13. Impulses enter nerve cells by way of the
   (a) dendrites
   (b) cytons
   (c) axons
   (d) terminal branches

14. Myelin is a nerve cell covering composed of
   (a) $CaCO_3$
   (b) collagen
   (c) fat
   (d) carbohydrate

15. The fact that a nerve cell transmits an impulse totally or not at all is a type of response known as
    (a) "take it or leave it"
    (b) "all or none"
    (c) "make it and take it"
    (d) "all or some"

16. The width of a synapse is approximately
    (a) 80 nm
    (b) 60 nm
    (c) 40 nm
    (d) 20 nm

17. The part of the brain that is necessary for life is the
    (a) cerebrum
    (b) cerebellum
    (c) fissure of Rolando
    (d) medulla oblongata

18. The sympathetic and parasympathetic systems function
    (a) antagonistically to each other
    (b) to reinforce the activities of each other
    (c) without relationship to the autonomic system
    (d) to counteract the control of the cerebrum

19. The hypothalamus is a part of the brain where
    (a) integration of sensory information occurs
    (b) integration of memory and higher thought processes occur
    (c) follicle-stimulating hormone is released into the bloodstream
    (d) the nervous and hormonal systems interact

20. Blood in the pulmonary artery
    (a) lacks oxygen
    (b) lacks carbon dioxide
    (c) contains nitrogen
    (d) contains all three gases

21. The most important function of erythrocytes is to
    (a) carry nutrients from cell to cell
    (b) protect the body against disease
    (c) carry oxygen to all cells
    (d) remove carbon from all cells

22. The lymph nodes are glands that
    (a) secrete hormones and neurohumors
    (b) propel tissue fluids through the body
    (c) control the production of red blood cells
    (d) filter bacteria from the tissue fluids

23. Lymphocytes are white blood cells that
    (a) phagocytize bacteria
    (b) detoxify histamines
    (c) produce anticoagulants
    (d) participate in immune responses

24. Breathing is controlled by the
    (a) diaphragm
    (b) respiratory centers in the brain
    (c) level of carbon dioxide in the blood
    (d) all three of the above

25. The lungs are enclosed in a set of double membranes known as the
    (a) pericardium
    (b) periosteum
    (c) pleural sac
    (d) peritoneum

26. In humans the organs of excretion are the
    (a) kidneys, lungs, rectum
    (b) large intestine, sweat glands, lungs
    (c) kidneys, lungs, sweat glands
    (d) sweat glands, kidneys, anus

27. The function of the urinary bladder is to
    (a) store urine
    (b) detoxify urea
    (c) add CO_2 to ammonia
    (d) filter out glucose

28. Blood is transported to the kidney from the dorsal aorta by the
    (a) renal vein
    (b) renal artery
    (c) arterioles
    (d) glomerulus

29. The cup-shaped portion of the nephron is the
    (a) loop of Henle
    (b) glomerulus
    (c) Bowman's capsule
    (d) proximal convoluted tubule

30. The widest organ in the alimentary canal is the
    (a) stomach
    (b) large intestine
    (c) colon
    (d) gallbladder

31. Two glands lying outside the alimentary canal but important to digestion are the
    (a) liver and kidney
    (b) pancreas and thoracic duct
    (c) liver and pancreas
    (d) liver and colon

32. The major work of digestion occurs in the
    (a) stomach
    (b) small intestine
    (c) large intestine
    (d) esophagus

33. Digestion begins in the
    (a) stomach
    (b) small intestine
    (c) esophagus
    (d) mouth

34. In the human male, sperm is stored in the mass of tubules known as
    (a) sperm ducts
    (b) Cowper's gland
    (c) the vas deferens
    (d) the epididymis

35. Follicles are sites where
    (a) fertilization occurs
    (b) chromosomes are halved
    (c) ova are produced
    (d) sperm receive nutrients

36. The primary germ layer that gives rise to the blood vessels is the
    (a) endoderm
    (b) mesoderm
    (c) ectoderm
    (d) protoderm

37. The membranous sac of water that surrounds the developing human fetus is the
    (a) amnion
    (b) chorion
    (c) allantois
    (d) placenta

38. Identical twins are produced from
    (a) two eggs fertilized by two sperm
    (b) two eggs fertilized by one sperm
    (c) one egg fertilized by one sperm
    (d) one egg fertilized by two sperm

39. The implanted embryo is attached to the uterus by means of the
    (a) placenta
    (b) umbilical cord
    (c) yolk stalk
    (d) allantois

40. The three major stocks into which modern anthropologists divide *Homo sapiens* are
    (a) Caucasoid, Mongoloid, African
    (b) Caucasoid, Bushman, Melanesian
    (c) Caucasoid, Hindu, Mongolian
    (d) Caucasoid, Nordic, Pygmy Black

PART C. Modified True-False. *If a statement is true, write "true" for your answer. If a statement is incorrect, change the* underlined *expression to one that will make the statement true.*

1. Scientist believe that at one time all primates lived in <u>savannas</u>.

2. Due to stereoscopic vision, primates can see in <u>two</u> dimensions.

3. The primate with the best developed brain is the <u>chimpanzee</u>.

4. The hammer, anvil, and stirrups are <u>sinuses</u> in the middle ear.

5. The humerus is the long bone in the <u>lower leg</u>.

6. Red blood cells are produced in the red marrow of long bones.

7. "Bone breakers" are cells known as osteoblasts.

8. Striated muscle is called involuntary muscle.

9. The sympathetic nervous system innervates striated muscle.

10. The type of muscle that makes up the heart is smooth muscle.

11. Z lines are bands of fat.

12. The unit of function and structure of the nervous system is the cyton.

13. Schwann cells are best associated with the myofibril sheath.

14. Nervous impulses are conducted by efferent neurons to the brain or spinal cord.

15. Motor neurons conduct impulses to the effectors.

16. Interneurons are located outside of the brain and spinal cord.

17. The reversal of polarity on a nerve cell is known as the resting potential.

18. The brain and the vertebral column make up the central nervous system.

19. The membranes that cover the brain are known as the phalanges.

20. The parasympathetic and sympathetic systems are divisions of the somatic nervous system.

21. Glands of internal secretion release chemicals collectively known as enzymes.

22. Chemicals known as releasing factors are secreted by the anterior pituitary.

23. The presence of estrogen in the blood signals the uterus to secrete luteinizing hormone.

24. The coronary arteries are positioned inside of the heart.

25. The liquid part of the blood is called serum.

26. The life of a red blood cell lasts for 30 days.

27. White blood cells function to protect the body against invasion by foreign worms.

28. Platelets initiate the clumping of blood.

29. Another name for the voice box is the pharynx.

30. Each bronchiole ends in an air sac known as a syrinx.

31. Rennin digests protein in milk.

32. The gallbladder synthesizes bile.

33. The gallbladder stores bile.

34. Digested fat molecules are absorbed into diverticuli and then enter the bloodstream through the thoracic duct.

35. Vibrations are carried to the brain by the optic nerve.

36. The back chamber of the eye is filled with aqueous humor.

37. The three small bones in the middle ear are the anvil, club, and stirrup.

38. Each identical twin has its own chorionic membrane and its own placenta.

39. The human female has one ovary (ovaries).

40. In the male reproductive system some organs are located outside the body and some are positioned internally.

CONNECTING TO CONCEPTS

1. Why are human beings classified as primates?

2. How does the skeletal structure enable the human being to stand upright?

3. Explain what is meant by the "all or none response" of the nerve cell.

4. Why is the pituitary gland considered to be the master gland of the body?

ANSWERS TO SELF-TEST CONNECTION

PART A

1. *Homo sapiens*
2. grasping
3. bipedalism
4. culture
5. heart
6. weight
7. rib cage
8. girdles
9. osteoblasts
10. osteoclasts
11. squeeze
12. striated or striped
13. neuromuscular
14. contractility
15. pacemaker
16. muscle

17. thalamus
18. chemical
19. cerebellum
20. autonomic
21. ducts
22. feedback
23. adrenals
24. systole
25. pulse
26. thoracic duct
27. nephron
28. lens
29. cones
30. Eustachian
31. morula
32. ductless or endocrine
33. parturition
34. peristalsis
35. pyloric sphincter
36. gallbladder
37. colon
38. ureters
39. optic
40. prostate

PART B

1. **(d)**
2. **(b)**
3. **(c)**
4. **(a)**
5. **(a)**
6. **(c)**
7. **(c)**
8. **(d)**
9. **(d)**
10. **(a)**
11. **(b)**
12. **(c)**
13. **(a)**
14. **(c)**
15. **(b)**
16. **(d)**
17. **(d)**
18. **(a)**
19. **(d)**
20. **(a)**
21. **(c)**
22. **(d)**
23. **(d)**
24. **(d)**
25. **(c)**
26. **(c)**
27. **(a)**
28. **(b)**
29. **(c)**
30. **(a)**
31. **(c)**
32. **(b)**
33. **(d)**
34. **(d)**
35. **(c)**
36. **(b)**
37. **(a)**
38. **(c)**
39. **(b)**
40. **(a)**

PART C

1. trees
2. three
3. human
4. bones
5. upper arm
6. true
7. osteoclasts
8. voluntary
9. smooth
10. cardiac
11. protein
12. neuron
13. myelin
14. afferent
15. true
16. inside
17. action
18. spinal cord
19. meninges
20. autonomic
21. hormones
22. hypothalamus
23. anterior pituitary
24. outside
25. plasma
26. 120
27. proteins
28. clotting
29. larynx
30. alveolus
31. true
32. liver
33. true
34. lacteals
35. auditory
36. vitreous
37. hammer
38. fraternal
39. two
40. true

CONNECTING TO LIFE/JOB SKILLS

Many careers in the biological sciences specialize in researching the systems of the human body. **Research biologists** with Ph.D. degrees and **research physicians** with M.D. degrees channel their life work into sorting out and discovering intricacies of tissues, organs, and systems that will improve understanding of the physiology of the body. The decade of the 1990s was known as the "decade of the brain." Renewed interest and effort has stimulated research workers to discover in more detail how the brain works. Some phase of human physiology may interest you. Why not pursue this interest by finding out more about it? Your local college or library media center will be of help.

Chronology of Famous Names in Biology

1637 **William Harvey** (England)—discovered how blood circulates in the body.

1759 **Philibert Gueneau de Montbeillard** (France)—made the first systematic measurements of the growth of a child.

1771 **John Hunter** (England)—dissected the body of the Irish giant, Charles Byrne, and discovered a much enlarged pituitary gland.

1790 **Luigi Galvani** (Italy)—discovered that electrical stimuli can cause muscles to contract.

1830 **Thomas Addison** (England)—discovered the effect of damaged adrenals on health: mottled skin, weight loss, anorexia, irritability.

1830 **Jan Evangelista Purkinje** (Czechoslovakia)—discovered the sweat glands in the skin and the fiber network in cardiac muscle; demonstrated the importance of fingerprints.

1849 **Arnold Adolph Gerthold** (Germany)—discovered the endocrine function of testes.

1853 **Thomas Curling** (England)—first reported myxedema.

1855 **Claude Bernard** (France)—discovered the glycogenic function of the liver.

1860 **Anders Retzius** (Sweden)—devised and named the cephalic index: maximum length to maximum breadth of the skull.

1885 **Paul Langerhans** (Germany)—discovered the islets in the pancreas.

1886 **Pierre Marie** (France)—discovered that an oversecretion of growth hormone causes acromegaly.

1873 **Camillo Golgi** (Italy)—devised a method of impregnating metallic salts into nerve cells to better determine their structure. He discovered the dendrites and axons of nerve cells.

1882 **Richard Owen** (England)—discovered the existence of parathyroid glands by dissecting an Indian rhinoceros.

1904 **Ernest Starling** and **William Bayliss** (England)—devised the term *hormone*.

1914 **Edward C. Kendall** (United States)—isolated thyroxin from cattle thyroid.

1915 **David Marine** (United States)—discovered the link between iodine and goiter.

1921 **Frederick Barting** and **Charles Best** (Canada)—isolated insulin.

1925 **David Marine** (United States)—associated iodine deficiency with the high incidence of goiter in Cleveland.

1929 **Edward Doisy** (United States)—isolated the ovarian hormone estrone.

1930 **Karl Landsteiner** (United States)—discovered human blood groups.

1942 **Charles Drew** (United States)—devised a more effective method for preserving blood for transfusions.

1944 **Herbert Evans** (United States)—discovered growth hormone.

1944 **Choh Hao Li** (United States)—isolated growth hormone and other pituitary secretions.

1948 **Otto Loewi** (France)—discovered the presence and actions of neurohumors.

1950 **Andrew F. Huxley** and **R. Niedegarde** (England)—found that two lines of skeletal muscle fibers move close together when a muscle contracts.

1950 **Gregory Pincus** (United States)—developed a steroid to suppress ovulation from the roots of a wild Mexican yam.

1955 **Alan L. Hodgkin** and **Andrew F. Huxley** (England)—won the Nobel Prize for showing how a difference of potential contributes to the functioning of a neuron. They studied the giant axons of the squid.

1958 **Hans Selye** (Canada)—discovered that stress affects the endocrine system.

1959 **Morris Goodman** (United States)—prepared animal serums with antibodies to measure the degrees of relationship among the various species of primates.

1965 **Hugh E. Huxley** and **Jean Hanson** (England)—developed the sliding filament theory of muscle contraction.

1980s **Candace Pert** (United States)—identified the specific receptor sites in the brain for opiates and other pain-deadening drugs and chemicals.

1990s **Antonio Damasio** (United States)—stimulated research in the neurological disorders of the mind and behavior.

1998 **Robert Furchgott** (United States)—won the Nobel Prize in Physiology and Medicine for his discovery of the role of nitric acid as an intercellular messenger that initiates cellular events that help to dilate blood vessels.

1999 **Gunter Blobel** (United States)—won the Nobel prize in Physiology and Medicine for his discovery of how proteins move across cell membranes. This work has particular significance in stemming the severity of human disorders such as cystic fibrosis and hypercholesterolemia, an inherited disorder of dangerously high cholesterol.

14

Nutrition: Eating for Health

WHAT YOU WILL LEARN

In this chapter you will examine why a balanced diet is necessary for maintaining good health. The food we eat each day constitutes our diet.

SECTIONS IN THIS CHAPTER

- Macronutrients: Carbohydrates, Proteins, Lipids
- Micronutrients: Vitamins, Minerals
- Eating Disorders
- Review Exercises for Chapter 14
- Conneccting to Life/Job Skills
- Chronology of Famous Names in Biology

NUTRITION AND DIET

Nutrition is the totality of methods by which an organism satisfies the energy, fuel, and regulatory needs of its body cells. Substances that contribute to the nutritional needs of cells are called **nutrients**. Animals take nutrients into the body by the **ingestion** of food. The term *food*, therefore, refers to edible materials that supply the nutrients needed by the body.

Nutrients required in large amounts are classified as *macronutrients*: carbohydrates, proteins, and lipids (fats). *Micronutrients*—vitamins and minerals—are needed in smaller amounts. Vitamins are organic compounds; minerals are inorganic. **Malnutrition** results from the improper intake of nutrients. This may be due to a person's eating too little food or to the intake of too much of one nutrient and not enough of others.

Nutritionists urge the eating of a *balanced diet*. A balanced diet is a good mixed diet that includes choices from the four major groups of food: the milk group, the meat group, the vegetable and fruit group and the breads and cereals group. Three or four servings from each of these groups each day will ensure a nutritionally useful diet.

Macronutrients: Carbohydrates, Proteins, Lipids

CARBOHYDRATES

Carbohydrates include all sugars and starches. You learned previously that carbohydrates are built from the basic unit $C_6H_{12}O_6$, **glucose**, a *monosaccharide* or *single sugar*. Table sugar, **sucrose**, is a *disaccharide* or *double sugar*. Different kinds of foods contain different kinds of sugars. The sugar in fruit is **fructose**, and the sugar in milk is **lactose**. Starches are also carbohydrates, formed from long chains of glucose molecules. Starches are classified as *polysaccharides*. Carbohydrates are used by body cells for fuel to provide energy so that cells can carry out the many biochemical activities necessary to sustain life.

All carbohydrates must be broken down into glucose or fructose by digestion before they can be used by cells as a source of fuel. If body cells receive more simple sugar than they can use as energy, some of the excess sugar is stored in the liver and muscles as *glycogen*, commonly called *animal starch*. If, however, carbohydrates are ingested in much larger quantities than the body needs, they are converted into fat and stored under the skin and around body organs.

Carbohydrates help form the structures of some important biological compounds, including parts of the cell membrane, and they assist the body in the manufacture of biotin and other B-complex vitamins.

Food sources of carbohydrates include potatoes, fruits, vegetables, cereal grains, beans, peas, sugar cane, beets, milk, baked goods, and pasta (Figure 14.1).

FIGURE 14.1 Some foods containing carbohydrates

PROTEINS

All proteins contain the elements carbon, hydrogen, oxygen, and nitrogen. In addition to these elements, some proteins contain sulfur. You know that protein is the only nutrient that supplies nitrogen to the body. Nitrogen is necessary for the growth and repair of body cells and tissues. Without enough protein in the body, a person slowly starves to death.

A protein molecule is constructed from building block units known as **amino acids**. Twenty amino acids are found in living cells; these are called **essential** amino acids. However, some proteins are composed of special amino acids, usually formed by a change in a common amino acid, that supplement the basic set of twenty essential amino acids.

The number of different proteins is enormous. Variety in proteins is made possible by variability in structure. Although proteins have a basic structural similarity, each kind is different because of the number, types, and order of amino acids that compose it.

Proteins are the most abundant of the organic compounds in body cells. They compose all of the fibrous structures in the body, including hair, nails, ligaments, the microfilaments in cells, and the myofibrils of muscles. They also form part of hemoglobin, certain hormones such as insulin, and thousands of enzymes that control biochemical processes of cells. Proteins are assimilated into protoplasm and are vital to the formation of DNA molecules. Proteins are also necessary in forming antibodies, molecules that constitute an important part of the immune system which functions to ward off disease, and in regulating the water balance and acid-base balance in the body.

Both plant and animal food sources supply protein. Figure 14.2 shows plant food sources of protein. Plant sources include vegetables, nuts, peas, and beans. Figure 14.3 shows some animal food sources of protein. Meat, fish, poultry, and eggs are common animal protein food sources. Vegetables, meat, and eggs are needed to supply the body with essential amino acids.

FIGURE 14.2
Plant food sources of protein

FIGURE 14.3 Animal food sources of protein

In the digestive system, proteins are broken down into their "building block" molecules, amino acids, which are then discharged into the blood. The bloodstream carries amino acid molecules to body cells, where they are used for repair and growth. If the body's store of carbohydrate is too low, amino acids can be changed to glucose, fatty acids, and glycerol to be used for energy. If proteins are used for energy, however, they will not be available to cells for the building of tissue. Therefore, the daily diet must include carbohydrate and protein in proper proportions.

Twelve of the twenty essential amino acids can be manufactured by the body; eight cannot be synthesized in the body from other amino acids and must be supplied by the diet. Protein that contains all the essential amino acids is called **complete** protein. Animal protein is complete protein; vegetable protein is **incomplete** protein.

A serious protein **deficiency disease** is **kwashiorkor**. This disease, which threatens the lives of many children in Africa, causes misshapen heads, barrel chests, bloated stomachs, spindly legs and arms, decreased mental ability, and poor vision.

LIPIDS (FATS)

Lipids include fats and oils and are a major group of biological compounds. Like carbohydrates, lipids contain the elements carbon, hydrogen, and oxygen. In lipids, however, the ratio of hydrogen to oxygen is much greater than 2:1 and varies from one lipid to another.

Fats, like carbohydrates, are fuel foods, supplying the cells with energy. Certain fats are essential to the structure and function of body cells, to the building of cell membranes, and to the synthesis of certain hormones. Fats also aid in the transport of fat-soluble vitamins. Foods rich in fats include butter, bacon, egg yolk, cream, and certain cheeses.

CALORIES

The energy potential of food is measured in Calories. (The Calorie, that is, the large calorie, is always written with a capital C.) One **Calorie** is the quantity of heat necessary to raise the temperature of 1 kilogram of water 1 degree Celsius. Fats are concentrated sources of energy. One gram of fat provides nine Calories, while one gram of protein or carbohydrate provides only four Calories. Therefore, foods rich in fats add to the caloric content of the human diet. Some body fat is necessary to cushion body organs and to prevent heat loss through the body's surface. Excessive intake of fat causes a person to gain weight.

FIBER

The **fiber** in the human diet comes only from plant sources. Fiber is not a nutrient, but it is important in the diet to stimulate the normal action of the intestines in the elimination of wastes. Fiber absorbs many times its weight in water and aids in the formation of softer stools. It also provides bulk, which promotes regularity and more frequent elimination.

Currently, it is suggested that dietary fiber may contribute protection against many noninfectious diseases of the large intestine, such as cancer of the colon, hemorrhoids, appendicitis, colitis, and diverticulosis. Incidences of these diseases seem to be much lower in countries where the diets are high in fiber. It is also believed that increased dietary fiber reduces blood cholesterol levels and helps to prevent the formation of fatty deposits on the inner walls of the arteries.

Raw fruits and vegetables, whole cereals and bread, and fruits with seeds (strawberries, figs, raspberries) are excellent sources of fiber.

Micronutrients: Vitamins, Minerals

VITAMINS

The study of micronutrients and their effects on the body was begun in 1906 by Dr. Frederick Gowland Hopkins, a physiology professor at Cambridge University, England. Dr. Hopkins did not isolate these microfactors in food, but he was able to demonstrate serious effects of deficiency diets on white laboratory rats. These illnesses had been noted many times over in human populations. In 1921, Casimir Funk, a Polish scientist, attached the name *vitamine* to these micro food substances which, when missing, cause human illness and body disorders.

Vitamins are organic compounds. They are classified as *water soluble* or *fat soluble*. In general, the water soluble vitamins are coenzymes necessary to the proper sequence of biochemical events that occur during cellular respiration. It is interesting to note that the primates (*Homo sapiens* included) and guinea pigs are the only vertebrate animals that cannot synthesize their own vitamin C from carbohydrates. Therefore, the daily requirements of ascorbic acid must be met through food intake. The functions of the fat soluble vitamins are not clearly understood.

Table 14.1 reviews the major vitamins, their functions and food sources, and the symptoms that commonly result from a deficiency.

TABLE 14.1
VITAMINS AND THEIR USES

Vitamin	Necessary for	Deficiency Symptoms	Food Sources
A—Retinol (fat soluble)	Healthy visual pigments in eye; healthy skin membranes	Dryness of membranes; poor growth; night blindness; inflamed eyelids	Fish liver oil, butter, cream, milk, margarine, brightly colored fruits and vegetables, leafy vegetables

TABLE 14.1
VITAMINS AND THEIR USES (continued)

Vitamin	Necessary for	Deficiency Symptoms	Food Sources
C—Ascorbic Acid* (water soluble)	Intercellular cement for teeth and bones; healthy capillary walls; resistance to infection	Sore gums, tendency to bruise easily, painful joints, loss of weight—all these symptoms are associated with scurvy.	Citrus fruits, tomatoes, cabbage, green leafy vegetables, green peppers
D—Calciferol† (fat soluble)	Regulates calcium and phosphorus metabolism and growth, building strong bones and teeth	Soft bones, poor tooth development and dental decay—rickets	Fish liver oil, irradiated feed, egg yolk, salmon
E—Tocopherol	Prevention of oxidation by red blood cells; muscle tone	Hemolysis of red blood cells	Wheat germ, green leafy vegetables
K—Menadione	Synthesis of prothrombin, clotting of blood	Prolonged bleeding from wound	Green vegetables, tomatoes
All the vitamins below belong to vitamin B complex:			
Thiamin and Niacin‡	Growth; healthy digestion; normal nerve function; good appetite; carbohydrate metabolism	Poor digestion; depression; nerve disorders; loss of appetite	Yeast, wheat germ, liver, enriched foods, bread, green vegetables
Riboflavin (water soluble)	Growth; health of skin and mouth; functioning of eyes; carbohydrate metabolism	Retarded growth; sores at corner of mouth; disturbances of vision; inflammation of the tongue	Same as for thiamin and niacin; meat

* Vitamin C is not stored by the body, oxidizes rapidly, and is readily destroyed by exposure to air.

† Humans make their own vitamin D when their skin is exposed to sunlight.

‡ Prolonged deficiency of thiamin results in beri-beri, a nerve disease that may result in paralysis. Prolonged deficiency of niacin results in pellagra, a disease characterized by a rash, graying and falling out of hair, depression, loss of weight, and digestive disturbances.

MINERALS

Minerals are inorganic compounds. Some, such as calcium and sodium, are needed in relatively large amounts. Calcium, together with phosphorus, is used in building bones and teeth. Calcium is also a regulator of muscle activity. Nerve cells could not carry impulses, nor could muscles contract, without the assistance of sodium and potassium. Sodium also functions in the regulation of body temperature, since large amounts of its salts are excreted by the sweat glands. Other minerals are needed by the body in only small amounts. These are known as the **trace** minerals. In general, the functions of trace minerals are regulatory in that they enable the enzymes of metabolism to work.

Table 14.2 reviews the minerals, their functions and major food sources, and the symptoms that commonly result from a deficiency.

TABLE 14.2
MINERALS AND THEIR USES

Vitamin	Necessary for	Deficiency Symptoms	Food Sources
Magnesium	Healthy bones and teeth; involved in protein metabolism	Weakening of bones and teeth; faulty metabolism	Green leafy vegetables
Sodium	Functioning of sodium-potassium pump; regulates water balance in cells; regulates nerve impulse; maintains acid-base balance of tissue fluids and blood	Leads to cardiovascular diseases and disorders of the nervous system	Salt
Iron	Synthesis of hemoglobin, myoglobin and the cytochromes	Anemia, difficulties in cellular respiration	Liver, red meats, egg yolk, whole grain cereals
Iodine	Synthesis of thyroxin	Goiter, sluggish metabolism	Marine fish, iodized salts
Fluorine	Aids in resistance to tooth decay	Breakdown of tooth enamel	Water treatment

TABLE 14.2
MINERALS AND THEIR USES (continued)

Vitamin	Necessary for	Deficiency Symptoms	Food Sources
Calcium	Building of bones and teeth; muscle contraction; nerve impulse transmission; permeability of cell membrane; activation of ATP enzymes	Loss of minerals from bone; anemia; nerve and muscle disorders	Milk and dairy products, eggs, whole grain cereals, green leafy vegetables
Potassium	Functioning of sodium-potassium pump; regulation of nerve impulse; muscle function; glycogen formation; protein synthesis	Nerve and muscle disorders; irregular heartbeat	Beans and peas, fruits, vegetables
Phosphorus	Building of bones and teeth; phosphorylation of glucose; building of ATP molecules; functions in cellular respiration; present in nucleic acids	Malfunctions of basic cell processes	Milk and dairy products

Eating Disorders

An incredible number (literally, millions) of teenage girls put themselves on starvation diets to prevent weight gain or to lose weight. These girls perceive themselves as being fat. Obsessive dieting can lead to the eating disorder **anorexia nervosa** (an-o-reck-see-a ner-voh-sah). Another eating disorder, **bulimia** (byoulee-me-a), is characterized by binge eating. Persons so afflicted stuff themselves with vast quantities of food in a short time and then purge their bodies by vomiting or taking laxatives.

Anorexia nervosa and bulimia are serious conditions. Anorexics lose muscle mass, bone mass, hair, and teeth. Many girls stop menstruating and fail to mature sexually. Bulimics destroy the linings of the mouth, esophagus, and stomach. Fatigue and muscle cramps occur in persons with either of these disorders.

Medical treatment, counseling, and family support are necessary to treat and, it is hoped, to cure these eating disorders.

REVIEW EXERCISES FOR CHAPTER 14

WORD-STUDY CONNECTION

amino acid
anorexia
balanced diet
bulimia
deficiency disease
dehydration
disaccharide
essential amino acids
fatty acid
fiber
nervosa
glucose
glycogen
kwashiorkor
synthesis lactose
lipid
macronutrient
malnutrition
micronutrient
fructose monosaccharide
nutrient
nutrition
polysaccharide
roughage
vitamin

SELF-TEST CONNECTION

PART A. Completion. *Write in the word that correctly completes each statement.*

1. The molecules in food that are used by cells for energy, growth and repair of tissue are _____.

2. The three major types of macronutrients are carbohydrates, proteins and _____.

3. Fibrous structures such as hair and nails, enzymes, and hemoglobin all contain the macronutrient _____.

4. The energy potential of food is measured in _____.

5. Fiber stimulates the action of the _____.

6. Some fatty acids are used to synthesize chemical messengers known as _____.

7. Vitamins and minerals belong to a class of substances known as _____.

8. Muscle activity is regulated by the mineral _____.

9. In general the water soluble vitamins are necessary for the events of cellular _____.

10. Minerals necessary only in small amounts are known as _____ minerals.

11. Carbohydrates include all sugars and _____.

12. A disaccharide is a _____ sugar.

13. Huge starch molecules composed of many units of simple sugar are known as _____.

14. Glucose is composed of the elements carbon, oxygen, and _____.

15. The number of carbon atoms in a molecule of glucose is _____.

PART B. Multiple Choice. *Circle the letter of the item that correctly completes each statement.*

1. Of the following pairs, the one in which both items are micronutrients is
   (a) calcium and ascorbic acid
   (b) protein and sucrose
   (c) maltose and calcium
   (d) glycerol and copper

2. Another name for animal starch is
   (a) glycerol
   (b) glycogen
   (c) glucose
   (d) glucagon

3. Two protein molecules are
   (a) levulose and vitamin E
   (b) copper and mannose
   (c) nitrogen and starch
   (d) insulin and antibodies

4. The role of fats in the body is
   (a) functional but not structural
   (b) structural but not functional
   (c) functional and structural
   (d) only structural

5. Fiber in the diet
   (a) causes diverticulosis
   (b) prevents infectious diseases
   (c) increases blood cholesterol
   (d) absorbs large amounts of water

6. The name Casimir Funk is correctly associated with
   (a) the discovery of rickets
   (b) inventing the name *vitamine*
   (c) finding the cause of beriberi
   (d) inventing the term *roughage*

7. The mineral sodium plays a role in all of the following except
   (a) regulation of body temperature
   (b) regulation of water balance
   (c) transmission of nerve impulses
   (d) synthesis of hemoglobin

8. Prolonged bleeding from a wound can result from a deficiency of
   (a) vitamin A
   (b) vitamin D
   (c) vitamin K
   (d) vitamin E

9. Vitamins belonging to the B complex group include
   (a) tocopherol and niacin
   (b) thiamin and niacin
   (c) ascorbic acid and menadione
   (d) calciferol and thiamin

10. Goiter is caused by a deficiency of
    (a) fluorine
    (b) zinc
    (c) iodine
    (d) magnesium

11. All proteins contain the elements carbon, hydrogen, oxygen, and
    (a) copper
    (b) phosphorus
    (c) sulfur
    (d) nitrogen

12. The number of amino acids essential to all cells is
    (a) 11
    (b) 20
    (c) 33
    (d) 44

13. The carboxyl or acid group present in amino acids is correctly represented as
    (a) CHOO
    (b) COHO
    (c) COOH
    (d) COHH

14. Anorexia nervosa refers to a (an)
    (a) eating disorder
    (b) genetic disease
    (c) starvation diet
    (d) fiber-like protein

15. The two major classes of micronutrients are
    (a) fiber and vitamins
    (b) water and minerals
    (c) vitamins and minerals
    (d) water and vitamins

PART C. Modified True-False. *If a statement is true, write "true" for your answer. If a statement is incorrect, change the <u>underlined</u> expression to one that will make the statement true.*

1. Kwashiorkor is a <u>carbohydrate</u> deficiency disease.

2. Nightblindness is caused by lack of vitamin <u>D</u>.

3. The mineral essential to the formation of hemoglobin is <u>zinc</u>.

4. Guinea pigs and *Homo sapiens* <u>can</u> synthesize vitamin C.

5. <u>Fats</u> are the most abundant of the organic compounds in body cells.

6. The major fuel foods are carbohydrates and <u>proteins</u>.

7. Raw fruits and vegetables are excellent sources of <u>fiber</u>.

8. Rickets is associated with a lack of vitamin <u>C</u>.

9. The mineral <u>phosphorus</u> is necessary for the synthesis of the hormone thyroxin.

10. <u>Fluorine</u> aids in resistance to tooth decay.

11. During dehydration synthesis one or more molecules of <u>hydrogen</u> form.

12. Sucrose is an example of a <u>monosaccharide</u>.

13. Enzymes are composed of <u>fats</u>.

14. Lipids are composed of glycerol and <u>amino</u> acids.

15. The three principal elements from which lipids are composed are carbon, <u>nitrogen</u>, and oxygen.

CONNECTING TO CONCEPTS

1. Why do body cells require carbohydrate foods?

2. What are the functions of proteins in the body?

3. If fats were not available to body cells, what processes or functions would be lost?

4. Why is fiber necessary in the human diet?

5. How may an eating disorder lower the quality of an afflicted person's life?

ANSWERS TO SELF-TEST CONNECTION

PART A

1. nutrients
2. fats
3. protein
4. Calories
5. intestine
6. hormones
7. micronutrients
8. calcium
9. respiration
10. trace
11. starches
12. double
13. polysaccharides
14. hydrogen
15. six

PART B

1. **(a)**
2. **(b)**
3. **(d)**
4. **(c)**
5. **(d)**
6. **(b)**
7. **(d)**
8. **(c)**
9. **(b)**
10. **(c)**
11. **(d)**
12. **(b)**
13. **(c)**
14. **(a)**
15. **(c)**

PART C

1. protein
2. A
3. iron
4. cannot
5. Proteins
6. fats
7. true
8. D
9. iodine
10. true
11. water
12. disaccharide
13. proteins
14. fatty
15. hydrogen

CONNECTING TO LIFE/JOB SKILLS

Maintaining a balanced diet each day is important to health and therefore the quality of life. In most households, family members eat lunch outside the home, so the noon meal may not be as nutritious as it should be. Plan menus for two meals a day for a family of five for one week. Your daily menus should include foods from the four food groups. You may wish to start walking through your local food markets to note the kinds of fruits, vegetables, and meats that are available. Your two-meal-a-day menus must include the essential macronutrients and micronutrients.

Chronology of Famous Names in Biology

1890 **Theodore Palm** (England)—wrote a treatise on the absence of bone deformity (rickets) in poor Japanese children.

1897 **Christian Eijkman** (India)—discovered that chickens and people kept on a diet of polished rice developed the disease beri-beri.

1898 **J. Lind** (England)—discovered that a diet containing citrus fruits prevents scurvy.

1906 **Frederick G. Hopkins** (England)—applied research methods in inducing and curing deficiency diseases in rats.

1919 **Kurt Huldschinsky** (Germany)—discovered that ultraviolet radiation and the hormone calciferol can cure rickets.

1920 **E. V. McCollum** and **Lafayette Mendel** (United States)—identified the first vitamin and named it A.

1920 **Joseph Goldberger** (United States)—discovered that pellagra is caused by a vitamin deficiency.

1921 **Casimir Funk** (Poland)—invented the name vitamine to describe a special group of nutrients.

1922 **E. V. McCollum** (United States)—discovered vitamin D.

1924 **Harry Steenbock** and **A. F. Hess** (United States)—discovered that irradiation of milk increases the vitamin D content.

1950 **D. M. Hadjimarkos** (United States)—researched the roles of micronutrients such as selenium, vanadium, and barium in the formation of dental caries.

1963 **Marcel El Conrad** (United States)—completed original research that elucidated the pathways of iron metabolism in the human body.

1964 **Reginald R. W. Townley** (Tasmania)—researched carbohydrate mal-absorptive disorders in children.

1965 **David Paige** (United States)—published a major study on lactose-intolerant populations, showing lactose enzyme deficiency as the genetic cause of milk intolerance.

1968 **Jana Parizkova** (Czechoslovakia)—completed research on problems of obesity that included studies of fat deposition and metabolism.

1994 **Jeffrey Friedman** (United States)—discovered that the hormone *leptin* controls appetites in lab mice. Mice lacking leptin become obese.

1996 **Stephen O'Rahilly** (England)—discovered that when the appetite control gene fails to code for the receptor protein melanocortin-4, obesity occurs in children.

15

Diseases of Homo sapiens and the Immune System

WHAT YOU WILL LEARN

In this chapter you will examine conditions and organisms that cause diseases in humans.

SECTIONS IN THIS CHAPTER

- Infectious Diseases
- The Immune System
- Noninfectious Diseases
- Review Exercises for Chapter 15
- Connecting to Life/Job Skills
- Chronology of Famous Names in Biology

OVERVIEW

A **disease** is a disorder that prevents the body organs from working as they should. In general, diseases can be classified as being *infectious* or *noninfectious*. **Infectious diseases** are caused by organisms that invade the body and do harm to the cells, tissues, and organs. As a rule, there is disease specificity whereby a specific disease-producing organism causes a particular disease. Disease-producing organisms are said to be **pathogens** and are described as being *pathogenic*. Usually, pathogens—or germs, as they are often called—are microorganisms, organisms microscopic in size. Pathogenic microorganisms include certain bacteria, protozoans, spirochetes, richettsias, mycoplasmas, and fungi. Parasitic worms and viruses also often produce disease in humans. Most infectious diseases are *contagious*—capable of being passed from one person to another by means of body contact or by droplet infection.

Noninfectious diseases are caused by factors other than pathogenic organisms. Among the factors that are responsible for noninfectious diseases are genetic causes, malnutrition, exposure to radiation, emotional disturbances, organ failure, poisoning, endocrine malfunctioning, and immunological disorders. Whatever the cause, a disease works counter to the well-being of a human.

Infectious Diseases

METHODS OF SPREADING

Contagious or **communicable** diseases are spread from one person to another in a number of ways. One such way is **direct contact**; that is, the infected person touches a well person. This may be accomplished by handshake or kissing, or through sexual intercourse (**sexually transmitted diseases**—see Table 15.1). Microorganisms are also spread **indirectly** when a noninfected person handles objects that have been in contact with the infected person. Eating from the same plate, drinking from the same glass, handling bed clothes or towels—there are any number of means by which germs are passed. *Droplet infection* is another common method of passing germs along. Disease germs are present in droplets of water that escape from the nose and mouth when sneezing, coughing, and talking. If these infected droplets are inhaled or taken in by mouth, the germs then enter the body of another person (Figure 15.1).

TABLE 15.1
SEXUALLY TRANSMITTED DISEASES

Disease	Causative Organism	Symptoms
Syphilis	*Treponema pallidum* (spirochete)	Body lesions, tumors, dementia
Gonorrhea	*Neisseria gonorrhoeae* (coccus)	Infection in genital/ reproductive system
Venereal herpes	Herpes simplex-2 (virus)	Sores at infection site, pains in joints
AIDS	HIV (retrovirus)	Syndrome of killer infections
Trichomoniasis	*Trichomonas vaginalis* (protozoan)	Venereal warts
Crab lice	*Phthirus pubis* (insect)	External parasite in pubic hair

TABLE 15.1
SEXUALLY TRANSMITTED DISEASES (continued)

Disease	Causative Organism	Symptoms
Monilial vaginitis	*Candida albicans* (fungus)	Vaginal and penile infections
Chancroid	*Hemophilus ducreyi* (bacterium)	Genital system infection
Hepatitis A, B	Virus	Infection of liver
Cervical cancer, warts	HPV (human papilloma virus)	Genital and body warts

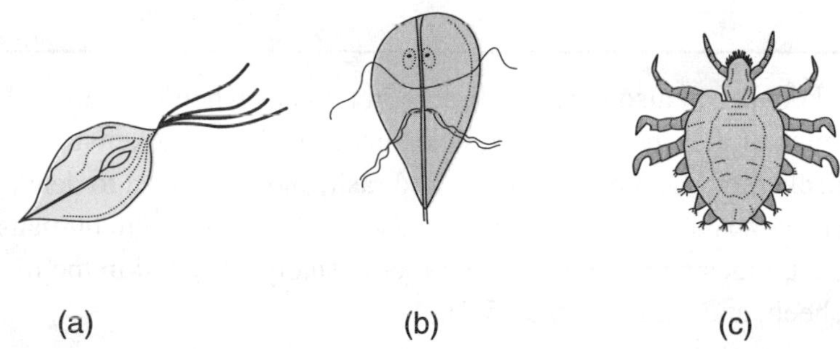

FIGURE 15.1 Some organisms that cause sexually transmitted diseases (a) *Trichomonas vaginalis*, protozoan (b) *Giardia enterica*, protozoan (c) *Phthirus pubis*, arthropod (body louse)

A current example of a highly infectious and ever-spreading disease is tuberculosis (TB). Figure 15.2 shows *Mycobacterium tuberculosis*, the bacterium that caused 2.9 million deaths in 1997. The bacterium travels through the air in droplets from infected people. Ninety-five percent of TB victims live in Third World countries. With increased immigration to the United States and within homeless shelters, however, there has been an alarming increase in tuberculosis in this country. Although tuberculosis is usually treatable with antibiotics, many resistant strains of the bacterium have emerged.

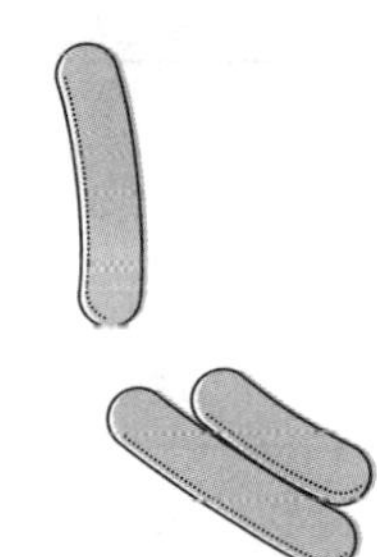

FIGURE 15.2 The bacterium *Mycobacterium tuberculosis*, causative agent of tuberculosis

Animals may be vectors, or carriers, of diseases. Biting insects or mammals may carry disease germs in their salivary glands and pass them on to a human who is bitten. Animal hair carries insects that may be disease carriers. Contaminated food and water spread diseases throughout human populations. Finally, human carriers of disease organisms (who remain unaffected by the germs that they carry) can spread pathogenic organisms to noncarriers.

COMMON INFECTIOUS DISEASES AND THEIR CAUSES

Bacteria are responsible for many human diseases, as outlined in Table 15.2.

TABLE 15.2
DISEASES CAUSED BY BACTERIA

Diseases Caused by Bacilli (Rods)	Diseases Caused by Cocci (Spheres)	Diseases Caused by Spirellae (Spirals)
Tuberculosis	Pneumonia (some forms)	Syphilis
Diphtheria	Gonorrhea	Asiatic cholera
Tetanus	Scarlet fever	Yaws
Typhoid fever	Rheumatic fever	Lyme disease
Bubonic plague	Streptococcus sore throat	
Whooping cough	Meningitis (some forms)	
Tuleremia	Childbed fever	
Leprosy		

Certain rickettsia can also produce disease in humans when they enter the body through the bites of their hosts—mites, ticks, lice, and fleas. Rickettsia diseases are often serious, characterized by high fever and rash, and often lead to death (Table 15.3). Some protozoa and some worms can also produce disease in humans. Most parasitic worms are ingested in contaminated food, usually encysted in the muscles of cows, pigs, sheep, and snails (Table 15.4).

TABLE 15.3
RICKETTSIA DISEASES

Human Disease	Disease Vector
Typhus	Lice
Trench fever	Lice
Rocky Mountain spotted fever	Ticks
Rickettsial pox	Mites

TABLE 15.4
WORM-CAUSED DISEASES

Diseases Caused by Flatworms	Diseases Caused by Roundworms
Tapeworm infection	Hookworm
Sheep liver fluke infection	Trichinosis
Chinese liver fluke infection	Pinworm infection
	Ascaris infection
	Filariasis

A few species of yeasts and molds are pathogenic for humans. They attack the skin, mucous membranes, and the lungs. Ringworm and athlete's foot are two such diseases. Although irritating and inconvenient, these diseases are not generally serious and can usually be readily treated.

Viruses—nonliving particles that come "alive" when they invade a cell—are major disease-producers in humans. Some viruses cause disease that affects the entire body; others cause disorders that affect a particular organ. Table 15.5 lists some common viral diseases.

TABLE 15.5
COMMON VIRUS DISEASES

Virus Disease	Organ Affected
Poliomyelitis	Nervous system and muscles
Rabies	Nervous system
Encephalitis	Nervous system
Viral pneumonia	Lungs
Common cold	Respiratory system
Influenza	Lungs and respiratory system
Fever blisters	Skin around the lips
Genital herpes	Genital organs
Mumps	Salivary glands
Viral hepatitis	Liver
Trachoma	Eyes
Measles	Spread throughout the body
Smallpox	Spread throughout the body
Chicken pox	Spread throughout the body
Yellow fever	Spread throughout the body
Dengue fever	Spread throughout the body
Psittacosis (parrot fever)	Spread throughout the body
Warts	Hands and feet
Cervical cancer	Cervix
AIDS	Immune system

HOW PATHOGENS DAMAGE THE BODY

Once pathogenic organisms enter the body there is interaction between the body and the germs that results in disease. The type of disease is determined by the type of pathogen that invades the body. The severity of the disease depends on the ability of the body to ward off infection (*resistance*) and the strength or *virulence* of the infecting germ.

Pathogens may affect body tissues and functions in a number of ways. Some pathogens produce enzymes that dissolve the materials that hold cells together, creating pathways for germs to enter tissues. Other germs produce substances that kill certain body cells. Pathogens frequently damage only certain cells and tissues. The rickettsia of Rocky Mountain spotted fever damages the liver, as does the Chinese liver fluke. The polio virus destroys nerve cells. The spirochete of syphilis (Figure 15.3) often destroys brain tissue. The typhoid bacillus attacks the lymph tissue of the intestinal wall.

Some disease-producing microorganisms secrete **toxins** that interfere with the metabolic activities of cells. Certain germs block vital passageways of the body and thus prevent normal functioning; for example, the organisms that cause diphtheria seal the throat with membranes and prevent breathing. Worm parasites frequently compete with the host for nutrients and produce malnutrition in the host.

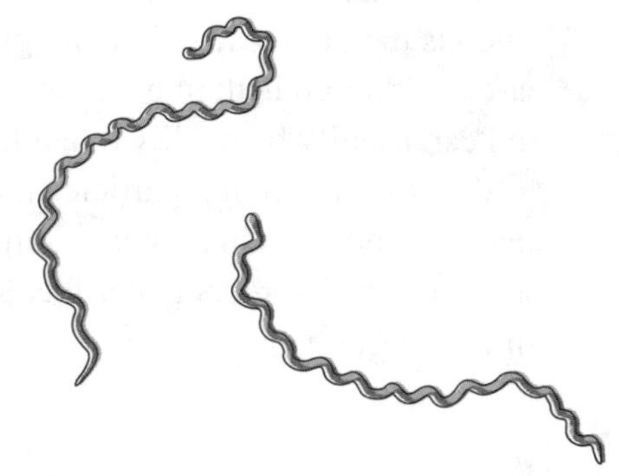

FIGURE 15.3 The spirochete *Treponema pallidum*, causative agent of syphilis, as seen through a dark-field microscope

NONSPECIFIC DEFENSES AGAINST DISEASE

When the body is attacked by pathogenic organisms, it puts up a series of defenses designed to destroy the enemy and maintain health.

SKIN

The first line of defense against invasion of the body by germs is the skin. The clean, unbroken skin is thick enough and tough enough to prevent most germs from penetrating. As a rule, germs that land on the skin do not live long enough to cause trouble because the skin itself has a germicidal quality that inhibits the growth of germs on its surface.

OTHER DEFENSES

The eyes, nose, and mouth are in effect breaks in the skin that can permit the penetration of germs into the body. Most germs entering the eye do not live long enough to cause distress. They are dissolved by **lysozyme**, an enzyme in tears. Nevertheless, some virulent germs survive and produce eye infections such as **conjunctivitis** (pink eye) or **tracoma**. Tracoma, a virus disease, is especially dangerous because it often causes blindness.

Germs numbering in the thousands enter the mouth daily with food and drink. Few of these survive to reach the intestines. The saliva in the mouth is able to kill many of the invaders. Those that reach the stomach face the killing action of hydrochloric acid and the digesting power of pepsin. However, some germs manage to survive. The germs of Asiatic cholera, thyphoid and paratyphoid fever, and other serious intestinal diseases are able to resist these body defenses and cause illness.

Uncountable numbers of disease germs are breathed in through the nose from the surrounding air. However, few of these ever reach the lungs. The nasal passages present a complicated maze of filters guarded by hairs that trap many germs. In addition, the mucous membranes lining the air passages secrete sticky mucus, which traps disease germs, rendering them inactive. Also, sneezing expels germs to the outside.

Germs that manage to reach the breathing tubes are for the most part trapped in mucous secretions from the cells that compose these tubes. In addition, the cilia of the cells that line the air tubes sweep the mucus-trapped germs back to the throat where they are swallowed and then destroyed in the stomach by hydrochloric acid and pepsin. Special "dust cells" in the air sacs of the lungs pick up some germs and carry them out. Despite these active defenses, some germs survive and cause respiratory diseases such as colds, influenza, and pneumonia.

The Immune System

The purpose of the **immune system** is to protect the body against infection. When microorganisms (germs) infect the body, the immune system is called into action. Its function is to destroy the invading germs or foreign proteins.

The organs and tissues of the immune system are distributed throughout the body. The immune system is made up of lymphoid tissue, fluid called **lymph**, and **white blood cells**, which act against foreign matter, germs and proteins, that enter the body. The immune system is closely involved with the blood circulatory system.

The cells of the immune system cluster in the lymphoid tissues. Lymphoid tissues include the adenoids and tonsils in the head region, the thymus gland in the chest cavity, bone marrow in the center of long bones, the spleen just below the heart, lymph nodes positioned under the arms and in the groin, and Peyer's patches in the small intestine.

CELLS OF THE IMMUNE SYSTEM

The cells of the immune system are able to recognize and act upon microorganisms and foreign proteins that enter the body. Any foreign substance or organism that causes the immune system to react is called an **antigen**. Antigens are usually proteins, glycoproteins (carbohydrate-protein molecules), or carbohydrates. These molecules may be carried on the cell membranes of invading microorganisms. Fats are usually not antigenic.

PHAGOCYTES

Macrophages and **neutrophils** are two types of white blood cells that engulf (phagocitize) invading microorganisms. These cells provide the human body with a very effective first line of defense.

Huge, amoeboid macrophages and the smaller neutrophils are specialized for attacking invading microorganisms. Some macrophages stay within the spleen and lymph nodes, where they engulf any microorganism invaders that pass their way. Other macrophages and the neutrophils travel through the body searching for invaders.

Chemicals released from damaged blood platelets attract the traveling macrophages and neutrophils. These cells gather at the site of infection and ingest foreign bacteria—hence the name **phagocytes**. The phagocytes ingest large numbers of bacteria and are themselves killed by the bacterial toxins (poisons). The accumulated dead bodies of macrophages and neutrophils form *pus*.

LYMPHOCYTES

The second line of defense provided by the immune system is made possible by a type of white blood cell known as a **lymphocyte**. There are two kinds of lymphocytes: the **B-cell lymphocyte** and the **T-cell lymphocyte**. Working together, these cells carry out a complex series of events known as the *immune response*.

The human immune response embraces two major biochemical events: the **humoral immune response** and the **cell-mediated response**. The humoral immune response involves the production of protein molecules called **antibodies**, and the cell-mediated response demands direct cellular action. Each of these responses is driven by a specific type of lymphocyte: either the B-cell or the T-cell.

Lymphocytes are produced by the **stem cells**, which reside in the marrow of long bones. Stem cells produce all of the blood cells. In some way still unknown, changes (differentiation) take place in the primary blood cells, enabling them to carry out specific functions. The B-cell lymphocytes mature in the bone marrow (thus the name B-cell); the T-cell lymphocytes, in the thymus gland (hence the name T-cell).

The B-cells and T-cells look alike in the inactive state. When activated, however, they differ significantly. Activated B-cells have a very rough endoplasmic reticulum caused by the enormous number of ribosomes that are attached to these membranes. Activated T-cells have large concentrations of free ribosomes in the cytoplasm.

Primary Immune Response

On the outer surface of their cell membranes, B-cell lymphocytes carry specific antigen-recognition proteins. Each cell is a specialist, carrying only one kind of recognition protein. When a newly produced B-cell meets with a matching antigen, the B-cell is activated and the antigen attaches to the recognition site on the membrane of the B-cell. This B-cell with its attached antigen becomes much larger and undergoes mitosis and differentiation, producing two new kinds of cells—**plasma cells** and **memory cells**.

The activated plasma cells produce an **antibody** that matches the captured antigen. Antibody production by the plasma cells is enormous—about 2000 molecules of antibody per minute! The antibodies attach to the foreign antigen and immobilize it. Now the macrophages and neutrophils are called into action. They ingest the antigen-antibody complexes. This early reaction of the immune system is called the **primary immune response**.

Secondary Immune Response

The memory cells store all the information needed to build the same kind of antibodies that the plasma cells make during the primary immune response. Plasma cells have very short lives. In contrast, memory cells have very long lives and can remain in existence for many years. Should a second attack by the same antigen occur, the memory cells go into action, producing huge numbers of antibodies through very rapid reproduction of plasma cells. The plasma cells are now produced in greater numbers than during the primary immune response. Thus the **secondary immune response** is much more effective than the primary immune response.

SUBPOPULATIONS OF THE T-CELL LYMPHOCYTE

Antibodies can inactivate viruses and prevent them from entering cells. Once a virus has entered a cell, however, it is safe from antibody. Therefore, for virus-infected cells, the immune system makes use of the cytotoxic T-cell.

Usually, a cell that is infected with a virus has viral antigens on its cell surface. The cytotoxic T-cell recognizes these foreign antigens on the surface of the cell membrane and kills that cell as it comes in contact with it. The ability of cytotoxic T-cells to destroy antigens without the aid of antibodies is known as *cell-mediated immunity*. Another type of T-cell forms a capsule around foreign cells, thereby localizing infection and preventing its spread.

A different subpopulation of T-cell lymphocytes is called **regulatory T-cells**. Of these, there are helper T-cells and suppressor T-cells. The helper T-cells assist other T-cells and B-cells in responding to antigens. The helper T-cells also activate some macrophages.

AIDS AND THE IMMUNE SYSTEM

Acquired immune deficiency syndrome (AIDS) is a killer disease that reached epidemic proportions in the 1980s through the mid 1990s. Newer drugs, often used in combination, and methods of treatment have been able to prolong the lives of some people afflicted with full-blown AIDS. Infection by the human immunodeficiency virus (**HIV**) causes the immune system to collapse, leaving the body open to devastating infections. HIV (Figure 15.4) attacks the CD-4 lymphocytes, which function as helper cells. When made inactive, the T-helper cells cannot stimulate reproduction of B-cell lymphocytes. The B-cells give rise to the antibody-producing plasma and memory cells. The result is that AIDS shuts down the patient's entire immune system. The infected person becomes host to a number of infectious organisms that cause diseases such as Kaposi's sarcoma, pneumocystic pneumonia, and yeast infections.

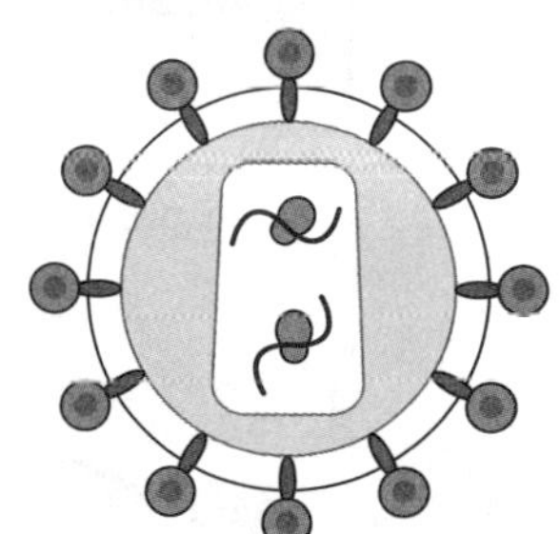

FIGURE 15.4
HIV in a human cell

AIDS is not a single disease but is a syndrome of symptoms caused by the various invading microorganisms that take advantage of an immune system that cannot function. Chapter 16 presents a discussion of the way in which the HIV retrovirus works.

PROTECTION AGAINST DISEASE

IMMUNIZATION

Immunity is the ability to resist the attack of a particular disease-producing organism. Immunity to one kind of disease germ does not automatically make a person immune to other types of disease germs. **Active immunity** is brought about by antibody production by a person's own body cells. Active immunity can be stimulated in either of two ways: by getting the disease and recovering from it or by being immunized against the disease. Immunization that produces active immunity involves the injection of weakened disease agents that stimulate antibody production but produce only mild symptoms or none at all. Active immunity is long-lasting because the body cells continue to produce the antibodies.

An injection of gamma globulins can give a person temporary immunity against certain specific diseases. This means that a person has borrowed antibodies in the blood and not those made by his (her) own cells. This kind of immunity is called **passive immunity**. It lasts only as long as the antibodies last; when they are used up, the immunity ceases.

Smallpox is a disease that has been almost entirely eradicated from even remote corners of Earth. The fight against this disease began in the 18th century when Edward Jenner vaccinated people with cowpox. Vaccination with cowpox stimulates the body to build its own antibodies and give lasting immunity to the dread disease of smallpox. Cowpox is a mild disease in humans and stimulates the cells to produce antibodies that happen to be effective against smallpox.

Successful methods of immunization have been developed against diphtheria, polio, whooping cough, lockjaw, plague, typhoid, yellow fever, cholera, measles, mumps, rubella, and typhus. In every community in the United States children are automatically immunized against diphtheria, tetanus, whooping cough, polio, and smallpox. Many of the states have laws that prevent children from attending public school without the prescribed immunizations.

SAFE DRINKING WATER

At one time diseases such as typhoid fever and cholera devastated the population of New York. Today not one death can be attributed to either of these diseases. The marked decrease in the incidences of these diseases is attributed to the development of a safe water supply.

Sanitary water engineers are employed to protect the community against the dangers of contaminated water. They use many water purification techniques. **Settling** is a process in which water is held in large tanks until suspended solids settle out. **Filtering** is accomplished by allowing water to trickle through sand beds several feet deep. This removes 90–95 percent of all bacteria as well as fine particles of solid matter. **Aeration** is the process in which water is sprayed up into the air. This technique kills some bacteria and allows more air to dissolve into the water. **Chlorination** is the chemical purification of water; chlorine gas or hypochlorite is added to the water in small quantities to kill any remaining pathogens.

Sanitary engineers and bacteriologists watch the community water supply very closely. They test it constantly for *Escherichia coli*. The presence of this bacterium is known as the **index of fecal contamination**. Since *E. coli* is normally present in the human intestines, its presence in drinking water indicates that the water supply must be contaminated with sewage. When *E. coli* is not present, the water is free of human wastes and is probably free of organisms that cause typhoid, cholera, and other diseases of the human intestines.

SAFE MILK AND FOOD

At one time diseases such as tuberculosis, Q fever, septic sore throat, brucellosis, and stomach upsets were transmitted to people from contaminated milk. To prevent the spread of disease through milk most local laws require that milk be **pasteurized** before sales. Commercial dairies chill the milk as quickly as possible after collection. This temporarily inhibits the growth of bacteria which may thrive at the body temperature of the cow. Then the milk is heated for a certain time and at a certain temperature that will kill the toughest disease germs present. After pasteurization, milk still contains thousands of living bacteria, but the disease-producing germs have been destroyed.

Methods of food preservation are used to prevent food from being spoiled and contaminated by bacteria. Canning sterilizes food and seals it up so that no bacteria can get in. Other methods are used that prevent the growth of bacteria in food products and prevent rapid spoilage. Some of these methods were devised by the ancients; some are relatively new. They include drying, salting, sugaring, pickling, fermenting, smoking, refrigeration, fast freezing, sterilization by heat or radiation, and the addition of chemical preservatives.

Noninfectious Diseases

The noninfectious diseases are not communicable because they are not caused by infectious organisms. These diseases have various causes other than germs. Table 15.6 provides a listing of some of the noncommunicable diseases.

Not all of the diseases listed in the table will be discussed in this chapter. Diseases resulting from endocrine insufficiency or oversecretion were summarized in Chapter 13, and some disorders that result from nutritional deficiency were summarized in Chapter 14.

TABLE 15.6
SOME NONINFECTIOUS DISEASES

Disease Type	Malfunction
Cardiovascular disease	Heart and blood vessels
Allergy	Hypersensitivity to certain substances
Genetic	Inborn defects
Emotional including drug-related psychoses	Psychological problems and mental stress
Occupational	Physical problems caused by work conditions
Poisoning	Illness caused by tissue toxins
Nutritional	Nutrient deficiencies in diet
Hormonal	Imbalance in endocrine secretions
Cancer	Wild reproduction of cells

CARDIOVASCULAR DISEASES

Problems of the heart and blood vessels are known as **cardiovascular diseases**. Because diseases of the heart and the circulatory system are leading causes of death in the United States, a great deal of research effort has been put forth for many years to determine the causes of heart defects and to find ways to prevent and/or cure these disorders. Heart defects trouble people of all ages.

Many infants are born with **congenital** heart defects. Sometimes the heart defect is a hole in the heart wall which divides the left and right ventricles. In some cases, the large arteries leaving the heart are in the wrong place. It also happens that a blood vessel that should have closed at birth failed to do so. Today many of these defects in children's hearts are corrected through procedures of **open heart surgery**. During the operation, the blood is sent through a machine that serves as a mechanical heart and lungs. After the real heart is repaired, the blood is returned to its normal pathways through the body.

In young and old alike, defective heart valves may cause trouble. Valves that are beyond self-repair are replaced by plastic ones.

Many things may go wrong with the adult heart. Sometimes the Purkinje fibers in the heart lose their ability to contract rhythmically and thus the pacemaking ability of the heart is impaired. This defect is treated by implanting in the patient's chest an artificial *pacemaker*, which makes possible appropriate heart stimulation.

Coronary artery disease prevents the proper blood supply from reaching the heart. The cells in heart muscle become damaged when not enough oxygen and nutrients reach them. Damaged heart cells cause heart attacks. If the damage to the heart is not too severe, a person will recover. Severe heart attacks cause death. High blood pressure (**hypertension**) is another major cause of heart attacks and strokes.

In recent years, some highly skilled heart surgeons have attempted exotic methods for dealing with patients who have badly diseased hearts. One of these methods has been heart transplant in which a diseased heart is exchanged for a healthy one. This technique has met with limited success in which the lives of transplant patients have been prolonged for a few months to several years. Recently, a patient was kept alive for about three months by means of an artificial heart. Neither of these methods has proven practical for large numbers of cases.

ALLERGY

Some people are very sensitive to substances that are quite harmless to most other people. The cells of a sensitive person produce antibodies to ward off whatever substance affects him/her. The antibodies become attached to the tissue cells, rendering the person *sensitized*.

Whenever that particular substance enters the body again, it reacts with the attached antibodies and damages the cells. These damaged cells prompt certain symptoms such as itching, sneezing, tearing eyes, red welts, large hives, fever, and a general feeling of not being well. The injured cells may release chemicals called **histamines**, which are responsible for the symptoms. **Anti-histamines** are substances that may neutralize the histamines and relieve the symptoms.

People may be sensitive to things that they inhale: pollen, dust, or powders. Other people may be sensitive to certain foods, such as wheat, eggs, milk, fish, nuts, chocolate, strawberries, or bananas. Still others react violently to drugs such as penicillin, aspirin, and streptomycin. A person may become *allergic* to a given substance at any time in life.

ASTHMA

Special attention must be given to the respiratory disorder **asthma**. The number of cases of asthma among children has risen so dramatically within the past two decades that research biologists and physicians worldwide have joined the investigation. Traditionally, asthma has been considered a form of allergic response to the more than 250 particles identified as asthma triggers: pollen, dust mite feces, cockroach scales, animal dander, smog, smoke, gases, molds, household cleaners, and many more!

In asthmatic persons, the release of histamine from cells in the nasal passages causes narrowing of air passages in the lungs. The narrowing of the bronchioles prevents air from entering the lungs, making it difficult for the person to breathe. Very severe attacks of asthma cause death.

Research during the 1990s (and continuing into this decade) has uncovered some new information about the possible causes of asthma. Dr. William Cookson in England started the asthma-research ball rolling in 1989, when he identified the gene on chromosome 11 that controls the immunoglobulin E response, a component of asthma. This was the first of several genes to be associated with the disorder. Dr. Cookson's work aroused a great deal of interest.

Worldwide, the results of research have shed new light on possible causes of asthma. Of particular note is the work done by research scientists Graham Cook and John Stanford. Their research indicates, rather unexpectedly, that too much cleanliness may be responsible for the increase in asthma among city-dwelling children. These scientists believe that the immune system has to be "primed" with small doses of germs throughout childhood. Children in rural areas gain exposure to certain bacteria when they play in the soil. Children living in overheated apartments, however, experience limited soil contact, but have more exposure to dust mites. This research indicates that soil bacteria may suppress the effect of triggers that stimulate the T-cell response that summons the release of histamine.

Figure 15.5 depicts the immune response. A type of immune cell found in tissue and blood is the **mast cell**. Mast cells are most significant in allergic reactions. When a substance that causes allergy enters the body, antibodies known as immunoglobulin E bind to it. These IgE antibodies also bind to mast cells. When the binding is complete, the mast cells produce a massive discharge of histamines that produce allergic symptoms. If bacteria or some other agents can prevent the binding of IgE to the mast cells, histamines are not released and therefore no allergic reaction can occur.

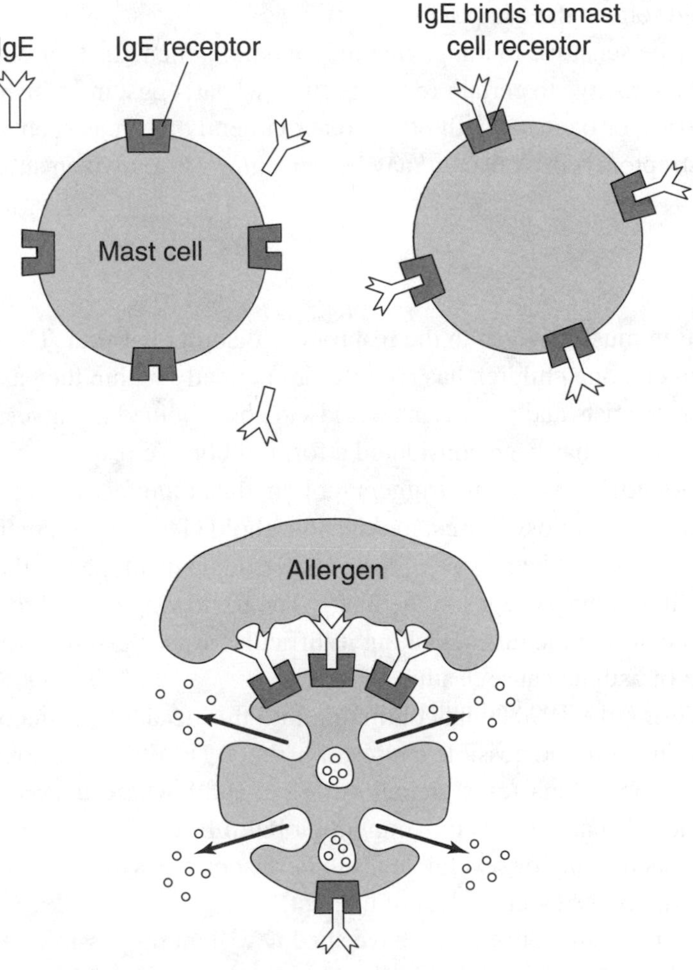

FIGURE 15.5 Immune cell response

OCCUPATIONAL DISEASES

The jobs of many people put them in contact with substances that present hazards to health. Fumes, dust, gases, vapors, fibers, and chemicals may have devastating effects on the health of a large number of people. Not too long ago it was discovered that asbestos causes lung cancer in those who have regular exposure to its fibers. Similarly, people who inhale the fibers of cotton or sugar cane may develop debilitating lung conditions and breathing problems. *Silicosis* is a type of lung disease occurring in miners and sandblasters who are in constant contact with rock dust. *Asbestosis* is a progressive inflammation of the lungs that results from breathing in fine fibers of asbestos. The disease occurs in workers in the construction trades and in miners.

Some occupations put workers in contact with poisons. Painters used to be subjected to lead poisoning from lead-base paints. (Today, most household paints are made without lead.) Lead is toxic to the body, settles in the brain and damages brain tissue. It is not unusual for farmers and exterminators to become poisoned by pesticides which they may absorb into the body by contact or by breathing. Recently, the pesticide Chlordane was shown to have adverse effects on the health of persons whose homes were contaminated by misuse of the product.

People in the medical professions are subject to occupational hazards. The ionizing radiation produced by X-ray machines can cause a wide variety of unfavorable conditions.

CANCER

ABNORMAL CELL GROWTH IN CANCER

Cancer is not a single disease. It is a whole constellation of different diseases that share a common characteristic. In every form of cancer there is an abnormal, uncontrolled growth of body cells. The cells that grow so wildly are not foreign cells that have invaded the body, but regular body cells that somehow have gone wrong (Figure 15.6). They grow in a disorganized fashion and compete with normal cells for space and nutrition.

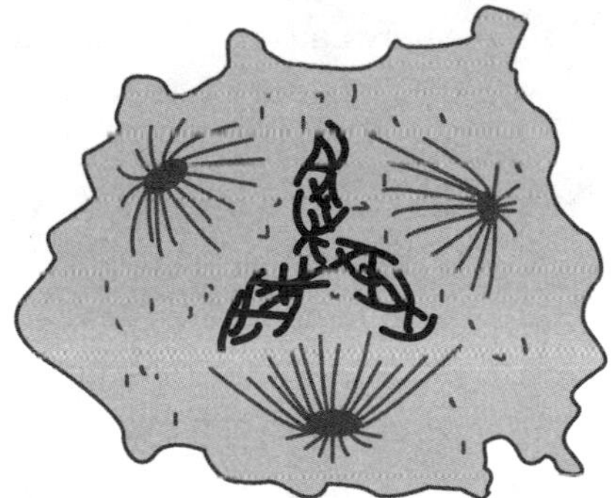

FIGURE 15.6 Cancer cell undergoing abnormal mitotic division

These wildly growing cells form a malignant mass called a **neoplasm** or a **cancer**. The malignant mass sends out fingers of cancerous cells that burrow into the normal tissues around it. As the invasion progresses, the normal tissue is gradually destroyed.

In the beginning cancer is localized in a particular part of the body. As time passes, fragments of malignant tissue separate away from the original malignant mass. They are carried off in the blood and lymph, reaching many new sites in the body. Wherever they remain, a new colony is established and grows like the original mass. This process in which cancer spreads is called **metastasis**.

There is significant evidence that the development of cancer involves changes both in genes and in cell signaling controls. Normally, cells have two types of genes that keep cellular reproduction (the cell cycle) under control. These genes are called proto-oncogenes and tumor suppressor genes; both cause cells to reproduce only when new cells are needed. If a proto-oncogene is mutated, it becomes an oncogene and triggers the cells' entry into a cycle of continual reproduction. If tumor suppressor genes are mutated, they become deactivated, allowing relentless reproduction. If these two genes are both altered, there is nothing to prevent tumor formation. There are additional cellular changes that have been found to contribute to cancer formation. These include altered apoptosis (cellular suicide), allowing cells to live long past their effective lifetime. Increased angiogenesis triggers blood vessel growth toward tumors, giving them plenty of nourishment. Binding proteins, such as cadherin, become less effective during the development of cancer, allowing cancerous cells to break away and metastasize to other organs.

All cancerous cells do not grow at the same rate. Some types of cancer grow quickly; others grow slowly. All cancer ultimately destroys normal tissue and kills the host. Every living cell has the potential to turn cancerous. Every variety of vertebrate animal is subject to the disease. So are humans of every race, nation, or habitat. Cancer cannot be blamed on modern living. Cancers have been found in Egyptian mummies and in fossils of ancient dinosaurs.

POSSIBLE CAUSES OF CANCER

Many causes of cancer have been investigated, based on statistical studies of the incidences of disease. A case in point is cigarette smoking. Extensive research has shown a very definite link between smoking and lung cancer (heart disease, too). Smokers also show a high incidence of cancer of the lips, mouth, and throat.

There is strong evidence that many types of chemicals are **carcinogens** (cancer-producing agents) or mutagens (which cause genetic mutations). Contact over a period of time with certain defoliants, insecticides, and chemical wastes has caused cancers of multiple varieties in large segments of the population. It is known that radium workers tend to get bone cancer. Workers in factories that produce aniline dyes often get cancer of the bladder.

Homo sapiens are constantly bombarded by various kinds of radiation: ultraviolet rays, X rays, gamma rays, cosmic rays, radioactive fallout, and the like. Any form of radiation, natural or human-made, can produce leukemia, bone cancer, skin cancer, or other types of cancer, such as lung, breast, and thyroid cancers. The effects of the atomic bomb explosions on Hiroshima and Nagasaki in Japan leave no doubt that a single exposure to a high dose of radiation can produce leukemia.

Research has also indicated that some forms of cancer are inherited. Retinoblastoma, a cancer of the eye found in young children, has been established as a type of cancer often passed on through the parents' genes. Genes have been found for specific types of breast, ovarian, and colon cancer; relatives of victims of these hereditary cancers can be genetically screened for the presence of cancer-causing genes.

There is strong evidence that certain forms of cancer are communicable, transmitted from one person to another by way of a virus. For example, cancer of the cervix has been found to be associated with certain types of HPV infection. The spread of Kaposi's sarcoma (a rare form of cancer associated with AIDS) indicates viral transmission. Electron microscope examination has revealed that some leukemia cells contain virus particles.

The best protection against any type of disease is to avoid contact with causative agents or chemicals whenever possible.

REVIEW EXERCISES FOR CHAPTER 15

WORD-STUDY CONNECTION

active immunity
aeration
AIDS
allergy
antibody
antigen
asthma
anti-histamine
carcinogen
cardiovascular diseases
contagious
coronary artery disease
cytotoxic T-cell
droplet infection
histamine
HIV
hypertension
immune system
immunization
index of fecal contamination
infections
lymph
lymphocyte
macrophage
mast cells
memory cells
metastasis
neoplasm
neutrophil
oncogene
passive immunity
pasteurization
phagocyte
plasma cells
regulatory T-cell
resistance
settling
sexually transmitted disease
tracoma
tubercle
vaccine
virulence
virus

SELF-TEST CONNECTION

PART A. Completion. *Write in the word that correctly completes each statement.*

1. Diseases that are caused by viruses, bacteria or other pathogens are known collectively as _____ diseases.

2. A disease that is spread from one person to another is said to be _____.

3. Rocky Mountain spotted fever is caused by an organism called a _____.

4. Polio is caused by _____ infection.

5. Most parasitic worms enter the body by way of contaminated _____.

6. The disease-producing ability of a pathogen is summed up by the term _____.

7. Tracoma is a disease of the eye caused by a _____.

8. Phagocytosis is best associated with _____ blood cells.

9. Fluid that bathes the body spaces is called _____.

10. Immune blood proteins are the _____ globulins.

11. The ability to resist disease is known as _____.

12. Cholera is spread through unclean _____.

13. Milk is _____ to prevent the spread of tuberculosis and Q fever.

14. The work of the Purkinje fibers can be taken over by an artificial _____.

15. During allergic attacks, cells release chemicals known as _____.

16. Lymphoid tissues in the head region include the tonsils and the _____.

17. The fluid of the immune system is known as _____.

18. Foreign proteins that cause an immune system to react are known as _____.

19. The purpose of the immune system is to protect the body against _____.

20. Lymphatic tissue known as Peyer's patches is located in the _____.

21. The HIV agent is best described as a (an) _____.

22. The bacterium that causes tuberculosis belongs to the genus _____.

23. When drinking water becomes contaminated with sewage, large numbers of the bacteria named _____ appear in the water.

24. The common name for *Phthirus pubis* is the _____.

25. Syphilis is a _____ transmitted disease.

PART B. Multiple Choice. *Circle the letter of the item that correctly completes each statement.*

1. The pathogens that are not microscopic in size are
   (a) virus particles
   (b) bacteria
   (c) worms
   (d) mycoplasmas

2. Transmission of germs by direct contact may be accomplished by
   (a) droplet infection
   (b) handshake
   (c) handling clothing
   (d) sneezing

3. An example of a disease caused by a bacillus is
   (a) meningitis
   (b) yaws
   (c) syphilis
   (d) tetanus

4. Viruses reproduce
   (a) in living cells
   (b) on dead organic matter
   (c) in quiet waters
   (d) in blood plasma

5. Ringworm is a disease of the skin that is caused by infection with a
   (a) protozoan
   (b) fungus
   (c) hookworm
   (d) bacterium

6. When germs break through the skin, certain chemicals are released from body cells that
   (a) kill the germs immediately
   (b) seal up the wound
   (c) cause the capillaries to expand
   (d) prevent pain and tenderness

7. Microorganisms that are enclosed in capsules are usually
   (a) phagocytic
   (b) anaerobic
   (c) harmless
   (d) pathogenic

8. By the time lymph leaves the lymph vessels and is returned to the blood, the lymph is
   (a) absolutely sterile
   (b) able to destroy bacteria
   (c) almost free of bacteria
   (d) crowded with bacteria

9. An injection of gamma globulins can give a person the type of immunity best described as
   (a) partial
   (b) temporary
   (c) lasting
   (d) active

10. The purpose of the aeration of water is to
    (a) remove debris
    (b) improve the color
    (c) prevent tooth decay
    (d) kill anaerobes

11. Allergy is caused by sensitizing
    (a) lymphocytes
    (b) antibodies
    (c) leucocytes
    (d) antitoxins

12. A malignant mass of cells is known as a (an)
    (a) metastasis
    (b) protoplasm
    (c) ectoplasm
    (d) neoplasm

13. Carcinogens are
    (a) virulent pathogens
    (b) immune proteins
    (c) contact poisons
    (d) cancer producers

14. A disease vector
    (a) carries the disease
    (c) cures the disease
    (b) causes the disease
    (d) controls the disease

15. The phagocytic activities of white blood cells were discovered by
    (a) Jenner
    (b) Metchnikoff
    (c) Salk
    (d) Sabin

16. Human body cells that ingest dead bacteria are the
    (a) pathogens
    (b) phagocytes
    (c) bacteriophages
    (d) antigens

17. Huge, amoeboid cells that engulf microorganisms that invade the body are the
    (a) eosinophils
    (b) stem cells
    (c) erythrocytes
    (d) macrophages

18. B-cells and T-cells are best classified as
    (a) lymphocytes
    (b) scavenger cells
    (c) neutrophils
    (d) macrophages

19. The humoral immune response produces
    (a) auxins
    (b) antigens
    (c) antibodies
    (d) autosomes

20. Lymphocytes are produced by
    (a) T-cells
    (b) stem cells
    (c) plasma cells
    (d) neutrophils

21. The organism *Treponema pallidum* is associated with the disease
    (a) asthma
    (b) scabies
    (c) syphilis
    (d) hepatitis

22. IgE antibodies bind to the immune cells known as
    (a) T-cells
    (b) mast cells
    (c) platelets
    (d) scavenger cells

23. Recent evidence indicates that asthma has a connection to a
    (a) gene
    (b) vacuole
    (c) mite
    (d) mosquito

24. Histamines are closely associated with responses known as
    (a) infectious
    (b) allergic
    (c) occupational
    (d) pathogenic

25. Cancer cells are markedly abnormal in that they
    (a) are generally infectious
    (b) are smaller than normal cells
    (c) are carried by ticks
    (d) reproduce wildly

PART C. Modified True-False. *If a statement is true, write "true" for your answer. If a statement is incorrect, change the* <u>*underlined*</u> *expression to one that will make the statement true.*

1. Malnutrition is an example of a <u>contagious</u> disease.

2. Human carriers <u>are</u> affected by the germs they carry.

3. Leprosy is a disease that is caused by a <u>coccus</u>.

4. Rickettsiae are microorganisms that live in the bodies of <u>protozoa</u>.

5. Smallpox is a <u>bacterial</u> disease that affects the whole body.

6. Hepatitis is an infection of the liver.
7. The ability of the body to ward off infection by disease organisms is known as virulence.
8. The virus of polio destroys lung tissue.
9. Lysozyme is a hormone present in tears.
10. Phagocytic white cells egest bacteria.
11. A boil is the result of a spreading infection.
12. Macrophages are very small white blood cells.
13. The name Edward Jenner is best associated with the disease polio.
14. Cowpox provides passive immunity.
15. Methods of food preservation include salting, freezing, and drying.
16. The organism that causes syphilis is classified as a rickettsia.
17. Antibodies are produced in enormous quantities by the T-cells.
18. Lymphocytes that have very long life spans are the macrophages.
19. Infection by the AIDS virus causes a shutdown of the circulatory system.
20. Cytotoxic T-cells kill cells that have foreign antibodies on their cell membranes.
21. *Giardia enterica* is best classified as a bacterium.
22. The HIV virus is noninfectious.
23. Animal dander may be an asthma inhibitor.
24. An IgE molecule is an isotope.
25. The virus of AIDS shuts down the circulatory system.

CONNECTING TO CONCEPTS

1. Why is Lyme disease classified as an infectious disease?
2. Thousands of germs enter the mouth daily, but few survive. Explain.
3. Describe the functions of the memory and plasma cells of the immune system.

4. When the T-cell lymphocytes fail to function, why does the entire immune system break down?

ANSWERS TO SELF-TEST CONNECTION

PART A

1. infectious
2. communicable or contagious
3. rickettsia
4. virus (viral)
5. food
6. virulence
7. virus
8. white
9. lymph
10. gamma
11. immunity (resistance)
12. water
13. pasteurized
14. pacemaker
15. histamines
16. adenoids
17. lymph
18. antigens
19. infection
20. small intestine
21. virus
22. *Mycobacterium*
23. *E. coli*
24. body louse
25. sexually

PART B

1. **(c)**
2. **(b)**
3. **(d)**
4. **(a)**
5. **(b)**
6. **(c)**
7. **(d)**
8. **(c)**
9. **(b)**
10. **(d)**
11. **(b)**
12. **(d)**
13. **(d)**
14. **(a)**
15. **(b)**
16. **(b)**
17. **(d)**
18. **(a)**
19. **(c)**
20. **(b)**
21. **(c)**
22. **(b)**
23. **(a)**
24. **(b)**
25. **(d)**

PART C

1. noninfectious
2. are not
3. bacillus
4. ticks, mites, lice, fleas
5. virus
6. true
7. resistance
8. nerve or muscle
9. an enzyme
10. engulf or ingest
11. local (contained)
12. large
13. smallpox
14. active
15. true
16. spirochete
17. plasma cells
18. memory cells
19. immune
20. antigens
21. protozoan
22. infectious
23. trigger
24. antibody
25. immune

CONNECTING TO LIFE/JOB SKILLS

Research work in biology is on-going and embraces innumerable areas of specialization. It is estimated that, beginning with the year 2000 and for each of the next 10 years, 11,200 graduates in the biological sciences will be needed to fill all of the available positions. If you are in high school and are considering a career in science, you should take a year of biology, a year of chemistry, a year of physics, several math courses, and an advanced placement science course in biology, chemistry, or physics. In addition, you must prepare yourself for higher level learning by developing your skills in reading, writing, and spelling. Furthermore, it is essential that you become skilled in the use of the computer. Most careers in science require advanced degrees.

Chronology of Famous Names in Biology

460 B.C. **Hippocrates** (Greece)—called the "father of medicine" because he was the first of the ancients to attempt scientific explanations of disease.

1527 **Jacques de Bothencourt** (France)—gave the name *venereal* to diseases transmitted by sexual intercourse. (This name has now been replaced by *sexually transmitted diseases.*)

1530 **Girolamo Fracastoro** (Italy)—gave the name *syphilis* to the heretofore unnamed sexually transmitted disease.

1796 **Edward Jenner** (England)—discovered that vaccination with cowpox renders immunity against smallpox.

1854 **Louis Pasteur** (France)—proved that microorganisms caused fermentation. In 1885 he developed the treatment for rabies.

1860 **Philippe Ricord** (United States)—determined that gonorrhea and syphilis are two separate diseases.

1872 **Moritz Kaposi** (Russia)—described the symptoms of Kaposi's sarcoma, which is now sometimes associated with AIDS.

1879 **Albert Neisser** (Germany)—isolated the organism that causes gonorrhea.

1880 **Charles Laveran** (France)—described the protozoan parasite of malaria.

1882 **Elie Metchnikoff** (Russia)—discovered the phagocytic activities of white blood cells.

1883 **Robert Koch** (Germany)—discovered the bacterium that causes tuberculosis.

1892 **Walter Reed** (United States)—discovered that yellow fever is transmitted by the *Aedes* mosquito.

1896 **Joseph Lister** (England)—developed aseptic techniques to prevent infection during and after surgery.

1906 **Howard T. Ricketts** (United States)—discovered that ticks infected with microorganisms (*Rickettsia*) cause Rocky Mountain spotted fever.

1908 **Paul Ehrlich** (Germany)—developed the first chemical substance—salvarsan—to inhibit the growth of the syphilis spirochete.

1911 **Emil von Behring** (Germany)—discovered that immunity to diphtheria can be given by injecting a person with antitoxin.

1929 **Alexander Fleming** (England)—discovered the growth-inhibition effect of penicillin on staphylococci.

1932 **Gerhard Domagk** (Germany)—discovered that prontosil is effective against *Streptococcus* infections.

1933 **George Dick** (United States)—devised a test to determine susceptibility to scarlet fever and discovered the germ that causes that disease.

1933 **Bela Schick** (United States)—developed the Schick test for susceptibility to diphtheria.

1935 **Wendell Stanley** (United States)—isolated the virus that causes tobacco mosaic disease.

1943 **Howard Florey** (United States)—developed an efficient method of mass producing penicillin from the mold *Penicillium notatum*.

1944 **Peter Medawar** (England)—did the basic research which brought to light the problems of tissue and organ transplant techniques and immunology. **1967**—discovered acquired immune tolerance factors that oppose tissue transplants.

1948 **Selman Waksman** (United States)—was the first to isolate streptomycin from *Streptomyces griseus*.

1954 **Jonas Salk** (United States)—developed a vaccine effective against poliomyelitis.

1963 **Albert Sabin** (United States)—developed oral polio vaccine.

1965 **K. Ishizaka** (United States)—discovered gamma globulin E (IgE), which is present in normal serum in small amounts.

1980 **Michael Bishop** and **Harold Varmus** (United States)—discovered that virus-encoded oncogenes originate in eukaryotic cells

1985 **Flossie Wong-Staal** (United States)—did pioneering research on the structure of the AIDS virus.

1989 **Samuel Broder** (United States)—played a key role in developing AZT as the first effective treatment for AIDS.

1989 **William Cookson** (England)—was the first to identify a gene that functions in the immune response to asthma.

1997 **Stanley Prusiner** (United States)—won the Nobel Prize in Physiology and Medicine for the discovery of prions, infectious proteins.

1999 **Ian Lipkin** (United States) and colleagues—identified the total genome of the West Nile virus that caused the encephalitis outbreak in New York. (His colleagues are Xi-Yu Jie, Thomas Brieze, Ingo Jordan, Andrew Rambadt, Hen Chang Chi, John S. Mackenzie, Roy A. Hall, and Jacqui Scharret.)

2007 **Mario R. Capecchi**, **Oliver Smithies** (United States), and **Sir Martin J. Evans** (United Kingdom)—together won the Nobel Prize in Physiology and Medicine for their development of gene targeting techniques that introduced embryonic stem cells to initiate specific genetic modifications ("knockout" genes). Experimental models using knockout mice have become indispensable tools in medical research.

16

Heredity and Genetics

WHAT YOU WILL LEARN

In this chapter you will review how traits are passed from one generation to another by way of genes and nucleic acids.

SECTIONS IN THIS CHAPTER

- Classical Principles of Heredity
- Modern Genetics
- Some New Directions in Genetics
- The Retrovirus of AIDS—Genetic Coding in RNA
- Importance of the Heredity Concept to Humans
- Review Exercises for Chapter 16
- Connecting to Life/Job Skills
- Chronology of Famous Names in Biology

Classical Principles of Heredity

The science of **genetics** has made tremendous advances since the 1960s, but the foundations were laid at the beginning of the 20th century. Before discussing the modern findings of genetics, we should review the classical principles upon which the science is founded.

MENDEL'S LAWS OF HEREDITY

Gregor Mendel, an Austrian monk, began the first organized and mathematical study of how traits are inherited. Using the garden pea as the test organism, Mendel identified seven different traits that were easily recognizable in this self-pollinating plant. He called each of these traits a **unit character**. Mendel began his work in 1856 when little was known about chromosomes and their functions in cell division and the concept of the gene had not yet been developed.

Mendel not only identified characteristics that seemed to be inherited, but for each unit character he identified an opposite trait. For example, if the unit character was height, the opposite traits were short and tall. If the unit character was seed coat color, the opposite traits were yellow and green.

Based on his observations of crosses that he made in pea plants by means of hand pollination, Mendel formulated three major laws or principles of inheritance. It should be noted that he kept very careful records of his experiments and converted his results into mathematical ratios.

THE LAW OF DOMINANCE

If two organisms that exhibit contrasting traits are crossed, the trait that shows up in the first filial generation (F_1) is the dominant trait. For example, when a purebred tall pea plant is crossed with a short pea plant, all of the offspring will be tall. The offspring will not be pure tall, however, and are therefore known as **hybrids**. The factor for shortness is hidden. We say today that the **phenotype** of the F_1 plants is tall. The term *phenotype* refers to traits that we can see. The **genotype** or genetic makeup of these plants is said to be hybrid or **heterozygous**, meaning "mixed."

When organisms with contrasting traits are crossed, the trait that shows up in the F_1 generation is called the *dominant* trait. The trait that is hidden is called the *recessive* trait. Since Mendel did not have the concept of the gene, he called the conditions that make for dominance and recessiveness **factors**. From the results of crosses made with the garden pea, he concluded that (a) two factors determine a characteristic, (b) two recessive factors are needed in order for the recessive characteristic to appear, and (c) one dominant factor and one recessive factor result in offspring that have the dominant trait.

THE LAW OF SEGREGATION

When hybrids are crossed, the recessive trait segregates out at a ratio of three individuals with the dominant trait to one individual with the recessive trait. The 3:1 ratio is known as the *phenotypic* ratio, because it refers to the traits that can be seen and not those factors hidden in the germplasm. The hybrid cross is also known as the F_1 cross and the offspring produced by this cross are known as the second filial generation, or F_2. In terms of modern knowledge, the F_2 generation also produces another type of ratio called the *genotypic ratio*, which refers to *gene* makeup. The genotypic ratio is 1:2:1, translated into 1 *homozygous* dominant (pure) individual: 2 *heterozygous* (hybrid) individuals: 1 *homozygous* recessive. Only the homozygous recessive shows the recessive trait.

THE LAW OF INDEPENDENT ASSORTMENT

Mendel believed that each trait is inherited independently of others and remains unaltered throughout all generations. We now know that Mendel's "factors" are genes that are linked together on chromosomes and that if genes are on the same chromosome, they are inherited together.

THE CONCEPT OF THE GENE

Around 1911, Thomas Hunt Morgan introduced the tiny fruit fly *Drosophila melanogaster* (Figure 16.1) as the new experimental organism for work in the field of heredity. The experimental work of Morgan resulted in the discovery that the chromosome is the means by which hereditary traits are transmitted from one generation to another. Morgan's chromosome theory of inheritance includes the concept that chromosomes are composed of discrete units called *genes*. Genes are the actual carriers of specific traits and move with the chromosomes in mitotic and meiotic cell divisions. Morgan further proposed that genes control the development of traits in each organism. When genes change, or *mutate*, the traits they control also change.

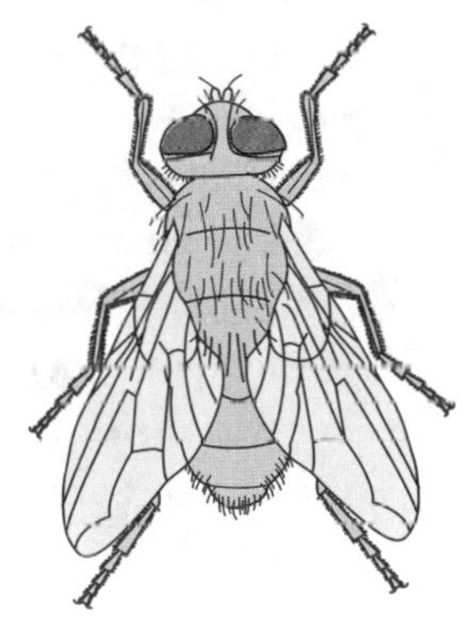

FIGURE 16.1 *Drosophila melanogaster*: fruit fly

SOME GENETIC SHORTHAND

A combination of the theories of Mendel and Morgan led to the development of a system of genetic "shorthand," which is used to represent the genetic makeup of a trait and to show rather simply what happens when organisms with specific traits are crossed. In this shorthand capital letters are used to indicate dominant genes, lowercase letters for recessive genes. A **Punnett square** is a diagrammatic device used to predict the genotypic and phenotypic ratios that will result when certain gametes fuse. Remember that as a result of meiosis each gamete has only one half the number of chromosomes that are in the somatic cells.

PROBLEM:

In fruit flies, long wing (*L*) is dominant over vestigial wing (*l*). What is the result of a cross between two flies that are heterozygous (*Ll*) for wing length?

SOLUTION:

	Male	×	Female
Parents:	*Ll*		*Ll*
Gametes:	Ⓛ ⓛ		Ⓛ ⓛ

Punnett square

	L	*l*
L	*LL*	*Ll*
l	*Ll*	*ll*

F_2

LL — 1 homozygous dominant long-winged fly
Ll — 2 heterozygous dominant long-winged flies
ll — 1 homozygous recessive short-winged fly

INTERMEDIATE INHERITANCE

Geneticists have discovered that in many cases a trait is not controlled by a single gene, but rather by the cooperative action of two or more genes. There are many instances in which Mendel's "law of dominance" does not hold true. A case in point is what was once called **blending inheritance**, or **incomplete dominance**. It is now known as **codominance**. When red-flowered evening primroses are crossed with white-flowered primroses, the hybrids are pink. Neither red nor white color is dominant and therefore the result is a blend. In sweet peas, the expression of red or white flowers is dependent upon two genes: a *C* gene for color and an *R* gene for enzyme. If *C* and *R* are inherited together, the flower color is red. If the dominant *C* is missing, the flower is white; if the dominant *R* is missing, the flower is also white. Therefore white flowers are the result of several different genotypes: *ccrr, ccRR, CCrr, Ccrr*.

CROSSING OVER

Genes are linked on chromosomes and are inherited in a group on a particular chromosome. However, linkage groups are broken by *crossing over*, a phenomenon that may occur during meiosis when homologous chromosomes are intertwined during synapsis. It is at this time that chromosomes may exchange homologous parts and thus assort linkage groups.

MUTATIONS

It was stated before that genes can change and that changes in genes are known as **mutations**. As a rule, mutations are usually recessive and they are usually harmful. Mutations usually occur at random and spontaneously. However, mutations may be induced by radiation or by chemical contamination.

There are several types of mutations. A loss of a piece of a chromosome is known as a **deletion**. The genes on the broken off piece of chromosome are lost. Sometimes a broken piece of chromosome sticks on to another chromosome, thus adding too many genes; this type of mutation is known as **duplication**. Sometimes a piece of chromosome becomes rearranged in the chromosome where it belongs, thus changing the sequence of the genes on that chromosome; this is known as an **inversion**, and it prevents gene-for-gene matching when chromosomes line up during meiosis. *Point mutations* are changes in individual genes.

Polyploidy is a condition in which cells develop extra sets of chromosomes: 3*N*, 4*N*. These extra sets of chromosomes change the characteristics of organisms. Usually polyploidy occurs in plants. Breeders may treat special plants with a chemical such as colchicine, which prevents the division of the cell after the nucleus has divided. Polyploid fruits and flowers are quite large.

SEX DETERMINATION

Among the discoveries made by Morgan while working with fruit flies was the existence of paired **sex chromosomes**. In males the two chromosomes of the pair are different in size and shape. One of these chromosomes is large and rod-like. The other chromosome is small and hook-shaped. The larger, rod-shaped chromosome is called the **X chromosome**; the smaller, hook-shaped one, the **Y chromosome**. Sex chromosomes determine the sex, not only of fruit flies, but also of all animal species (Figure 16.2). In human beings, there are 22 pairs of **autosomes**, chromosomes that affect all characteristics except sex determination. One pair of chromosomes determines the sex of an individual (Table 16.1). In females, the sex chromosomes are designated as *XX*. In males, the sex chromosomes are *XY*.

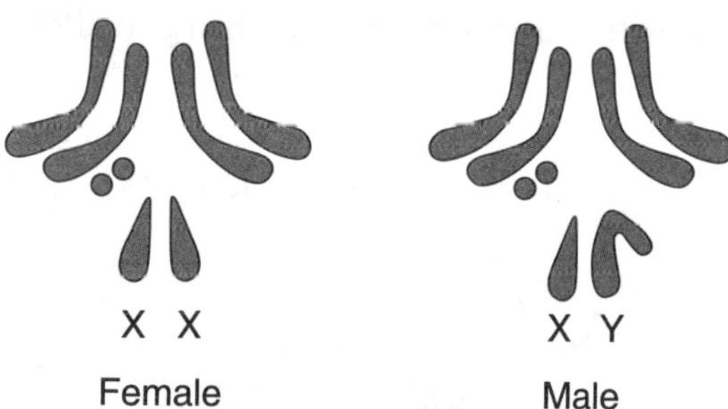

FIGURE 16.2 Chromosomes in the female and male fruit fly. Note the shapes of the sex chromosomes.

TABLE 16.1
SEX DETERMINATION

From mother (egg)	From father (sperm)	Offspring (zygote)
X chromosome	X chromosome	XX=female
X chromosome	Y chromosome	XY=male

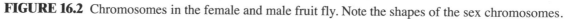

SEX-LINKED TRAITS

Certain disorders are sex-linked, usually passed from mother to son by a defective gene on the *X* chromosome; red-green color blindness is one such sex-linked trait that is found fairly frequently in males and hardly ever in females.

Hemophilia, another sex-linked trait that directly affects males, is a disorder of the clotting mechanism in the blood. Hemophiliacs experience spontaneous internal bleeding and blood capillary breakdown. Duchenne **muscular dystrophy**, a disease in which there is gradual destruction of muscle cells, is a third sex-linked disease passed from mother to son. A **carrier** is a person who carries the defective gene on a sex chromosome but is not afflicted with the disease.

The sex-determining chromosomes in females are XX; in males, XY. If a woman has a defective gene on an X chromosome, there is a reasonable chance that the defective chromosome will be passed on to her son. A son having the XY chromosome has no counteracting gene for the defect on the Y chromosome. Females having XX chromosomes are more likely to carry on one of the X chromosomes a dominant allele that masks the expression of the recessive allele. An **allele** is an alternative form of a gene that occupies a given place on a chromosome.

MULTIPLE ALLELES

As mentioned above, *alleles* are two or more genes that have the same positions on homologous chromosomes. Alleles are separated from each other during meiosis and come together again at fertilization, when homologous alleles are paired, one from the sperm cell and one from the egg cell. Two or more alleles determine a trait.

The inheritance of some characteristics cannot be explained by the action of a single pair of alleles. It has been shown experimentally that multiple alleles determine certain traits. However, within each cell no more than two alleles are present.

The inheritance of the ABO blood group in humans is an example of the existence of multiple alleles: I^A , I^B, and i. In this example, alleles I^A and I^B are codominant with each other and i is recessive to both I^A and I^B. Table 16.2 shows blood types and possible genotypes.

TABLE 16.2
BLOOD GENOTYPES

Blood Type	Genotype
A	I^AI^A or I^Ai
B	I^BI^B or I^Bi
AB	I^AI^B
O	ii

Modern Genetics

THE ROLE OF NUCLEIC ACIDS

DNA is an important part of the chromosome structure of all cells. DNA is a **nucleic acid**, as is RNA. The unit of structure and function in the nucleic acid is called a **nucleotide**. A nucleotide is composed of a *phosphate group*, a five-carbon sugar, and a protein base. If the five carbon sugar is *ribose*, the nucleic acid is **ribonucleic acid (RNA)**. If the five-carbon sugar is deoxyribose, then the nucleic acid is *deoxyribose nucleic acid* (DNA). Figure 16.3 is a diagrammatic presentation of a nucleotide.

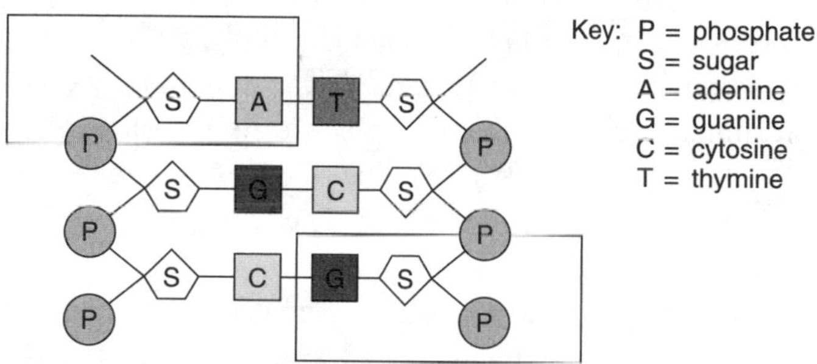

FIGURE 16.3 Nucleotides are the units on which DNA molecules are built

The protein bases in nucleic acids are ring compounds. Those bases with single rings are **pyrimidines**. Bases with double rings are **purines**. The pyrimidines in nucleic acid are *thymine*, *cytosine*, and *uracil*. The purines are *adenine* and *guanine*. The four bases that make up the DNA molecule are adenine (A), guanine (G), thymine (T), and cytosine (C). The four bases that make up the RNA molecule are adenine (A), guanine (G), cytosine (C), and uracil (U).

In the early 1950s, James Watson, an American, and Francis Crick, an English investigator, unraveled the structure of DNA. The Watson-Crick model of DNA, as it has come to be known, indicates that the DNA molecule is shaped like a double helix. It consists of two long chains of nucleotides turned around each other in the shape of a double spiral. Figure 16.4 shows the DNA molecule as a double helix. Notice that this diagram resembles a stepladder that has curved sides and straight rungs. The sides of the DNA molecule consist of alternate phosphate-sugar groups. The rungs of the "ladder" consist of protein bases joined by weak hydrogen bonds. It is known that adenine and thymine link together while guanine and cytosine link together.

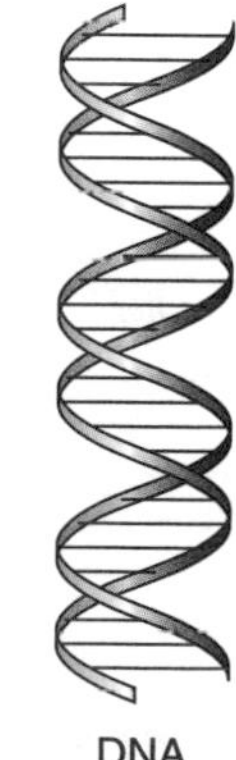

FIGURE 16.4
DNA: double helix

The RNA molecule is usually single stranded and much smaller than DNA. RNA is synthesized in the nucleus and functions in the cytoplasm where it controls the synthesis of proteins by the ribosomes.

HOW DNA FUNCTIONS

Deoxyribonucleic acid is able to function as genetic material because it has unusual characteristics. It is stable, can make more of itself, and can control the production of enzymes. DNA is stable as demonstrated by its ability to pass hereditary traits from one generation to another unchanged. However, DNA does change occasionally. Such changes are responsible for **mutations**.

DNA REPLICATION

DNA can make more of itself in a process called **replication**. Replication refers to a duplication of molecules. Prior to the onset of cell division (mitosis and meiosis) DNA molecules replicate in a way that is at the same time both simple and precise. The double-stranded helix unwinds, forming a structure that resembles a straight-sided ladder with horizontal rungs (Figure 16.5). The rungs of the ladder are formed by the joining of a pyrimidine molecule with a purine through a weak hydrogen bond. The bond breaks and the two strands "unzip" between the base pairs. Complementary nucleotide chains are joined to the free base ends in the unzipped strands. Figure 16.6 illustrates replication of DNA. As the result of the incorporation of free nucleotides into the unzipped strands, two new molecules of DNA are formed that are identical to each other and to the original molecule.

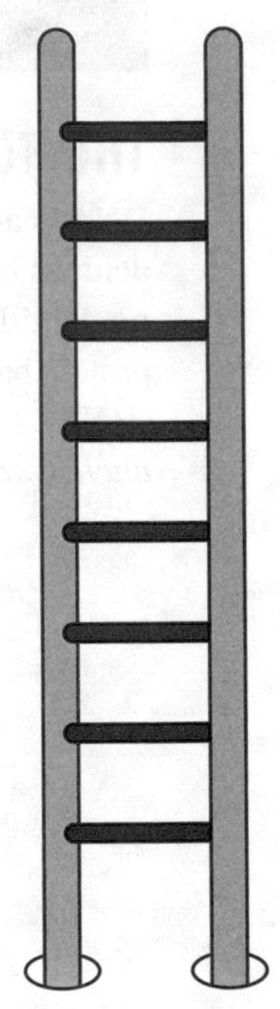

FIGURE 16.5 Straight-sided DNA

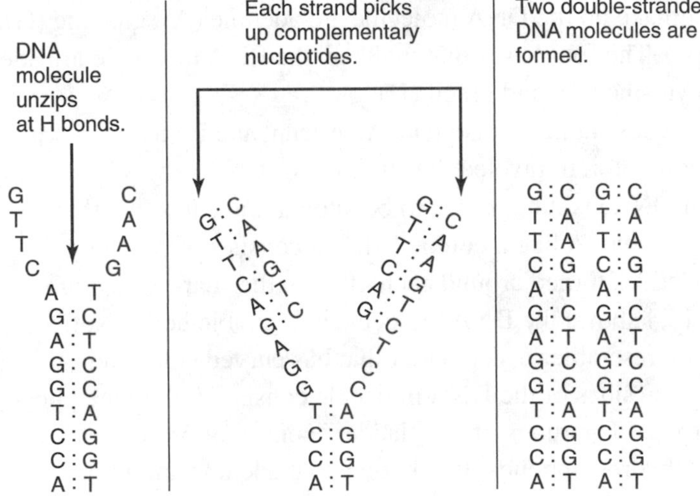

FIGURE 16.6 Replication of DNA

GENETIC CODE

The DNA molecule carries coded instructions for controlling all functions of the cell. At the present time, scientists know most about the functions of the **genetic code** that controls protein synthesis. They have determined that triplet combinations of bases code for each of 20 amino acids. The coded sequence of amino acids determines the formation of different types of proteins.

Transcription

The code for proteins is present in messenger RNA (mRNA) molecules, which are complementary to DNA molecules. For example, suppose that a portion of a DNA molecule carries a code such as AAC GGC AAA TTT. Its mRNA complement will be as follows:

```
AAC GGC AAA TTT
 |   |   |   |
UUG CCG UUU AAA.
```

The genetic code for a given polypeptide is contained in a DNA segment and is copied into a molecule of messenger RNA by a process known as **transcription**. Table 16.3 shows the genetic codes for 20 amino acid molecules. At the end of the table there is a set of three bases labeled "stop." These bases control the termination of transcription.

TABLE 16.3
GENETIC CODES FOR TWENTY AMINO ACIDS

Amino Acid	Code
Alanine	GCU, GCC, GCA, GCG
Arginine	CGC, CGU, CGA, CGG, AGA, AGG
Asparagine	AAU, AAC
Aspartic acid	GAU, GAC
Cysteine	UGU, UGC
Glutamic acid	GAA, GAG
Glutamine	CAA, CAG
Glycine	GGU, GGC, GGA, GGG
Histidine	CAU, CAC
Isoleucine	AUU, AUC, AUA
Leucine	UUA, UUG, CUU, CUC, CUA, CUG
Lysine	AAA, AAG
Methionine	AUG
Phenylalanine	UUU, UUC
Proline	CCU, CCC, CCA, CCG
Serine	UCU, UCC, UCA, UCG, AGU, AGC
Threonine	ACU, ACC, ACA, ACG
Tryptophan	UGG
Tyrosine	UAU, UAC
Valine	GUU, GUC, GUA, GUG
stop	UAA, UAG, UGA

Forms of RNA

There are three kinds of RNA. You just read that **messenger RNA**, or **mRNA**, is assembled from a strand of DNA. Its function is to carry the genetic code for a particular protein.

The second type of RNA, **transfer RNA** or **tRNA**, is also formed by DNA. The tRNA molecule is so shaped that it will pick up only one specific kind of amino acid. Figure 16.7 provides a diagrammatic representation of tRNA. The molecule is shaped like a clover leaf, with a short end (5') and a long end (3'). Amino acids are accepted at the 3' end. At the side opposite from the molecule's ends, there is a region labeled **anticodon**. The anticodon is the site where tRNA attaches to the mRNA codon. A **codon** is a triplet that specifies an amino acid base. Since there are 20 different amino acid molecules, there are 20 different forms of tRNA molecules, one for each amino acid. The function of tRNA is to bring a particular amino acid to a place specified by mRNA.

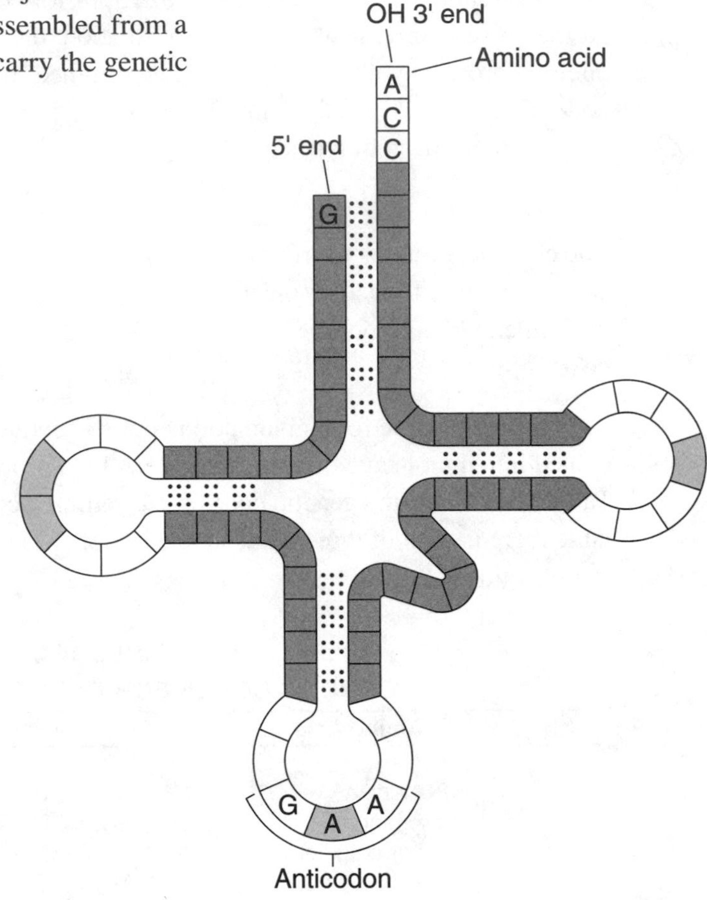

FIGURE 16.7 Diagrammatic representation of a tRNA molecule. The 3' end accepts a particular amino acid. The anticodon joins with the mRNA codon for which it is specified.

The third type of RNA, **ribosomal RNA** or **rRNA**, is assembled by DNA in the *nucleolus*, a small, rounded structure within the nucleus. Ribosomal protein is synthesized in the cytoplasm and then moves into the nucleolus. In the nucleolus, two subunits of rRNA join with the ribosomal protein to form a ribosome. The subunits of rRNA differ in size: one is larger than the other. After formation, the ribosomes move out of the nucleolus into the cytoplasm. The synthesis of each of the three kinds of RNA takes place in the nucleus of the cell. You can see that the nucleus is the site of biochemical activities necessary for the life of this cell and for the lives of cells in multicellular organisms.

The mRNA, carrying the polypeptide code, now moves from the nucleus to the cytoplasm. The mRNA attaches itself to several ribosomes, each having its own ribosomal RNA. Specific tRNA molecules bring to the ribosomes their own kinds of activated amino acids. Transfer RNA molecules that fit the active sites of mRNAs on the ribosomes temporarily attach to them. As a result, amino acids are lined up in the proper sequence.

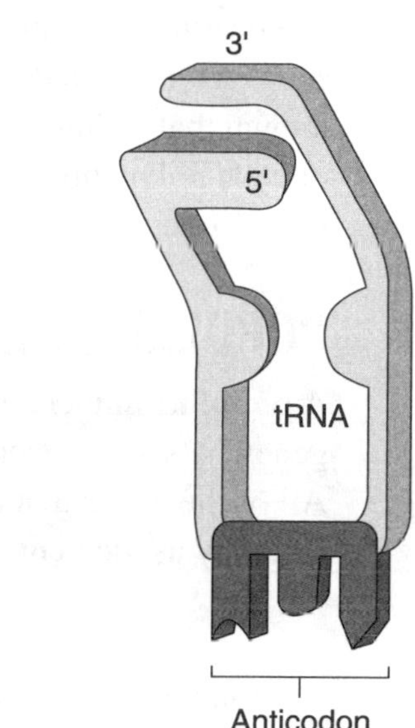

FIGURE 16.8 Graphic representation of a transfer RNA molecule

Translation: Assembly of a Polypeptide

A graphic representation of a tRNA molecule is given in Figure 16.8, which is a "close-up" view of this type of molecule. Note that the site on the 3' end picks up and carries a specified amino acid. Note also the anticodon, which is specific for an mRNA codon. Figure 16.9 is a schematic drawing showing how mRNA and tRNA fit into a ribosome during the making of a polypeptide. Messenger RNA brings the genetic code for a polypeptide to the ribosome. Transfer RNA reads the code. The ribosome puts it all together.

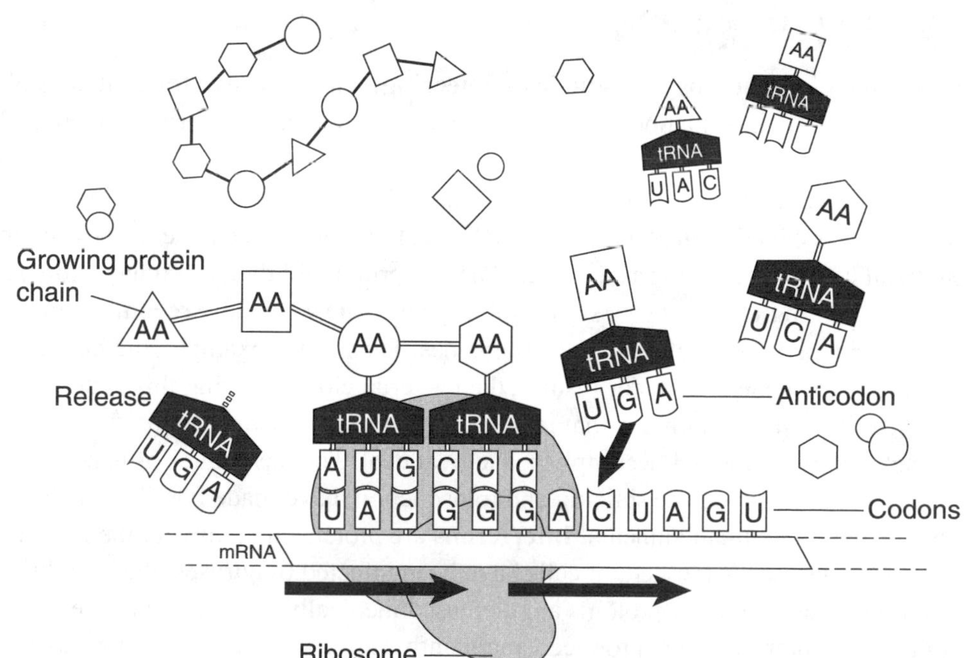

FIGURE 16.9 Representation of the way in which tRNA and mRNA fit into a ribosome

One Gene—One Polypeptide Hypothesis

In 1941 Beadle and Tatum used the red bread mold *Neurospora crassa* to find out how genes influence the synthesis of enzymes. As a result of their work the "one gene—one enzyme" hypothesis was formed. This hypothesis suggests that the synthesis of each enzyme in a cell is controlled by the action of a single gene. At present, it is known that a single enzyme may be composed of several polypeptides. The synthesis of each polypeptide is governed by a different gene. Hence, the new hypothesis, "one gene—one polypeptide," is considered to be more accurate.

POPULATION GENETICS

A **population** includes all members of a species that live in a given location. Modern geneticists are concerned about the factors in populations that affect gene frequencies. All of the genes that can be inherited (heritable genes) in a population are known collectively as the **gene pool**. The **Hardy-Weinberg principle** uses an algebraic equation

$$p^2 + 2pq + q^2 = 1$$

to compute the gene frequencies in human populations. The conditions set by the Hardy-Weinberg principle for determining the stability of a gene pool are as follows: large populations, random mating, no migration, and no mutation.

Some New Directions in Genetics

GENETIC ENGINEERING

Through the **biotechnology** of **genetic engineering**, scientists have found it possible to transfer genetic information from one organism to another. In a series of technical steps known as **gene splicing**, geneticists are able to remove a specific gene from the cells of a human and splice it within the circular DNA strand from a bacterium.

In a process called **transduction**, this spliced strand of DNA can be introduced into a bacterial cell, using a virus as a carrier. The bacterium will then reproduce **clones**, organisms bearing identical genes. The spliced ring of DNA is now **recombinant DNA**, bearing a human gene and bacterial genes. Then if, for example, the human gene carries the genetic code for insulin, the bacterial clones bearing this gene will synthesize insulin, the hormone that regulates sugar metabolism.

Genetic engineering has accomplished a great deal. The techniques of genetic engineering, which include gene splicing and transduction, have made possible medicinal drugs for treating human ailments. **Interferons** are proteins that prevent the replication of RNA viruses. After certain cells have been infected by viruses, they produce interferons. These immune proteins are then used medically to prevent and treat certain diseases. Interferons are produced abundantly and inexpensively by means of recombinant DNA.

The technology of producing recombinant DNA has proved very useful in the:

1. production of insulin, using bacterial cells;
2. production of interferons;
3. production of human growth hormone;
4. production of enzymes for the cheese industry;
5. development of processes for safer and more efficient cleanup of organic wastes generated in the food-processing industries;
6. treatment of hepatitis.

CLONING

A **clone** is a population of cells, or whole organisms, that has descended from an original parent cell, stimulated to reproduce by asexual means. Therefore, the clone is genetically identical to the original cell. Over the past four decades, *tissue culture* experiments have been carried out with plants in which single, non-embryonic cells have been induced to develop along a pathway of a fertilized egg. The results with such species as carrot, African violet, Boston fern, and Cape Cod sundew indicate that a whole plant can be propagated from a single nonreproductive cell.

During the 1990s, the techniques used in cloning experiments became more advanced. There are indications that frogs, mice, and rats were cloned in research laboratories. In 1997 Ian Wilmut, a research scientist working for PPL Therapeutics in Scotland, and his research team transplanted the nucleus from a mammary cell of a sheep into an unfertilized egg cell of another sheep. The experimental egg cell was then implanted in the uterus of a host sheep, where it underwent cleavage and gestation. A female sheep genetically identical to the mother, that is, a clone, was produced and affectionately named Dolly. In the year 2000, cloning technology took another step forward. Five identical piglets were produced through cloning (Figure 16.10). It may be possible one day to clone pigs that could provide organs for human transplants.

FIGURE 16.10 Five cloned piglets with identical genes *Used with permission of AFP*.

GENE THERAPY

Research geneticists are trying to cure genetic diseases by means of **gene therapy**. In theory, for any disease caused by a single defective gene, there should be a way to replace that gene with a normal, functional one by way of a normal cell using recombinant DNA. Thus far, however, gene therapy has not produced many positive results. After several discouraging episodes, a bone marrow transplant in a boy suffering from sickle-cell anemia has effected a remission of the disease.

The Retrovirus of AIDS–Genetic Coding in RNA

The human immunodeficiency virus (HIV), the virus that causes AIDS, is a *retrovirus*, that is, a virus that carries its genetic information coded in ribonucleic acid (RNA) instead of in deoxyribonucleic acid (DNA). It is usual for RNA to carry only the genetic codes that build proteins. Retroviruses, however, produce the enzyme **reverse** transcriptase, which enables the virus to carry out its **replication** processes. Through replication, HIV makes more of itself.

When HIV infects (enters) the type of white blood cell known as a CD-4 lymphocyte, the RNA of the virus captures the DNA of the invaded cell. Now reverse transcriptase goes to work, forcing the DNA of the CD-4 cell to follow the coded instructions contained in the viral RNA. This results in the production of HIV viruses within the CD-4 cell. When the CD-4 cell can no longer hold the number of viruses produced, the cell membrane bursts, releasing numerous HIV viruses into the blood—and so the process continues. These viruses infect other CD-4 lymphocytes, ultimately destroying them.

To summarize, under usual circumstances, the information of heredity is coded in DNA molecules. The genetic codes for protein synthesis are contained in DNA and then transferred to RNA. Through a series of steps involving three kinds of RNA, proteins useful to the cell are synthesized. The virus of AIDS, however, reverses the process. Genetic codes are transferred from RNA to DNA, forcing the captured cell to make HIV viruses instead of proteins.

Research workers have found that most retroviruses that infect animal cells have three genes, which have been named *pol*, *env*, and *gag*. *Env* contains the genetic blueprint that determines the structure of the viral outer coat. *Gag* holds the code for the internal structure of the virus. *Pol* provides instructions for the code for the characteristic enzymes of the virus.

Unlike other retroviruses, the virus of AIDS contains *eight* genes. The functions of four of these genes are unknown. However, a gene that research workers have named *tat* codes for the enormous production rate of the HIV virus, which accounts for the huge infecting power of the AIDS virus. At present virologists are hard at work trying to unravel the mysteries of the other four HIV genes.

Importance of the Heredity Concept to Humans

The concept of heredity has always been important to humans. Ancients living 5000 years ago in Babylonia and Assyria carried out the process of pollination to make their date palms produce fruit. Since, however, they did not understand the mechanisms of fertilization and genetics, they turned to superstitious practices to solve agricultural and other problems.

PLANT AND ANIMAL GENETICS

Plant and animal breeders throughout history have engaged in practices to improve their crops and their domesticated animals. Using the method of **selection**, they mated organisms based on observed characteristics that were desirable to the breeder. But selection has its limits. Species cannot be improved beyond their genetic capabilities. Now, species improvements are made through hybridization, in which varieties of organisms are crossed (mated) in the hope that the favorable characteristics of each will show up in the offspring. For example, a cross between Texas longhorned cattle and the Indian Brahman bull has resulted in hybrids that are resistant to infections and produce good-quality meat. Many varieties of hybrid sweet corn have been produced by selective breeding and carefully planned genetic crosses between strains having desired characteristics. These new strains of corn have improved texture, delicious taste, and resistance to certain fungus diseases that infect corn plants.

HUMAN GENETICS

On page 400 you read about three sex-linked disorders—color blindness, hemophilia, and Duchenne muscular dystrophy—that occur in humans. Table 16.4 summarizes three other human genetic diseases. Figure 16.11 shows a **genotypic pedigree** of Queen Victoria's descendants in regard to the sex-linked disease hemophilia. Notice how the gene for this disease passed from female carriers to sons.

In 1956 the work of Joe Hin Tjio and Albert Levan provided a method of counting human chromosomes accurately. They developed the technique of producing a *karyotype*, which is a photograph of matched chromosome pairs. Shortly after this work was done, geneticists were able to point to the cause of Down syndrome, a condition of mental retardation and physical handicap caused by a tripling of chromosome number 21. When chromosomes cause a medical issue, it is called a chromosomal disorder, and when individual genes cause the problem, it is a genetic disorder.

TABLE 16.4
SOME HUMAN GENETIC DISORDERS

Disorder	Description
Phenylketonuria	Results from the inability of a gene to synthesize a single enzyme needed for the normal metabolism of the amino acid phenylalanine. Phenylketonuria renders a child mentally retarded. Testing the urine of newborns allows for corrective dietary treatment to prevent the disease.
Sickle-Cell Anemia	Results from the inheritance of two mutant genes for abnormal hemoglobin, which causes the sickling of red blood cells. The afflicted person experiences severe pain caused by obstructed blood vessels and anemia caused by the decreased hemoglobin content of the sickle-shaped cells.
Tay-Sachs Disease	Results from the inheritance of two recessive genes that cause a malfunctioning of the nervous system whereby nerve cells are destroyed. The deterioration in nerve tissue is caused by the accumulation of fatty material due to inability to synthesize a specific enzyme. The disorder greatly shortens the life span; it is rare for afflicted children to live beyond the age of 6 years.

The study of human disease caused by genetic disorders has resulted in some remarkable findings. For example: a large number of human disorders (club foot, cleft lip and palate, spina bifida, and water on the brain—to name a few) are caused by the interaction of several genes. One mutated gene can cause a series of biochemical defects that are then translated into human abnormalities. The procedure of **amniocentesis**, in which a small amount of amniotic fluid is removed from a pregnant woman, is used to study cells of the embryo. In this way certain chromosomal defects can be determined before birth.

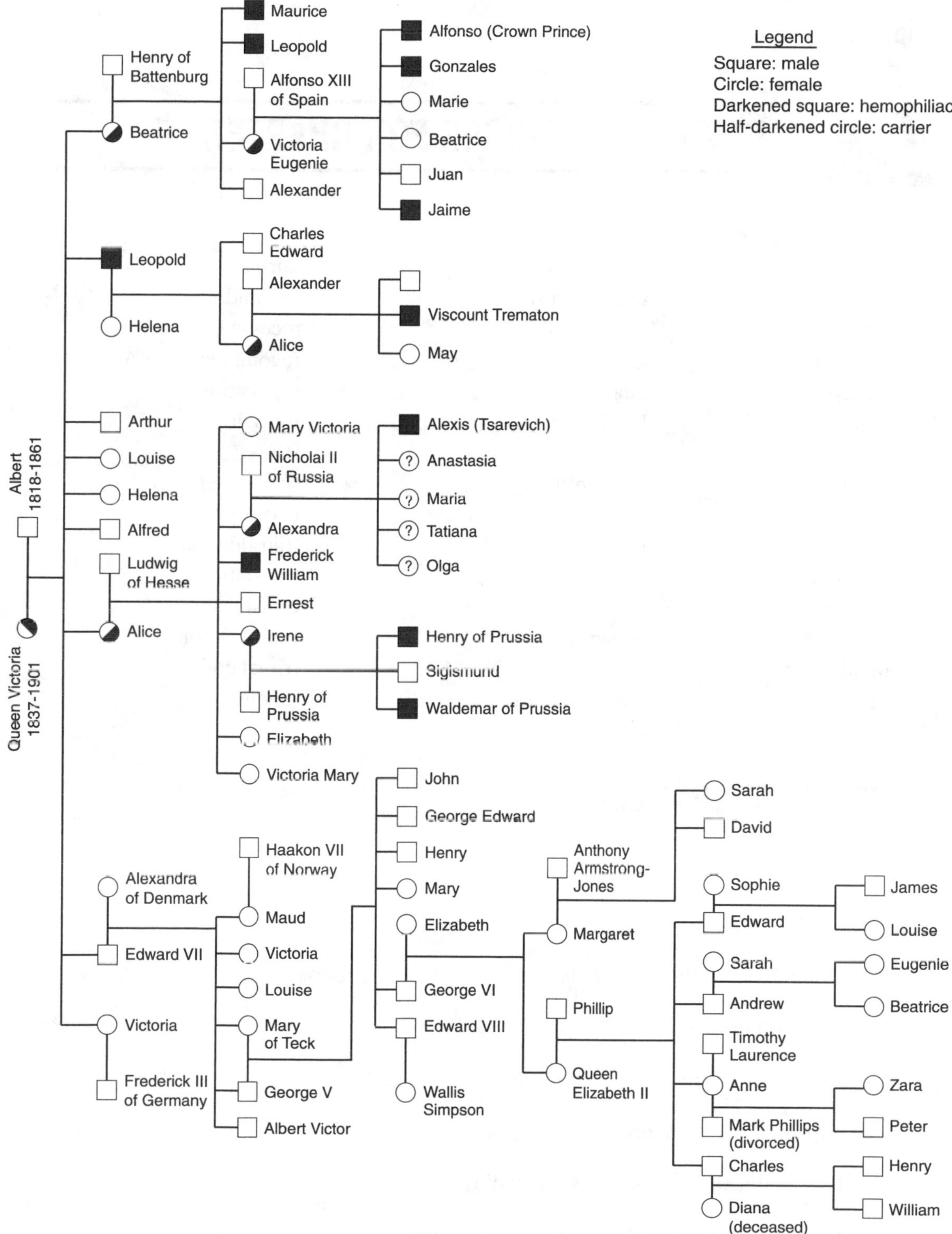

FIGURE 16.11 A genotypic pedigree of Queen Victoria's descendants, showing the inheritance of hemophilia

REVIEW EXERCISES FOR CHAPTER 16

WORD-STUDY CONNECTION

adenine
alleles
amniocentesis
anticodon
autosome
clone
codominance
codon
crossing over
cytosine
deletion
dominant
DNA
duplication
gene
gene pool
gene splicing
gene therapy
genetic code
genotype
guanine
Hardy-Weinberg principle
hybrid
interferon
intermediate inheritance
karyotype
mutation
nucleotide
pedigree
phenotype
point mutation
polyploidy
purine
pyrimidine
recessive
recombinant DNA
replication
retrovirus
RNA
sex-linked
thymine
transduction
translation
translocation
transcription
triplet code
uracil

SELF-TEST CONNECTION

PART A. Completion. *Write in the word that correctly completes each statement.*

1. The first organized study of heredity was made by _____.

2. Characteristics that can be observed are called _____.

3. Genes on the same chromosome are said to be _____.

4. Short wing in fruit flies is known as _____ wing.

5. Linkage groups are broken by _____.

6. The chemical _____ induces polyploidy.

7. Two or more genes that have the same positions on homologous chromosomes are said to be _____.

8. Pyrimidines are nucleic acids with _____ rings.

9. The complete name for DNA is _____.

10. Messenger RNA is synthesized in the _____.

11. DNA that has been changed by adding a gene from another species is known as _____ DNA.

12. Organisms arising from the same parent cell and bearing identical genes are called _____.

13. Bacterial cells bearing the human gene for insulin will synthesize _____.

14. The technique of adding a human gene to a circular strand of bacterial DNA is termed gene _____.

15. The overall technology of transferring genetic information from one organism to another has become known as genetic _____.

16. The shortened name of the virus of AIDS is _____.

17. The nucleic acid _____ codes the genetic information of the virus that causes AIDS.

18. Genetic information in cells is coded in the nucleic acid _____.

19. The process in which a virus makes more of itself is known as _____.

20. The genetic material of a T-4 lymphocyte is captured by a (an) _____ virus.

21. Each tRNA molecule is specialized for carrying a particular kind of _____.

22. The assembling of a polypeptide takes place inside the cell within an organelle known as a _____.

23. Messenger RNA is replicated from a segment of the genetic substance called _____.

24. The process of cloning requires the stimulation of a (an) _____ to initiate cleavage.

25. The cellular organelle that exerts master control over the synthesis of polypeptides is the _____.

PART B. Multiple Choice. *Circle the letter of the item that correctly completes each statement.*

1. If two organisms that exhibit contrasting traits are crossed, the trait that shows up in the F_1 generation is called
   (a) condominant
   (b) dominant
   (c) recessive
   (d) allelic

2. The trait that remains hidden in the F_1 generation is the
   (a) dominant
   (b) codominant
   (c) recessive
   (d) phenotype

3. The specimen used by Mendel in his work was the
   (a) fruit fly
   (b) snapdragon
   (c) firefly
   (d) garden pea

4. The law of segregation states that in the F_1 generation of a hybrid cross, three out of four offspring will exhibit the dominant trait. This is referred to as the
   (a) genotypic ratio
   (b) phenotypic ratio
   (c) homozygous ratio
   (d) heterozygous ratio

5. The genotypic ratio of a hybrid F_1 cross is
   (a) 3:1
   (b) 8:2
   (c) 1:2:1
   (d) 1:3:1

6. *Drosophila melanogaster* is a
   (a) fruit fly
   (b) gene
   (c) bread mold
   (d) cross over

7. Red cattle crossed with white cattle produce a red and white hybrid. This type of inheritance is known as
   (a) mutation
   (b) codominance
   (c) recessive
   (d) transformation

8. A loss of a piece of chromosome is known as a
   (a) translocation
   (b) transfiguration
   (c) deletion
   (d) duplication

9. Homologous chromosomes intertwine during
   (a) meiosis
   (b) mitosis
   (c) transduction
   (d) translocation

10. Human blood groups are formed by
   (a) a single pair of genes only
   (b) multiple alleles
   (c) several pairs of genes
   (d) no genes

11. Introducing recombinant DNA into a bacterial cell by means of a virus carrier is accomplished by
   (a) transference
   (b) translocation
   (c) transduction
   (d) translation

12. Certain cells produce interferons after having been infected by
   (a) viruses
   (b) fungi
   (c) bacteria
   (d) molds

13. Interferons are best classified as
   (a) hormones
   (b) toxins
   (c) catalytic enzymes
   (d) immune proteins

14. Cells of human embryos can be studied by the technique of
    (a) genetic engineering
    (b) amniocentesis
    (c) translocation
    (d) gene splicing

15. A photograph of matched chromosome pairs is a
    (a) micrograph
    (b) karyotype
    (c) daguerrotype
    (d) pictograph

16. Reverse transcriptase is associated with
    (a) a retrovirus
    (b) AIDS infection
    (c) DNA transcription
    (d) CD-4 lymphocytes

17. The human immunodeficiency virus carries its genetic code in molecules of
    (a) HIV
    (b) DNA
    (c) ATP
    (d) RNA

18. When a T-cell is infected with an HIV virus, the T-cell ultimately
    (a) makes more of its kind
    (b) makes more protein
    (c) increases its DNA content
    (d) is destroyed

19. The usual number of genes in an animal virus is
    (a) one
    (b) two
    (c) three
    (d) four

20. The function of RNA is to code for
    (a) carbohydrates
    (b) proteins
    (c) fats
    (d) minerals

21. The genetic disorder traced through the descendants of Queen Victoria is
    (a) color blindness
    (b) Tay-Sachs disease
    (c) hemophilia
    (d) Down syndrome

22. Tissue culture laboratory procedures were used to clone
    (a) carrots
    (b) piglets
    (c) sheep
    (d) elm trees

23. Research scientist Rosalind Franklin initiated the pioneering work on
    (a) color blindness
    (b) deoxyribonucleic acid
    (c) ribosomal structure
    (d) cloning techniques

24. Messenger RNA migrates
    (a) from the cytoplasm to the nucleus
    (b) from the nucleolus to the ribosomes
    (c) from the Golgi body to the cytoplasm
    (d) from the nucleus to the cytoplasm

25. The "building blocks" of polypeptides are
    (a) thymine and uracil
    (b) mRNA
    (c) amino acids
    (d) deactivated protein

PART C. Modified True-False. *If a statement is true, write "true" for your answer. If a statement is incorrect, change the* <u>underlined</u> *expression to one that will make the statement true.*

1. Point mutations are changes in single <u>cells</u>.

2. When a piece of broken chromosome sticks to another complete chromosome, the defect is known as <u>linkage</u>.

3. The five-carbon sugar in DNA is named <u>ribose</u>.

4. The sex chromosomes of human males are <u>XX</u>.

5. A person with genotype *ii* has blood type <u>AB</u>.

6. Thymine is always joined to guanine.

7. The DNA molecule is shaped like a double circle.

8. DNA molecules are composed of units called nucleic acids.

9. RNA does not contain the base uracil.

10. Pyrimidine molecules are joined to purines through weak oxygen bonds.

11. Down syndrome is a genetic disease caused by the doubling of chromosome number 21.

12. Recombinant DNA is carried into a bacterial cell by means of a catalyst.

13. Recombinant DNA is formed by using a virus to introduce a strand of DNA into a bacterial cell by a process called transplantation.

14. Interferons prevent the replication of DNA.

15. The names Tjio and Levan are correctly associated with counting genes.

16. The viral gene *env* codes for the environment of the virus.

17. Thus far, the number of genes identified in the AIDS virus is six.

18. A virus that passes its genetic code from RNA to DNA is a reverse virus.

19. Reverse transcriptase helps the HIV virus to carry out its metabolic processes.

20. A cell that is invaded by a virus is called a home cell.

21. Hemophilia and muscular dystrophy are genetic diseases that are classified as being trisomic.

22. Cloning is a form of sexual reproduction

23. Two forms of RNA are replicated in the nucleus of the cell.

24. In the tRNA molecule, the loop opposite the 3' and 5' ends is known as the codon.

25. Ribosomes are assembled from protein and RNA in the cytoplasm of the cell.

CONNECTING TO CONCEPTS

1. Although Gregor Mendel did not know about genes, he formulated three basic principles of heredity. Explain.

2. Cite an example in which Mendel's law of dominance does not hold true.

3. What does the "one gene-one enzyme" hypothesis suggest?

4. How does a retrovirus differ from a regular virus?

5. What ethical questions might be raised against the use of genetic engineering?

ANSWERS TO SELF-TEST CONNECTION

PART A

1. Mendel
2. phenotypes
3. linked
4. vestigial
5. crossing over
6. colchicine
7. alleles
8. single
9. deoxyribonucleic acid
10. nucleus
11. recombinant
12. clones
13. insulin
14. splicing
15. engineering
16. HIV
17. RNA
18. DNA
19. replication
20. invading
21. amino acid
22. ribosome
23. DNA
24. egg
25. nucleus or chromosome

PART B

1. **(b)**
2. **(c)**
3. **(d)**
4. **(b)**
5. **(c)**
6. **(a)**
7. **(b)**
8. **(c)**
9. **(a)**
10. **(b)**
11. **(c)**
12. **(a)**
13. **(d)**
14. **(b)**
15. **(b)**
16. **(a)**
17. **(d)**
18. **(d)**
19. **(c)**
20. **(b)**
21. **(c)**
22. **(a)**
23. **(b)**
24. **(d)**
25. **(c)**

PART C

1. genes
2. duplication
3. deoxyribose
4. XY
5. 0
6. adenine
7. helix
8. nucleotides
9. thymine
10. hydrogen
11. tripling
12. virus
13. transduction
14. RNA
15. chromosomes
16. outer coat
17. eight
18. retrovirus
19. replication
20. host
21. sex-linked
22. asexual
23. three
24. anticodon
25. nucleus

CONNECTING TO LIFE/JOB SKILLS

The field of genetics offers many career opportunities such as **research scientist**, **physician**, and **laboratory technologist**. Another career, for which there has been increasing demand for trained personnel, is **genetic counselor**. Genetic counselors work with a medical team. They give advice about genetic disorders and about the chances that couples face of having children who might inherit disorders. They also help people to understand their disorders and to cope with their fears. Genetic counselors must have an M.S. degree with a major in human genetics. In addition, clinical experience is required. To become licensed, they must pass an examination given by the American Board of Genetics. You can obtain additional information by contacting the American Genetic Association, 818 18th Street NW, Washington, DC 20060.

Chronology of Famous Names in Biology

1863 **Gregor Mendel** (Austria)—initiated the first mathematical study of inheritance.

1869 **Fredrich Meischer** (Germany)—extracted nucleic acid from cells and named the substance.

1900 **Hugo de Vries** (Netherlands)—discovered mutations in the evening primrose.

1903 **Walter S. Sutton** (United States)—discovered the mechanism of meiosis while studying the sperm cells of grasshoppers.

1908 **Sir Archibald Garron** (England)—discovered "inborn errors of metabolism."

1911 **Thomas Hunt Morgan** (United States)—developed the theory of the gene. He was the first investigator to use the fruit fly for genetic research.

1914 **Robert Feulgen** (Germany)—found that fuchsin-red dye is specific for DNA.

1928 **Frederick Griffith** (England)—accidentally discovered that harmless diplococci could be transformed into harmful bacteria that cause pneumonia.

1941 **George Beadle** and **Edward L. Tatum** (United States)—developed the "one gene—one enzyme" theory in work with *Neurospora crassa*.

1944 **Oswald T. Avery** (United States)—found the transforming factor in DNA.

1949 **P.A. Levene** (United States)—showed that nucleic acid could be broken down into four nitrogenous bases.

1950 **Edwin Chargaff** (United States)—found that nitrogenous bases in DNA do not occur in equal proportions.

1950 **E.C. Steward** (United States)—cloned carrot plants from differentiated cells.

1950 **Robert Briggs** and **Thomas King** (United States)—cloned tadpoles.

1951 **Rosalind Franklin** and **Maurice Wilkins** (England)—developed the technique of using X-ray crystallography to produce images of DNA. They proposed that DNA is a molecule in the shape of a helix.

1952 **Alfred Hershey** and **Martha Chase** (England)—performed experiments with labeled viral DNA, showing that it carries the complete hereditary message.

1952 **Alfred Mirsky** (United States)—showed that tissue cells contain equal amounts of DNA.

1952 **James Watson** (United States) and **Francis Crick** (England)—were the first to demonstrate the double-helix structure of DNA.

1952 **Norton Zinder** and **Joshua Lederberg** (United States)—discovered that DNA is the genetic material.

1958 **Linus Pauling** (United States)—found a difference in electrophoresis patterns between sickle-cell hemoglobin and normal hemoglobin.

1959 **Vernon Ingram** (United States)—showed that the difference between normal and sickle-cell hemoglobin is one amino acid in 300.

1960 **Hans Gruneberg** (India)—discovered that a single mutant gene in rats causes a complex of disorders.

1960 **Arthur Kornberg** (United States)—was the first to synthesize DNA *in vitro*.

1961 **Francis Crick** (England)—provided experimental evidence supporting the triplet code while working with the T_4 virus.

1985 **William A. Hazeltine** (United States)—discovered the function of the *tat* gene in the HIV virus.

1997 **Ian Wilmut** (Scotland)—successfully cloned a healthy sheep that continues to live through the adult stage.

1998 **Michael W.Young** (United States)—isolated and deciphered the functions of four of the genes linked to the biological clock of the fruit fly.

2000 **Andrew Yeager** (United States)—successfully transplanted bone marrow into the body of a boy suffering from sickle-cell anemia and cured the disease.

2000 **David Ayares** (United States) and colleagues—successfully cloned five identical piglets.

2002 **Sydney Brenner** (South Africa)—won the Nobel Prize in Physiology and Medicine for his explanation of apoptosis, or programmed cell death, and the genetic regulation of organ development.

17

Principles of Evolution

WHAT YOU WILL LEARN

In this chapter you will explore scientific evidences of evolution that support the doctrine of change in living things through the billions of years of Earth's history.

Evolution concerns the orderly changes that have shaped Earth and have modified the living species that inhabit it. Evolution is a fusion of biological and physical sciences that have provided supporting data that confirm the fact that over periods of time major changes have occurred in the interior of Earth and on its surface, accompanied by modifications in climate. All of the changes in Earth are classified as nonbiological or **inorganic evolution**. Changes that have taken place in living organisms are known as biological or **organic evolution**.

SECTIONS IN THIS CHAPTER

- Evidence of Evolution
- Theories of Evolution
- The Origin of Life
- Evolution of Humans
- Review Exercises for Chapter 17
- Connecting to Life/Job Skills
- Chronology of Famous Names in Biology

Evidence of Evolution

Evidence that **evolution**—gradual change over a period of time—has occurred in living things is provided by many sciences and includes facts from the geologic record, information gained by the study of fossils, and evidence from cell studies, biochemistry, comparative anatomy, and comparative embryology.

THE GEOLOGIC RECORD

Look at Table 17.1—The Geologic Time Scale. This scale is to be read from the bottom upward because it represents the age of Earth as determined by the rock layers of Earth. Scientists have established that Earth is 4.6 billion years old.

TABLE 17.1
THE GEOLOGIC TIME SCALE

Era	Period	Epoch	Millions of Years Ago	Climate and Life
PRECAMBRIAN			Over 570	Earth cooling; shallow seas; bacteria; cyanobacteria; eubacteria evolve.
PALEOZOIC	Cambrian		570	Primitive marine algae; marine invertebrates in great numbers.
	Ordovician		500	Mild climate; seas cover continents; plants invade land; marine algae abundant; jawless fish evolve.
	Silurian		430	Rise of mountains in Europe; continents flat; first vascular plants; arthropods appear on land.
	Devonian		395	U.S. covered by oceans, Europe mountainous; primitive tracheophytes; origin of first seed plants; first liverworts; age of fishes, sharks.
	Carboniferous	Pennsylvanian Mississippian	345	Subtropical climate; swamps; great coal forests; lycopsids, sphenopsids, ferns, gymnosperms; reptiles evolve; amphibians dominant; first insects; fungi.

TABLE 17.1
THE GEOLOGIC TIME SCALE (continued)

Era	Period	Epoch	Millions of Years Ago	Climate and Life
	Permian		280	Glaciers in southern hemisphere; Appalachians rising; first conifers, cycads, ginkos; expansion of reptiles; decline of amphibians; extinction of trilobites.
MESOZOIC	Triassic		225	Extensive arid and mountainous areas; dominance of land by gymnosperms; first dinosaurs; first mammals.
	Jurassic		190	Europe covered by ocean; last of seed ferns; gymnosperms dominant; reptiles dominant; origin of birds; dinosaurs abundant.
	Cretaceous		136	Rise of angiosperms; decline of gymnosperms; extinction of dinosaurs; second great radiation of insects.
CENOZOIC		Paleocene	65	First primates and carnivores.
		Eocene	54	Mild to tropical weather; small horses.
		Oligocene	38	Dominance on land by mammals (anthropoid apes, ungulates, whales); birds, insects.
		Miocene	26.0	Forests decrease; dominance of angiosperms.
	Tertiary	Pliocene	7.0	Cool; hominoid apes; first humans.
		Pleistocene	2.5	Increase in herb population; first *Homo* sp.
	Quarternary	Recent	0.01	Four ice ages; *Homo sapiens.*

Geologists have measured the age of Earth, of ancient events, and of many life forms through a process called **radioactive dating**. Scientists have determined that certain elements disintegrate by giving off radiation spontaneously and at a regular rate. Such elements are said to be **radioactive**. In the process of emitting radiation, the radioactive substance changes to something else. For example, uranium-238 changes to lead. The **half-life** of U-238, the rate at which one half of the uranium in a rock sample will change to lead, is 4.5 billion years. Uranium's rate of decay is not affected by any chemical or physical conditions. Therefore measuring the uranium-lead ratio in a sample of rock is a very reliable method for establishing the age of the rock.

FOSSIL EVIDENCE

Fossils are the preserved remains of plants and animals. They are often found in sedimentary rock, which is formed by the gradual settling of *sediments*. The age of fossils is estimated by the use of *carbon dating*. The sample plant or animal fossil is tested for the ratio of radioactive carbon (carbon 14) to nonradioactive carbon (carbon 12). By using the rate of decay of carbon 14 to carbon 12, the age of a fossil can be determined.

The fossil records contained in the layers of sedimentary rock provide reliable evidences of change in plant and animal species. The lower down the rock layer, the older the fossil. Top layers contain more recent fossil remains of more complex species. The hard parts of animals, such as a shell or a skeleton, become fossilized in the hardened sediments of rock. Imprints, casts, or molds are other types of fossil remains, produced by an organism leaving a footprint, a track, or form in the sediment.

In 1998, Wes Linster (14 years of age) uncovered a nearly perfect skeleton of a winged prehistoric animal as he was fossil hunting in Montana. Scientists estimate that this bird-like predator lived more than 75 million years ago. This fossil dromeosaur (Figure 17.1) has been given the scientific name *Bambiraptor feinbergi*, indicating that it was a juvenile and had not reached its full size. It had a large brain, a bird-like wishbone sternum, and wing-like "arms." This fossil was a rare find in that it is complete with no missing bone.

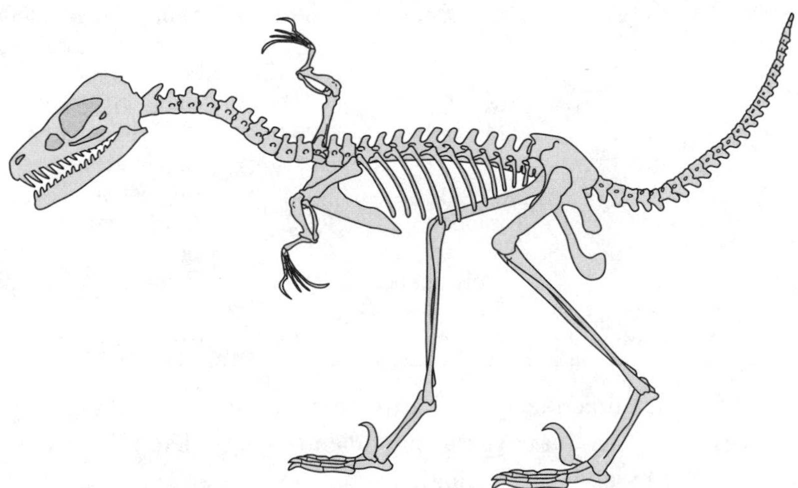

FIGURE 17.1 *Bambiraptor*, a prehistoric winged dinosaur that lived 75 million years ago

Other types of fossil remains have also provided evidence of ancient species. **Amber** is the hardened resin of trees. Insects trapped in the sticky gum remain preserved as the resin hardens into amber. Ice has preserved some rather huge animals, such as the woolly mammoth, which was probably caught in the glaciers of Siberia. Leaf imprints have been preserved in coal while it was being formed. Fossilization, the absorbing of mineral matter by dead plant and animals, preserves species in stone. Bogs have been the sources for preserved wood fossils. Since bacteria of decay cannot thrive in the acid environment of bogs, wood samples have been kept intact. Saber-toothed tiger fossils have been found in the La Brea tar pits in Los Angeles.

Dr. Dan Gebo and his research team uncovered the fossil remains of an extremely small primate—so small that it can fit into the palm of the hand. However, this fossil find, named Eosimias and called the "dawn monkey" (Figure 17.2), is the possible link between the prosimians (lower primates) and the anthropoids (higher primates). These bones were found in China. It is believed that early primates migrated from Asia to Africa, where they evolved into baboons and chimpanzees.

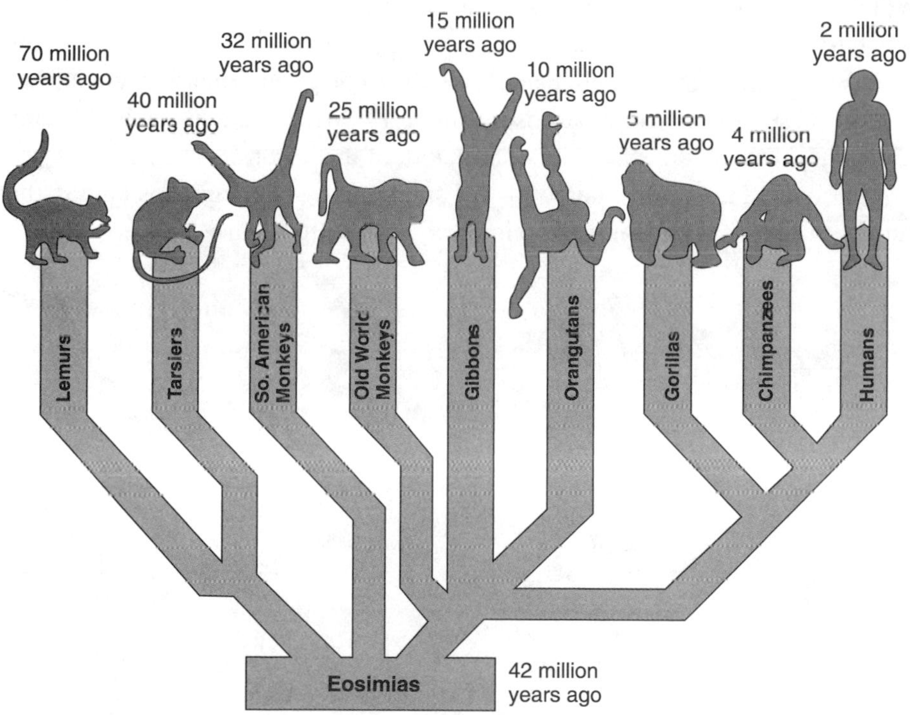

FIGURE 17.2 Representation of the link between Eosimias and the anthropoids (dates are approximate)

EVIDENCE FROM CELL STUDIES AND BIOCHEMISTRY

The cells of all living organisms have comparable structures that function in similar ways. All eukaryotic cells have a cell membrane, a nucleus (except mature red blood cells), cytoplasm with energy-producing mitochondria, and ribosomes where proteins are synthesized. The fact that the cells of all living things have similar structures that perform the same tasks indicates that there is evolutionary unity in all living things.

On the molecular level, there is similarity in genetic material of cells. Similar genes direct the formation of similar cell structures and similar proteins. For example, insulin produced in the hog pancreas is so similar to human insulin that hog pancreas is a source of insulin that is prescribed for diabetics. This means that the human and the hog have some very similar DNA molecules. It is not unusual for animals of various species to synthesize some proteins of similar natures. Interestingly enough, the primates (including humans and apes) and the guinea pigs show an unusual kind of biochemical relationship. These are the only vertebrate animals that cannot synthesize vitamin C from carbohydrates.

EVIDENCE FROM COMPARATIVE ANATOMY

A comparative study of the bone structures and body systems of animals from the various phyla reveals a great deal of similarity. A comparative study of the skeletal systems of vertebrates shows that many of the bones are very much alike. Much of our evidence for evolution comes from a study of homologous structures. **Homologous** structures are bones that look alike and have the same evolutionary origin although they may be used for different purposes. Figure 17.3 shows the bones in the forelimbs of vertebrates. Notice the similarity of structure of these bones. The flipper of a whale, the arm of a human, and the wing of a bird are homologous structures having the same evolutionary origin and maintaining similarity of structure.

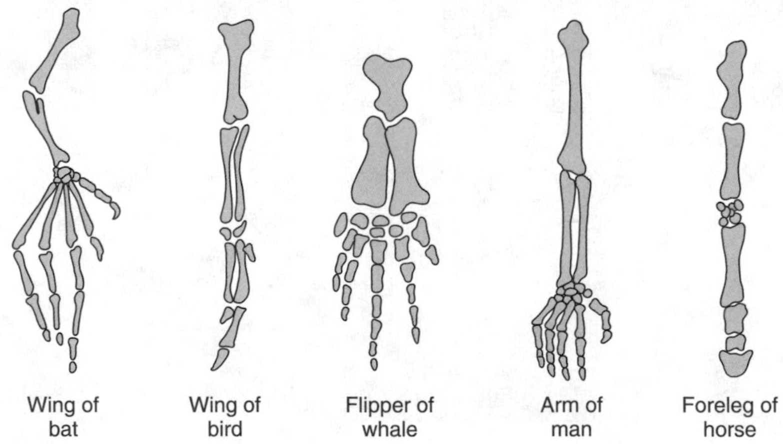

FIGURE 17.3 Homologous forelimbs

A comparison of the digestive, nervous, and circulatory systems of vertebrates indicates similar evolutionary origin. Biologists believe that there must have been a common ancestor from which the vertebrate line descended. Figure 17.4 shows the brains of some common vertebrate species. Notice their similarities and differences.

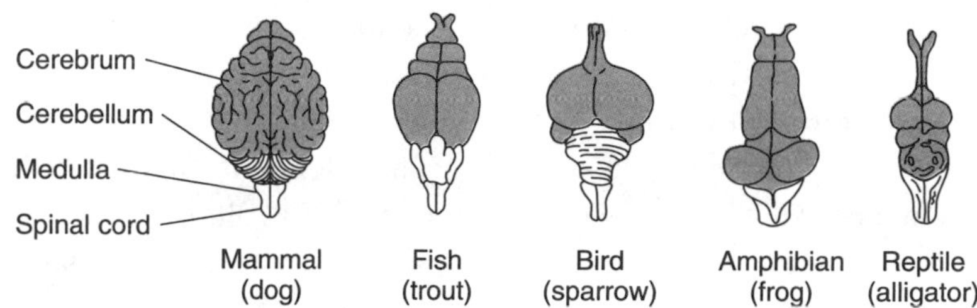

FIGURE 17.4 A comparison of vertebrate brains

EVIDENCE FROM COMPARATIVE EMBRYOLOGY

Embryology is the study of developing forms or embryos. A comparative study of the embryos of seemingly unrelated vertebrates indicates similar evolutionary origins. Figure 17.5 shows some of these embryos. Notice the marked similarity in structure. As the embryonic development continues, the distinctive traits of each species begin to take form.

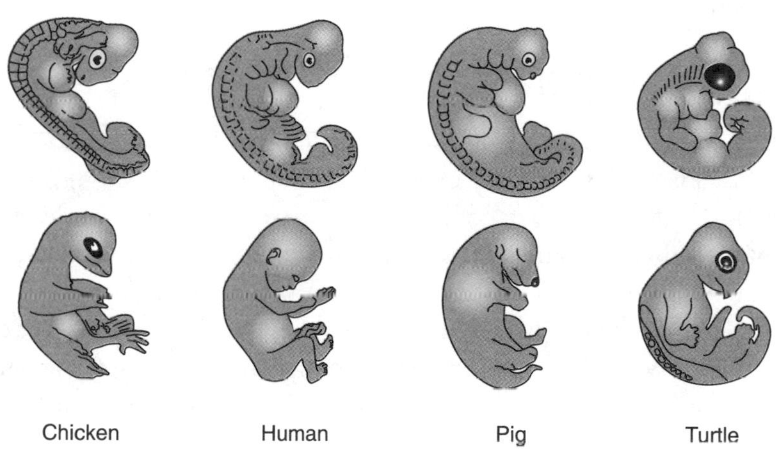

FIGURE 17.5 A comparison of vertebrate embryos

Theories of Evolution

Since the 18th century several theories have been proposed to explain evolution. Among these are the **use and disuse** theory of Lamarck and the theory of **natural selection** formulated by Charles Darwin. Recently advances in genetics and biochemistry have led to the formulation of a modern theory of evolution.

"USE AND DISUSE"

The French biologist Jean Baptiste Lamarck (1744–1829) presented the first well organized theory of evolution. Lamarck recognized that living animals and plants throughout the world were different. He proposed that these differences were due to changes that caused significant variations among species. He also believed that changes are going on all of the time, because evolution is a continuing process.

Lamarck believed that organisms evolved in straight line fashion from the "less perfect to the more perfect." We know today that branching evolution better explains the diversity of the many species. Lamarck also believed that the changes that take place in living organisms are stimulated by needs in the environment. He believed that systems and body structures developed in response to "**use**" and that those structures in "**disuse**" eventually disappeared. Lamarck also believed that **acquired characteristics** could be passed on from parent to offspring.

German biologist August Weismann (1834–1914) used a rather dramatic experiment to disprove Lamarck's theory of inheritance of acquired characteristics. Weismann cut off the tails of mice and then mated the mice. He carried out this process for 25 generations. In each generation he found that mice grew tails just as long as the tails of the mice in the first generation. Weismann proved that cutting off the tails of mice (resulting in an acquired characteristic) did not alter the "germplasm" of future generations.

NATURAL SELECTION

Charles Darwin (1809–1882) was an English naturalist who first developed a theory of evolution that laid the groundwork for modern biological thinking. At the age of 22 in the year 1831, Charles Darwin commenced on a trip around the world as the official naturalist of the ship *H.M.S. Beagle*. Darwin was commissioned to collect as many specimens of animal and plant life as possible and to record observations of the natural phenomena that he saw. Darwin's trip on the *Beagle* extended over a five-year period. The ship traveled around the coast of South America to the Galápagos Islands. From there the ship pursued a westerly course passing the southern coast of Australia, then going north to the southern coast of Asia and then south along the southern coast of Africa back to England.

In all of the places where the ship stopped, Darwin collected specimens of living organisms and recorded what he observed. For the next twenty-five years Darwin studied and organized his notes. Meanwhile, Darwin's cousin, Alfred Russel Wallace (1823–1915), had independently begun writing a book based on ideas very similar to Darwin's. When Wallace presented his manuscript to his cousin for review, Darwin realized that he needed to publish his own work immediately. In 1859, Darwin published *On the Origin of Species by Means of Natural Selection, or the Preservation of Favored Races in the Struggle for Life*. In this book, Darwin's theory of natural selection was clearly set forth. Today this work is referred to by a shorter title, *The Origin of Species*.

Darwin's theory of natural selection can be summed up thusly: large numbers of new plants and animals are produced by nature. Many of these do not survive because nature "weeds out" weak and feeble organisms by killing off those that cannot adapt to changing environmental conditions. Only the strongest and most efficient survive and produce progeny. Specific tenets of Darwin's theory of evolution follow.

OVERPRODUCTION

The theory of natural selection cites the fact that every organism produces more gametes and/or organisms than can possibly survive. If every gamete produced by a given species united in fertilization and developed into offspring, the world would become so overcrowded in a short period of time that there would be no room for successive generations. This does not happen. There is a balance that is maintained in the reproduction of all species and therefore natural populations remain fairly stable, unless upset by a change in conditions.

COMPETITION

As stated above, not all the offspring (and gametes) survive. There is competition for life among organisms: competition for food, room, and space. Therefore there is a **struggle for existence** in which some organisms die and the more hardy survive.

SURVIVAL OF THE FITTEST

Some organisms are better able to compete for survival than others. The differences that exist between organisms of the same species, making one more fit to survive than another, can be explained in terms of **variations**. Variations exist in every species and in every trait in members of a species. Therefore some organisms can compete more successfully for the available food or space in which to grow or can elude their enemies better. These variations are said to add survival value to an organism. Survival value traits are passed on to the offspring by those individuals that live long enough to reproduce. As weaker individuals are weeded out of the species, those individuals that remain do so because they are best adapted to live in their environment. As time goes on the special adaptations for survival are perpetuated and new species evolve from a common ancestral species.

The environment is the selecting agent in natural selection because it determines what variations are satisfactory for survival and which are not. Any change in an environment can affect the usefulness of a given variation. A change in the environment can render a once useful variation useless. In this way the direction of evolutionary development can change.

MODERN EVOLUTIONARY THEORY

The major weakness in Darwin's theory of natural selection is that he did not explain the source, or genetic basis, for variations. He did not distinguish between variations that are hereditary and those that are nonhereditary. He made the assumption that all variations that have survival value are passed on to the progeny. Like Lamarck, Darwin believed in the inheritance of acquired characteristics.

We now know that there are several ways in which variation is produced within a species. Variation that is inherited is an immediate, direct result of mutation.

Nonhereditary variation is a function of the impact of a mutation on a population through natural selection or genetic drift.

CAUSES OF VARIATION

Hugo De Vries (1845–1935), a Dutch botanist, explained variations in terms of **mutations**. His study of 50,000 plants belonging to the evening primrose species enabled him to identify changes in leaf shape and texture and in plant height that were passed on from parent plant to offspring. In 1901 De Vries offered his mutation theory to explain organic evolution. He did not know how mutations come about or where they occur. Today, we know that mutations are changes in genes that can come about spontaneously or can be induced by some mutagenic agent.

Spontaneous mutation rates are very low. In *Escherichia coli* one cell in every 10^9 will mutate from streptomycin sensitivity to streptomycin resistance. In human beings the rate of mutation in Huntington's disease occurs in 5 out of every 10^6 gametes. It is a known fact that different genes have different mutation rates. This variation is probably due to the different chemical composition of genes or perhaps to the different places that they occupy on chromosomes. Mutations alone do not affect major changes in the frequencies of alleles.

An important cause of variation within species is genetic recombination that results from sexual reproduction wherein the genes of two individuals are sorted out and recombined into a new combination, producing new traits—and thus variation.

Gene flow is another agent of evolution, responsible for the development of variations. It is the movement of new genes into a population. Gene flow often acts against the effects of natural selection. *Genetic drift* is a change in a gene pool that takes place in a population as a result of chance; this can result in geographic or reproductive isolation. Geographic isolation refers to genetic changes that occur within a species that has been physically separated by land or water barriers from other members of the same species. Reproductive isolation occurs when members of a small population reproduce exclusively among themselves; this is sometimes called inbreeding. If a mutation occurs in a gene of one person, and that person does not reproduce, the gene is lost to the population. Sometimes a small population breaks off from a larger one. Within that small population is a mutant gene. Because the mating within the small population is very close, the frequencies of the mutant gene will increase. In the Amish population, for example, where there is very little (or no) outbreeding, an increase in the homozygosity of the genes in the gene pool is evinced in the high frequencies of genetic dwarfism and polydactyly (six fingers). The isolated smaller population has a different gene frequency than the larger population from which it came. This is known as the **founder effect**.

Genetic drift and the random mutations that increase or decrease as the result of genetic drift are known as non-Darwinian evolution.

Speciation is the forming of one or more new species from a species already in existence. This can happen when a population becomes divided and part of the original species continues life in a new habitat. Islands cut off from the mainland by surrounding water, isolated mountain peaks, migration of a part of the species to a new area—all these are examples of geographic isolation from the main population. The separated populations cannot interbreed. Over evolutionary time, different environments present different selective pressures, and the change in gene pools will eventually produce new species. For example, Darwin identified 14 different species of finches living on the Galapagos Islands; all had descended from a single species of finch that lived on mainland Peru.

TIME FRAME

There is a time frame for evolution. Darwin's concept of **gradualism** supports the idea that evolutionary change is slow, gradual, and continuous. According to the more current concept of **punctuated equilibrium**, species have long periods of stability, lasting for four or five million years, and then suddenly change as a result of some geological or other environmental change.

The Origin of Life

Theories of evolution are concerned not only with how living organisms have changed through time but also with how life began—how living organisms first evolved on Earth. In the 1920s the Russian scientist A.I. Oparin began investigations into how life could have evolved from inorganic compounds under the conditions of early Earth. He found that he could produce *coacervate droplets* that had the ability to incorporate simple enzymes in their structure. This initial work opened the door to many studies of how life could have begun.

In the 1950s Stanley Miller set up an experiment in which he duplicated the chemical conditions and the temperature of the early seas. Into sterile water he put some methane gas, hydrogen, and ammonia. He sealed off the system so that nothing could leave or enter it. The water was heated to a temperature similar to that of the early seas. The water and gas mixture were subjected to electric sparks. Miller let the experiment run for about a week. At the end of that time, he tested the water and found present in it organic compounds that were not there before. Miller's experiments help to explain how organic matter appeared in the early seas.

In 1963, Carl Sagan, duplicating Miller's experiment, was able to produce ATP (adenosine triphosphate) from inorganic matter. ATP molecules are essential to all living cells and function as energy storage molecules.

Following the lead of Stanley Miller, other investigators demonstrated how organic molecules could have been formed in the early seas. Melvin Calvin demonstrated the polymerization of complex molecules. Sidney Fox produced microspheres, long chain peptides surrounded by something resembling a membrane. All of these experiments have contributed to the theory of the origin of life.

Scientists believe that the first organisms were anaerobic heterotrophs that obtained nutrition from the organic molecules in the "hot, thin soup," an expression used to describe the early oceans. As oxygen began to increase in the atmosphere, conditions changed for the heterotrophs. Those that could not adjust to the changing atmospheric conditions were destroyed. Eventually, organisms that could make their own organic compounds from inorganic materials in the presence of sunlight began to evolve. These early autotrophs were the blue-green algae.

The blue-green algae increased the oxygen content of the air—remember oxygen is a byproduct of photosynthesis—and this further threatened the continued existence of the heterotrophs. As a result of the high concentration of oxygen in the atmosphere, an ozone layer developed that further diminished the organic compounds available to the groups of existing heterotrophs. From this group there evolved heterotrophs that could utilize oxygen in cellular respiration during which energy was released from organic molecules taken in as nutrients. Thus the carbohydrates produced by the autotrophs and the oxygen in the air supplied the newly evolved heterotrophs with the nutrients necessary for survival. This sequence of events (anaerobic heterotrophs → simple autotrophs → aerobic heterotrophs) is known as the **heterotroph hypothesis**. It is an attempt to explain the origin of autotrophic and heterotrophic cells.

Evolution of Humans

The evolution of *Homo sapiens* has always been of interest to modern humans. People like to know their origins. In 1924 the first specimen of *Australopithecus africanus* was discovered in South Africa. It is estimated that this prehuman form lived between 2 and 3 million years ago. In 1974, a 3-million-year-old fossilized skeleton was found in Ethiopia; it is believed to be the remains of one of the earliest hominid ancestors, *Australopithicus afarensis*. This species lived between 3.9 and 2.9 million years ago. Studies of the several australopithecine fossils found indicate that these species may well have been forerunners of humans. The brain case indicates that the brain was within the size range of modern apes. The teeth were more human-like than those of modern apes. They were arranged in rows without gaps between the canines and the premolars, and the canine teeth did not stick out beyond the adjacent teeth. The bones of the pelvis show that the australophithecines were bipedal, walking on two feet. Because not all of the australopithecine fossils are alike, anthropologists believe that two forms of hominids lived at about the same time: *Australopithecus africanus* and *Paranthropus robustus*. Very little is known about the life patterns of these two species.

Fossil finds indicate that another species of prehuman lived on Earth probably for as long as 500,000 to 1,000,000 years. *Homo erectus*, as this species is called, lived on various continents. The brain case is smaller than that of modern humans, but the teeth are larger. The fossil teeth show traces of an enamel collar, a primitive trait in apes and in humans. The species became extinct (Figure 17.6).

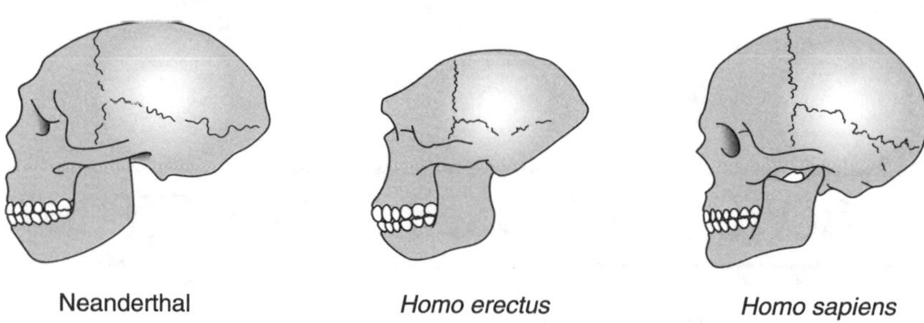

FIGURE 17.6 A comparison of skulls of three human species

In 1856 the fossil remnants of a human species called *Homo sapiens neanderthales* were found in a cave in Germany. It is believed that the Neanderthal appeared on Earth about 125,000 years ago. They lived through two glacial periods and then became extinct between 50,000 and 15,000 years ago. It is commonly accepted theory that over thousands of years, modern humans gradually replaced the Neanderthal grade of humans, until there remained only one species of human being—*Homo sapiens* (Figure 17.7).

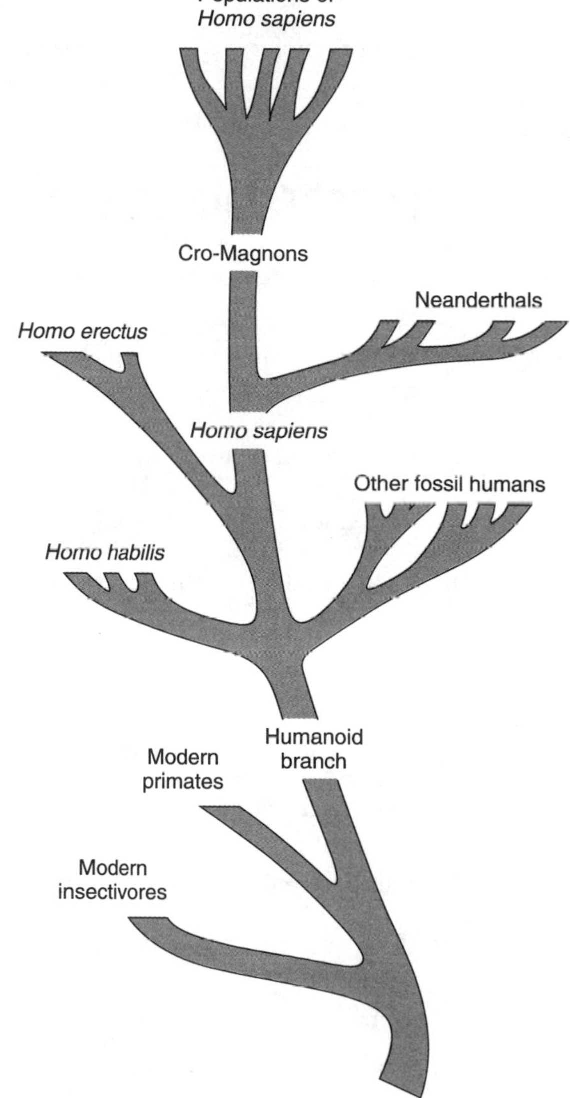

FIGURE 17.7 The tree of human evolution

REVIEW EXERCISES FOR CHAPTER 17

WORD-STUDY CONNECTION

amber
evolution
fossil
founder principle
gene flow
genetic drift
gradualism
heterotroph hypothesis
homologous
mutation
natural selection
petrification
punctuated equilibrium
radioactive dating
speciation
use and disuse

SELF-TEST CONNECTION

PART A. Completion. *Write in the word that correctly completes each statement.*

1. Changes that have taken place in the structure and energy of Earth are classified as _____ evolution.
2. Elements that emit atomic particles are said to be _____.
3. The kind of rock in which fossils are found is _____ rock.
4. Cells of all living things have _____ structures.
5. The flipper of a whale and the wing of a bird are _____ structures having the same evolutionary origin.
6. The theory of "use and disuse" was proposed by _____.
7. The theory of natural selection was proposed by _____.
8. Hugo De Vries explained variations in terms of _____.
9. Loss of genes from a population is known as _____ (2 words).
10. The experiments of _____ helped to explain how organic molecules appeared in the early seas.

PART B. Multiple Choice. *Circle the letter of the item that correctly completes each statement.*

1. *Australopithecus africanus* is considered to have been a
   (a) human hominid
   (b) human
   (c) ape
   (d) old world monkey

2. A species of early human, the fossils of which have been found all over the world, is
   (a) *Australopithecus*
   (b) Neanderthal
   (c) *Homo erectus*
   (d) *Homo habilis*

3. The early autotrophs were probably
   (a) bacteria
   (b) viruses
   (c) amoebae
   (d) blue-green algae

4. The first living cells were probably
   (a) autotrophs
   (b) heterotrophs
   (c) parasites
   (d) symbionts

5. The modern theory of the origin of life was initially developed by
   (a) Miller
   (b) Fox
   (c) Sagan
   (d) Oparin

6. The founder effect explains
   (a) the origin of life
   (b) abnormal gene frequencies in a small population
   (c) the methods by which genes leave a population
   (d) finding and securing of lost genes

7. Genes are sorted out and recombined by the process of
   (a) genetic drift
   (b) mutation
   (c) sexual reproduction
   (d) vegetative reproduction

8. The evening primrose was the experimental specimen used in the study of
   (a) genetic drift
   (b) mutation
   (c) gene flow
   (d) vegetative propagation

9. Darwin's theory of evolution is known as
   (a) natural selection
   (b) struggle for existence
   (c) survival of the fittest
   (d) competition

10. The "continuity of the germplasm" was demonstrated in an experiment using mice by the investigator named
   (a) Miller
   (b) Lamarck
   (c) Fox
   (d) Weismann

PART C. Modified True False. *If a statement is true, write "true" for your answer. If a statement is incorrect, change the <u>underlined</u> expression to one that will make the statement true.*

1. The age of Earth is measured by a process called <u>reactive</u> dating.

2. Uranium-238 ultimately turns into <u>boron</u>.

3. Earth is approximately <u>10</u> billion years old.

4. Fossil plants and animals are dated by measuring the <u>hydrogen</u> content.

5. Fossil insects are most likely to be found in <u>ambergris</u>.

6. The absorbing of mineral matter by dead plant and animal bodies which turns them into stone is known as <u>fossilization</u>.

7. Human beings can synthesize vitamin C from carbohydrates.

8. The study of developing forms is known as anatomy.

9. The *H.M.S. Beagle* is most closely associated with Wallace.

10. The concept of punctuated equilibrium helps to explain the time frame of reproduction.

CONNECTING TO CONCEPTS

1. Explain how radioactive dating is used to determine the age of Earth.

2. How do you know that dinosaurs did not live when humans appeared on Earth?

3. How does fossil evidence support the theory of evolution?

4. How can you prove that mutations are inherited?

ANSWERS TO SELF-TEST CONNECTION

PART A

1. inorganic
2. radioactive
3. sedimentary
4. similar
5. homologous
6. Lamarck
7. Darwin
8. mutations
9. genetic drift
10. Miller

PART B

1. **(a)**
2. **(c)**
3. **(d)**
4. **(b)**
5. **(d)**
6. **(b)**
7. **(c)**
8. **(b)**
9. **(a)**
10. **(d)**

PART C

1. radioactive
2. lead
3. 4.6
4. carbon
5. amber
6. true
7. cannot
8. embryology
9. Darwin
10. evolution

CONNECTING TO LIFE/JOB SKILLS

Teams of scientists work together to uncover the secrets and history of fossil remains. Scientists with different (and somewhat related) specialties form very successful fossil-hunting teams.

Field of Study	Provides Information About
Anthropology	Humans, including races of people
Physical anthropology	Human evolution, including that of primitive people
Archaeology	Ancient humans—culture and tools
Paleontology	Ancient bones and fossils
Paleoanthropology	Fossil human bones
Paleoecology	Ancient climates and vegetation

You may wish to find out more about one or more of the sciences listed above. All are related to investigating humans' history on Earth. You will find reading about the work of the scientists who are engaged in exploring the history of Earth and its plants and people fascinating. Your local library and museum of natural history are good places to start.

Chronology of Famous Names in Biology

1747 **Comte de Buffon** (France)—was the first scientist to state that living things can change.

1771 **Erasmus Darwin** (England)—was an early advocate of the idea of evolution.

1784 **Jean Baptiste Lamarck** (France)—presented the first organized theory of evolution, although his ideas are not accepted today. He proposed the theory of "use and disuse," which suggests that organs that are well developed from use are passed on to offspring in the well-developed stage.

1800 **Thomas Malthus** (England)—wrote the notable "Essay on Population," in which he said that food supplies cannot keep pace with rapid increases in human population.

1830 **Charles Darwin** (England)—developed the theory of natural selection on which the modern ideas of evolution are based.

1830 **Alfred Wallace** (England)—proposed a theory of evolution similar to that of Charles Darwin.

1874 **August Weismann** (Germany)—proposed that germplasm carries the hereditary material that affects successive generations of progeny.

1891 **Eugene Dubois** (Java)—found the braincase of the first known fossil of *Homo erectus*.

1924 **Raymond Dart** (South Africa)—identified the first recovered specimen of *Australopithecus*.

1935 **E. B. Ford** (England)—introduced the concept of industrial melanism to explain change in color of species of moths.

1940 **Trofim D. Lysenko** (Russia)—tried to change genes and inheritance by subjecting plants to periods of heat and cold. His methods in agriculture failed. Lysenko is remembered for his fallacious proposals concerning the "vernalization of wheat."

1947 **Robert Broom** (South Africa)—found fossil bones of *Australopithecus africanus*.

1950 **Bernard Kettlewell** (England)—provided an excellent investigative study of evolutionary selection in the peppered moth.

1960 **Louis Leakey** and **Mary Leakey** (Tanzania)—found fossils that predate humans.

1995 **Daniel Gebo** (United States)—identified features in Eosimias that indicate a link between the prosimians and the anthropoids.

1998 **Wes Linster** (United States)—discovered the preserved skeleton of a prehistoric winged predator.

18

Ecology

WHAT YOU WILL LEARN

In this chapter you will explore the interrelationships among living things and the factors in the nonliving environment that affect life. You will also learn how living things can change the environment.

SECTIONS IN THIS CHAPTER

OVERVIEW

Ecology is the science that studies the interrelationships between living species and their physical environment. The word *ecology* was coined in 1869 by the German zoologist Ernst Haeckel to emphasize the importance of the environment in which living things function. The environment includes living or **biotic** factors and nonliving factors, referred to as **abiotic** factors.

The abiotic factors consist of physical and chemical conditions that affect the ability of a given species to live and reproduce in a particular place. Included in the abiotic factors are temperature, light, water, oxygen, pH (acid-base balance) of soil, type of substrate, and the availability of minerals. Certain kinds of plants and animals will flourish in a natural community if the conditions are present that permit their survival. Species interact to influence the survival of one another. One important principle of ecology is that no living organism is independent of other organisms or of the physical environment, if they share the same community.

The Concept of the Ecosystem

Certain terms are used in ecology to provide a consistent description of conditions and events. A **population** refers to all of the members of a given species that live in a particular location. For example, a beech-maple forest will contain a population of maple trees, a population of beech trees, a population of deer, and populations of other species of plants and animals. All of the plant and animal populations living and interacting in a given environment are known as a **community**.

The living community and the nonliving environment work together in a cooperative ecological system known as an **ecosystem**. An ecosystem has no size requirement or set boundaries. A forest, a pond, and a field are examples of ecosystems. So is an unused city lot, a small aquarium, the lawn in front of a residential dwelling, or a crack in a sidewalk. All of these examples reflect areas where interaction is taking place between living organisms and the nonliving environment.

COMPONENTS OF THE ECOSYSTEM

Modern ecologists think of the ecosystem in terms of its interacting forces or components. One such component is the physical environment. This includes the air, which is made up of 21 percent oxygen, 78 percent nitrogen, 0.03 percent carbon dioxide, and the remainder inert gases. The soil is the source of minerals that supply plants with compounds of nitrogen, zinc, calcium, phosphorus, and other minerals.

The green plants in the ecosystem are another component. These are the **producers**, so-called because they are able to make their own food using the inorganic materials of carbon dioxide and water and minerals from the soil. Directly dependent upon the producers are the *primary* **consumers**—**herbivores**, or plant-eaters. Herbivores come in all sizes: crickets, leaf cutters, deer, and cattle. The **carnivores**, or flesh-eaters, such as snakes, frogs, hawks, and coyotes, are *secondary consumers* because they feed on the herbivores. The **tertiary consumers** are those that feed on the smaller carnivores and herbivores as well. There are also **scavengers** in the ecosystem. Earthworms and ants feed on particles of dead organic matter that have decayed in the soil. Vultures eat the bodies of dead animals.

The **decomposers** form another important part of ecosystems. Bacteria and fungi are organisms that break down dead organic matter and release from it organic compounds and minerals that are returned to the soil. Many of the materials returned to the soil are used by the producers in the process of food-making. Without the work of the decomposers the remains of dead plants and animals would pile up, not only occupying space needed by living organisms, but also keeping trapped within their dead bodies valuable minerals and compounds.

The structure of the ecosystem remains the same whether its location is on land or in water. A marine ecosystem illustrates this fact. The autotrophs are microscopic plants called *phytoplankton*, which float on the top of the water. The phytoplankton is composed of billions of single-celled green algae, which produce a vast quantity of food by means of photosynthesis just as the land plants do. Living in close association with the phytoplankton are heterotrophs such as protozoa and brine shrimp. Since these species feed on the algae, they are also herbivores. Other types of consumers are dependent upon the algae also. Among these are the secondary consumers, small fish that constitute the carnivores that feed on the herbivores. Then there are the larger fish that feed on the smaller species of fish. Dwelling on the bottom sediments are the scavenger worms and snails that feed on the dead plant and animal bodies that fall to the bottom. Living among the sediments are the decomposers—the fungi and bacteria—that break apart plant and animal remains.

THE ECOLOGICAL NICHE

An important concept of ecology is the **niche**. An ecological niche is a feeding pattern exhibited by species that compose a community, that is, a feeding way-of-life in relationship to other organisms. For example, small woodpeckers and nuthatches feed on grubs that are present in the crevices of trees. Although the woodpecker and the nuthatch feed on grubs present in the same tree, these species are not in competition with each other. The woodpecker searches for its food in the crevices at the bottom of the tree and works its way upward. The nuthatch functions best at the top of the tree and works its way downward. These two species may live close together and may feed on similar organisms on the same tree, but they select grubs from different locations on the tree. Therefore, it can be said that the woodpecker and the nuthatch occupy different ecological niches.

When two species live in the same place and feed in the same way, using the same food at the same time, *competition* results. The species that has special adaptations for reaching the food first or has greater reproductive potential will be the survivor. The other species will be eliminated. The process by which elimination establishes one species per niche in a particular habitat is known as the **competition-exclusion principle**, illustrated by the zebra mussel (Figure 18.1).

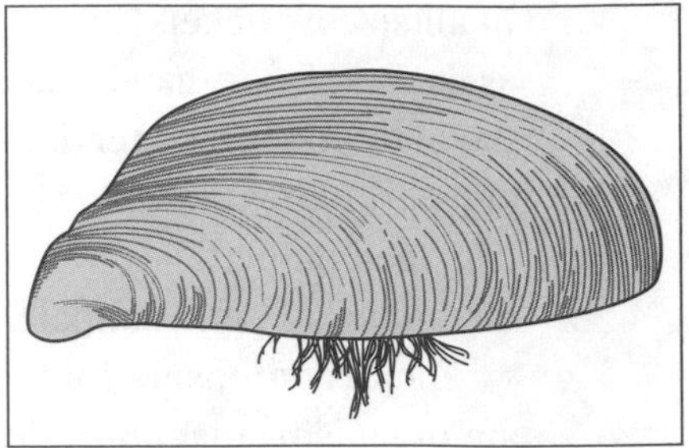

FIGURE 18.1 Zebra mussels eat up microscopic plants, depriving fish of food, reproduce prolifically, and clog up utility pipes in the Hudson River.

Energy Flow in an Ecosystem

The source of all energy in an ecosystem is the sun. Autotrophs are able to capture just a small portion of the sun's energy and use it to make food enough for all of the living organisms in the ecosystem. The green plant is able to store energy temporarily in ATP molecules and in the nutrients that it makes. Energy is then transferred from green plants into the animal body where it is used to power the vital functions necessary for life. Once energy is used to do work it is converted into heat, which then escapes the body and radiates into the atmosphere. The cycles of photosynthesis (energy trapping and conversion) and respiration (energy release and use) must be repeated *ad infinitum* if the ecosystem is to continue.

ENERGY AND FOOD CHAINS

The flow of energy through an ecosystem can be studied by way of **food chains**, which show how energy is transferred from one organism to another through feeding patterns. An example of a food chain on a cultivated field might be as follows:

Lettuce → Rabbit → Snake → Hawk

Each stage in the food chain represents a feeding or *trophic level*. Lettuce is the producer, the green plant that provides the food that supports the ecosystem. The rabbit is the primary consumer and represents the second trophic level. The carnivorous snake, the secondary consumer, feeds upon the rabbit and thus represents the third trophic level. This food chain ends with the hawk, the tertiary consumer, which occupies the smallest trophic level in terms of energy. Every food chain begins with an autotroph and ends with a carnivore that is not eaten by a larger animal. When carnivores die, their bodies usually serve as food for scavengers. The bacteria and fungi of decay decompose the remains not devoured by the scavengers.

The flow of energy in a food chain is in a straight line pattern. Most of the energy is concentrated in the level of the producer. At each succeeding level the energy is decreased. However, the feeding relationships among organisms in an ecosystem are usually, in actuality, more complex.

ENERGY AND FOOD WEBS

Let us look again at the same simple food chain:

$$\text{Lettuce} \rightarrow \text{Rabbit} \rightarrow \text{Snake} \rightarrow \text{Hawk}$$

Suppose all of the rabbits disappear from the cultivated field. Will the snakes die of starvation? The answer to this question can be seen in Figure 18.2, which illustrates a food web. Notice that the snake can feed upon a shrew, mouse, or owl. In other words, other **primary consumers** (herbivores) serve well as food for snakes. Figure 18.2 shows that there are several alternative energy pathways in a **food web**. It is the alternative pathways that enable an ecosystem to keep its stability. One species does not eradicate another in the quest for food.

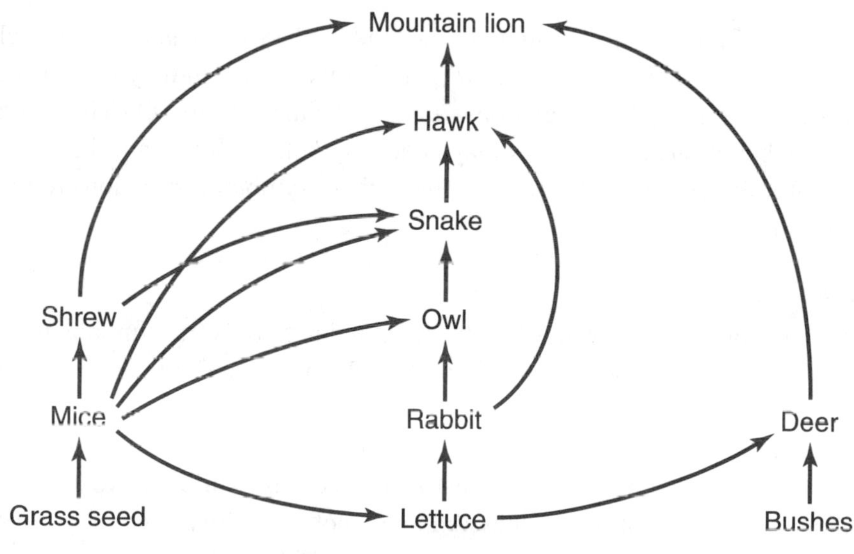

FIGURE 18.2 A food web

The relationships between predators and prey help to keep an ecological community stable. The predators help to keep the populations upon which they feed in check. For example, if rabbits were allowed to reproduce without their numbers being thinned by natural predators, the population of rabbits would become overwhelming. The burgeoning rabbit population would devour the available vegetation and then the species would experience starvation and death. Such imbalances in nature do occur when humans upset the natural stability of an ecosystem. Past experience has shown that ecosystems are upset when new species are introduced into an area where there are no predators or natural enemies of the species. Figure 18.3 shows a **food pyramid**, another way of illustrating energy flow in an ecosystem. Notice that the autotrophs at the base of the pyramid support all of the heterotrophs (consumers) that exist at each nutritional level and that there is a decrease of available energy at each level.

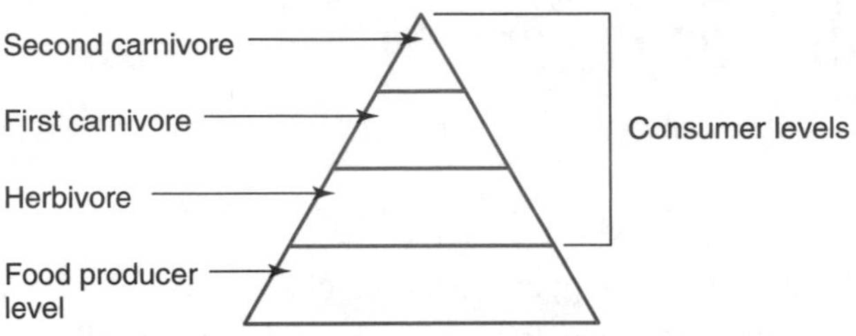

FIGURE 18.3 Food pyramid

NUTRITIONAL RELATIONSHIPS

SYMBIOSIS

Taken from the Greek word meaning "living together," **symbiosis** indicates a close association of *two organisms of different species*. One organism may be considered the host; the other, the *visitor*. The relationship may take three forms. (1) The visitor and the host both benefit. (2) The visitor may be neither helpful nor harmful to the host. (3) The visitor may harm the host. Each one of these symbiotic situations represents a special nutritional relationship.

Mutualism

In this relationship two organisms live together and both benefit. Mutualism is illustrated by the relationship between a crocodile and a bird, the Egyptian plover. When the crocodile opens its mouth, the plover jumps in and picks off the leeches that are attached to the gums of the crocodile. The crocodile keeps its mouth open while the bird is inside. The bird gets its food by eating the leeches, while the crocodile is relieved of the irritation of the leeches. Although both the plover and the crocodile benefit from this relationship, each species can live without the other.

The **lichen** (like-en) is a combination of two different species that live together so closely that the whole structure is mistaken for a single plant-like organism. A lichen consists of a fungus and an alga. The fungus gathers water through its hyphae, and the alga uses this water to make food through photosynthesis. The fungus obtains food, and the alga gets water and a place to live—anchored to the fungus. In this case, neither species can live without the other.

Commensalism

In this relationship, one species (the visitor) lives on another species (the host) without doing harm to it. A suckerfish (*remora*) attaches itself to the underside of a shark and is transported from place to place by the shark. The suckerfish also eats food droppings from the shark's mouth. However, the shark is not hurt or inconvenienced in any way by the suckerfish.

Parasitism

In a parasitic relationship, the host is always harmed by the visitor (a parasite). In nature, parasitism is widespread and is a successful way of life for the parasite. Bacteria that invade the body and cause disease are parasites, as are fleas or ticks that infest the external surfaces of animals. Infectious worms that inhabit the internal bodies of other animal species are also parasites. Whether external or internal, parasites benefit by absorbing food from the body of the host and doing harm to the host.

A vivid example of parasitism is the guinea worm (Figure 18.4), which spends the major part of its life cycle in the human body. Infestation runs high in Africa and India. The guinea worm is contracted by drinking pond water infested with almost invisible water fleas that have eaten worm larvae. Gastric juices kill the fleas but do not harm the worm larvae, which pass through the stomach lining. After mating, the males die, while the females, carrying new larvae, pass to other parts of the body. The worms become 2–3 feet long and form blisters on the feet and legs of the host. Worms slowly emerge from these blisters. When the victim enters the water, hundreds of thousands of larvae are released, continuing the cycle. Some parasites, as illustrated by the guinea worm, have extremely complicated life cycles, living in two hosts. The Chinese liver fluke divides its life cycle among three hosts: snail, fish, and human. The West Nile virus has two hosts: mosquitoes and humans, or mosquitoes and birds.

FIGURE 18.4 Life cycle of the guinea worm

PREDATOR-PREY RELATIONSHIPS

A **predator** is an animal that hunts another animal to use as food. Lions hunt and kill antelopes. Cats stalk birds and mice. A predator kills one animal at a time for food. The predator lives an independent life, however, not inside of or attached to another animal.

Biogeochemical Cycles

Certain compounds cycle through the abiotic portion and the biotic communities of ecosystems. These compounds contain elements that are necessary to the biochemical processes that are carried out in living cells. Among these elements are carbon, hydrogen, oxygen, and nitrogen. In elemental form, they are useless to cells and must be combined in chemical compounds. Let us trace the pathways of some of the vital compounds from Earth to living organisms to the atmosphere and then back to Earth. This cycle of events is best described by the term **biogeochemical** cycle (Figure 18.5).

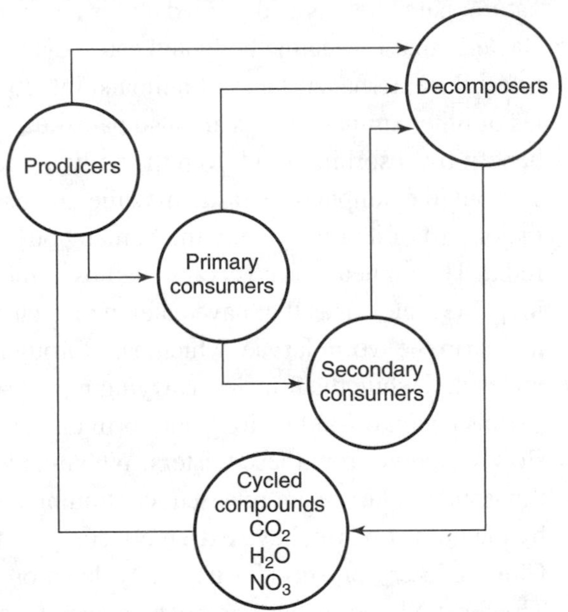

FIGURE 18.5 Biogeochemical cycle

THE WATER CYCLE

Water is the source of hydrogen, one of the elements necessary for the synthesizing of carbohydrate molecules by green plants. Water is necessary as the dissolving medium for substances that cross cell membranes and enter cells. Without water, there can be no life.

There are three ways in which water vapor enters the atmosphere. (1) Water evaporates from land surfaces and from the surfaces of all bodies of water. (2) Water vapor enters the air as a waste product of respiration of animals and plants. For example, every time you exhale, water vapor is released into the air. (3) Great amounts of water are lost from plants through the openings in the leaves; this water loss due to evaporation is called *transpiration*.

Water vapor in the air is carried to high altitudes where it is cooled and forms clouds by condensation. Clouds dissipate and their moisture falls to earth in the form of precipitation. Most of the precipitation returns to the oceans, lakes, or streams; less than 1 percent of it falls on land. Of the water that does fall on land, about 25 percent will evaporate from the various land surfaces before it can be absorbed by plants or used by animals. Water that does not evaporate enters the soil and becomes available to plant roots and soil organisms.

Soil water that is not absorbed by plants seeps down into the ground until it reaches an impervious layer of rock. The water moves along this rock as **groundwater** until it reaches an outlet into a larger body of water such as a lake or an ocean. The water cycle, illustrated in Figure 18.6, then repeats.

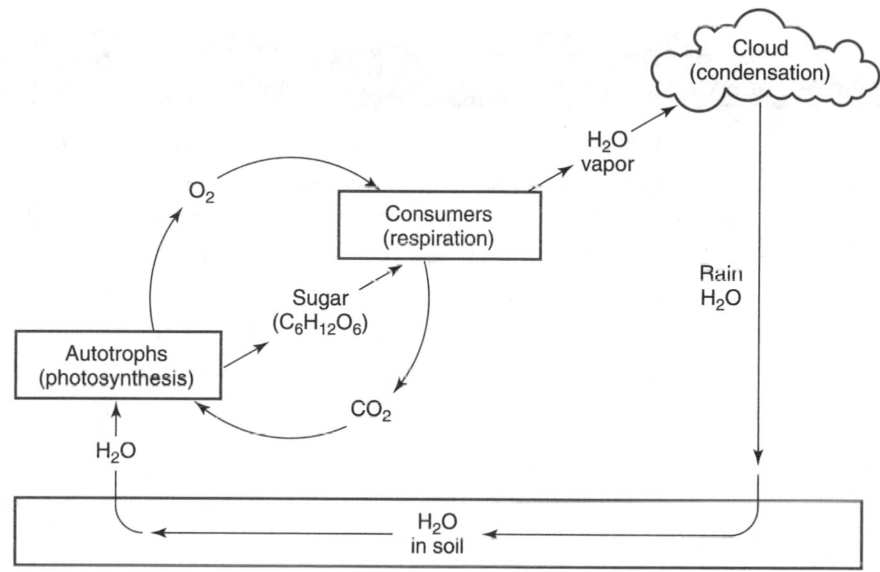

FIGURE 18.6 Water cycle

CARBON DIOXIDE/OXYGEN CYCLE

Carbon dioxide and oxygen cycle through the abiotic and biotic components of ecosystems. Oxygen is released from plants into the atmosphere during photosynthesis. Oxygen from the air is taken in by animals and used during cellular respiration. Carbon dioxide, a waste product of respiration, is released into the air by animals and plants. Plants take in the carbon dioxide and use it during photosynthesis. The cycle repeats.

NITROGEN CYCLE

About 80 percent of the volume of the air is elemental nitrogen. Although it is an important component of proteins, nitrogen must be in a combined form before it can be used by the cells of living things.

Nitrogen becomes available to plants through the action of nitrogen-fixing bacteria that change atmospheric nitrogen into nitrates, which are used by the plants for protein synthesis.

Nitrogen is also fixed by lightning. The energy produced by a discharge of lightning joins nitrogen with oxygen to form nitric oxides. These oxides combine with water vapor to form nitrous and nitric acids. When these acids fall on soil with rainwater, they are changed into nitrites and nitrates.

Denitrifying bacteria in the soil change nitrates back to atmospheric nitrogen. The cycle then repeats.

The Limiting Factor Concept

There are environmental factors such as light, temperature, and amount of rainfall that set the limit of conditions for species survival. Every species can live within a range of values for each factor in its physical environment but cannot survive beyond the low and high values called the **limits of tolerance**. For example, the eggs of the frog *Rana pipiens* exhibit a limit of tolerance for the temperature range between 0° and 30°C. The eggs of *Rana pipiens* will die in temperature ranges below or above these figures. However, at 22°C more *Rana pipiens* eggs hatch than at any other temperature. This means that 22°C is **optimum temperature** for the survival of this species.

LIGHT

The effect of light as a **limiting factor** was demonstrated in 1920 when W. W. Garner and H. A. Allard showed that day length was an important factor in the flowering of plants. The physiological response made by plants to changes in daylength is known as **photoperiodism**. Some plants (short-day) flower only if they are exposed to light for *less than* a certain amount of time each day; other plants (long-day) must have a certain minimum length of photoperiod. Researchers explain the ability of plants to measure time by the action of phytochrome, a light absorbing pigment that is associated with the cell membrane and with some of the cell's internal membranes.

TEMPERATURE

Between 0°C and 40°C is the range of temperature between which most species are active. In temperatures below 0°C there is a slowing down of biochemical processes. Freezing causes spicules of ice to form in cells; the results are lysing of the cells and disruption of cell processes. Above 40°C, the protein of most species denatures, rendering enzymes, hormones, and the structural proteins of cells useless. However, some forms of life are able to exist outside the range of 0° to 40°. The seeds of most deciduous trees must go through a period of cold, very often in temperatures far below the freezing point of water, before they can germinate. It is also a fact that certain microorganisms, such as bacteria and some species of algae, can live in hot springs and geysers, where temperatures may exceed 90°C.

Warm-blooded mammals that live in cold climates have developed physiological mechanisms for preserving body heat. The combination of large amounts of body fat and the ability to hibernate during the most forbidding of the cold months serves to maintain the lives of these organisms. On the other hand, cold-blooded animals have body temperatures that are quite similar to the temperature of the external environment. They too become inactive in cold weather and then require protective shade during the hot season.

Temperature is a factor that regulates the growth of plants and the activities of animals. There is no doubt that plant and animal hormones evoke physiological changes in response to the rise and fall of temperature.

RAINFALL

Rainfall is also a limiting factor. Land-dwelling plants and animals depend upon rainfall as the source of water. Species that live successfully in deserts or in areas where there is seasonal drought have developed adaptations for water conservation. The kangaroo rat lives in humid burrows under the desert sands where, huddled together with others of its kind, it licks the condensed drops of water vapor from the hair of its burrow mates. The lungfish (*Dipnoi*) estivates under the mud in dried ponds, awaiting the seasonal downpour of rain that will fill up its habitat and restore the species to active life. The seeds of several flowering plant species of the desert will germinate only when a heavy rain wets the substrate around them; the rainfall seems to remove from the seed coat a chemical compound that inhibits germination. Similarly, species in areas of heavy rainfall have special adaptations. Some plant species in the rain forests of Brazil have developed adaptations such as aerial roots that can absorb moisture from the air rather than becoming waterlogged in too wet soil.

Greenhouse Effect

The term **greenhouse effect** refers to the fact that the earth is surrounded by an atmosphere that acts similarly to the glass walls of a greenhouse. In the earth's atmosphere are a number of gases, both naturally occurring and human generated, which trap heat from the sun. These gases, like greenhouse walls, let in light and prevent heat from escaping. The greenhouse effect has been important in the evolution of life on earth. It has provided a favorable environment for the development of the human race, and in fact it is unlikely that humans would exist without the greenhouse effect because the earth would be about 60 degrees colder than it is, too cold to sustain human life.

Greenhouse gases include carbon dioxide (CO_2), methane (CH_4), nitrous oxide (N_2O), water vapor (H_2O), and a number of synthetic fluorinated gases. Greenhouse gases enter the atmosphere naturally as a result of naturally occurring water vapor and carbon dioxide production, weather and geological activity, as well as swamp and livestock methane emissions. Greenhouse gas levels have been tested using ice core samples and have been found to have remained relatively stable until the Industrial Revolution. Since the late 1800's, however, human activities have generated significant increases in greenhouse gas emissions as a result of deforestation, combustion of fossil fuels, use of fertilizers, other industrial and agricultural activities, decay of organic waste, and the use of some aerosol propellants, refrigeration, and fire suppression products.

Global Warming

The greenhouse effect is intimately related to **global warming**, a gradual and sustained increase in the earth's surface temperature. Scientists have expressed concern that increases in greenhouse gases as a result of human activity may be causing irreversible damage to the earth. Excessive increases in these gases traps solar radiation in the atmosphere and warms the earth. As a result, cloud cover and vegetation may be changed in ways that intensify the warming influence of gas pollutants. For example, increased carbon dioxide in the air will generate faster vegetative decay, releasing carbon dioxide back into the air more quickly. The resulting increased carbon dioxide in the air will speed up plant growth which will reduce airborne carbon dioxide. Faster plant growth will produce more vegetative waste to decay, producing even more carbon dioxide. This circular series of events is an example of **feedback**. Scientists have identified several feedback mechanisms that influence and are influenced by greenhouse gases and which, potentially, contribute to global warming.

While most scientists agree that some level of global warming exists, there is debate within the scientific community about the degree of risk that it presents. The threat of continued global warming has been associated with melting polar ice caps, fluctuations of sea levels, changes in weather patterns, altered land masses and ecosystems, and the potential extinctions of species. Scientists are trying to develop effective models that will demonstrate how human endeavors can reduce the greenhouse effect and the intensification of global warming.

Ecological Succession

It was stated before that an ecosystem consists of a living community and its nonliving environment. As physical conditions change in an ecosystem, so do the plants and animals that inhabit that system. The orderly change of the biotic community in an ecosystem is known as **ecological succession**. There are two levels of succession: primary and secondary.

As the name implies, primary succession is the introduction of living species into a region that was formerly barren. Let us suppose that a rocky cliff is devoid of soil. By the processes of erosion and weathering, the rock is pulverized. Lichens that inhabit the bare rock contribute to its chemical erosion. Dead lichens contribute organic matter to the pulverized rock material. Now the physical conditions have changed just enough to allow the growth of mosses. These too add organic matter to the substrate and also contribute some water-holding ability to the newly formed soil. The moss population is followed by the grasses and then the ferns. In orderly succession there follow the low-growing bushes, the highergrowing shrubs, and then trees. Each population inhabiting the region makes it better for another species.

Primary succession culminates in the existence of a stable community, which is known as a **climax community**, where one or two large trees predominate. Over a period of years secondary succession may occur in which one species of tree is gradually replaced by another. For example, white pine trees may replace gray birch trees. Limiting factors such as temperature, light, rainfall and mineral content of the soil influence the type of climax community. A climax community in a broad geographical area having one type of climate is known as a **biome**.

World Biomes

Earth is divided into several major regions called biomes, in which temperature, precipitation, populations, and geography are consistent and similar; these are illustrated in Figure 18.7.

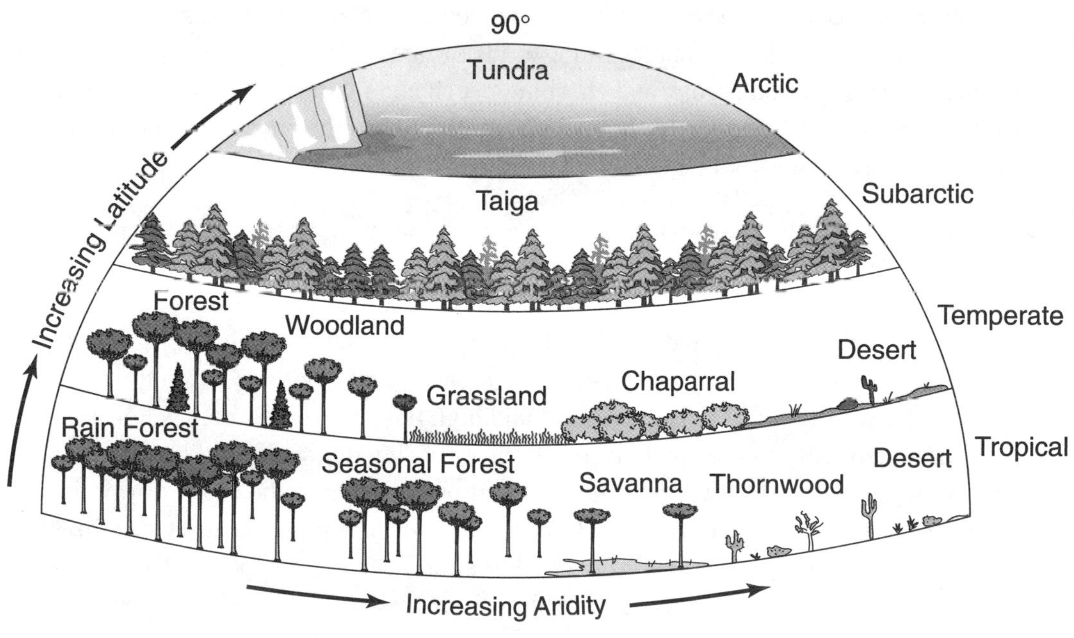

FIGURE 18.7 World biomes

THE TUNDRA

Vast stretches of treeless plains surrounding the Arctic Ocean where cold is the limiting factor (60°F in the summer to –30°F in the winter) are known as the arctic tundra. Here the ground is permanently frozen a few feet below the surface and is responsible for the many lakes and bogs that characterize the region. The kinds of organisms that inhabit the tundra have developed adaptations for survival in extremely cold temperatures. The plant species consist of lichens, mosses, grasses, and sedges. During the summer, the flowering herbs bloom in brilliant color for a very short time along with the dwarf willows. This seemingly meager plant life is able to support the food chains of the tundra. It feeds thousands of migrating birds and insect swarms that appear during the brief summer. Many mammals, such as musk oxen, caribou, polar bears, wolves, foxes, and marine mammals, remain active for most of the year.

THE TAIGA

The coniferous forests of Canada survive well in the long severe winters. The spruce and fir trees predominate. The kinds of mammals that inhabit the taiga are the black bear, the wolf, lynx, and squirrel.

THE DECIDUOUS FOREST

The forests of the temperate regions are dominated by broad-leaved trees that lose their leaves in the winter. Examples of the kinds of trees that compose these hardwood forests are oak and hickory, oak and chestnut, beech and maples and willows, cottonwood and sycamore. The types of animals inhabiting these forests are deer, fox, squirrel, skunk, woodchuck, and raccoon.

THE DESERT

Deserts form in regions where the annual rainfall is less than 6.5 centimeters, or where rain occurs unevenly during the year and the rate of evaporation is high. The temperature changes drastically from hot days to cold nights. Plants that survive in the desert have specific adaptations for low moisture and high temperature. Shrubs, such as creosote and sagebrush, shed their short, thick leaves during dry spells and become dormant as protection against wilting. Fleshy desert plants, such as the cacti of American deserts and the euphorbias of African deserts, store water in their tissues. Cheat grass and wild flowers are annuals that grow quickly after a desert downpour, bloom, produce seed, and die. Mosses, lichen, and blue-green algae lie dormant on the sand and become active when moisture is present. The animals of the desert include lizards, insects, kangaroo rats, and arachnids.

THE GRASSLANDS

Grasslands occur where the annual rainfall is low and irregular. Grasslands usually occupy large areas of interior continents that are sheltered from moisture-laden winds. In the United States the grasslands are known as the Great Plains; in Russia they are called the *steppes*; in South Africa, the *veldt*; and in South America, the *pampas*. Grasses have adaptations for living in soil where the rainfall is low and erratic. Grazing animals are suited for life on the grasslands. At one time the American grasslands supported huge herds of antelope and bison. The settlers replaced these natural herds with cattle and sheep. Besides the domesticated animals, a number of predator species inhabit the grasslands, including coyotes, bobcats, badgers, hawks, kit foxes, and owls. These feed primarily on burrowing rodents.

THE TROPICAL RAIN FOREST

The tropical rain forest is an incredibly complex biome, characterized by consistently high temperatures and constant rainfall. This type of biome is found in equatorial Central and South America, in Southeast Asia, and in West Africa. The trees are tall and the vegetation is stratified into four layers: emergent trees, upper canopy, understory (or lower canopy), and forest floor. Vegetation is so thick that the forest floor is shaded from light, and soil quality tends to be poor. This biome contains the greatest diversity of any terrestrial biome. In fact, it is estimated that while tropical rain forests make up only 7 percent of the earth's surface area they contain more than 50 percent of the world's species. One of the opportunities provided by this huge biodiversity is in pharmaceutical discovery and research; the search for new drugs is called bioprospecting. The animals of the rain forest include monkeys, lizards, snakes, insects, and birds.

THE MARINE BIOME

The marine biome, composed of oceans, estuaries, and coral reefs, is the world's most stable biome. This biome has tremendous impact on the rest of the earth; its algae supplies a large percentage of the earth's oxygen and takes in huge amounts of atmospheric carbon dioxide; it is a source of food and water; and it stabilizes temperatures and climates around the world. Ocean waters cover almost three fourths of Earth's surface and support the greatest abundance and diversity of organisms in the world. Averaging 3.5 to 4.5 kilometers in depth, a marine biome constitutes the thickest layer of living things in the biosphere. The dominating physical factors determine the type of living organisms that compose its communities. Temperature, light intensity, salinity, waves, tide currents, and pressures are the limiting factors that set the conditions for life.

Light in the ocean extends for a depth of 180 meters; beyond this there is total darkness. The upper layers of water are known collectively as the continental shelf, a region that supports a vast array of living things. Phytoplankton live at the surface and serve as food for primary consumers such as sardines and anchovies. These herbivores serve as food for larger fish such as salmon, mackerel, and tuna. Mud-burrowing animals such as clams, snails, worms, and shrimp are eaten by crabs, lobsters, starfish, and a number of different kinds of fish such as cod, halibut, haddock, ray, and flounder. In the dark region of the sea, phytoplankton are absent. The small animals that live at the bottom of the sea feed on organic debris that falls from above. The temperature of the sea is fairly constant. Its salinity lowers the freezing point of water.

Humans and the Biosphere

People influence the biotic and the abiotic environments in which they live.

NEGATIVE EFFECTS OF HUMAN ACTIVITIES

Unfortunately, the changes that humans have produced have, for the most part, been negative. These changes are summarized in the table below.

EFFORTS TO CORRECT HUMAN ERRORS

Increased awareness about the ecology of our environment has prompted attempts to stop the destruction. People have become aware of the life-threatening dangers of chemical pollution brought about by irresponsible dumping of chemical wastes. Powerful interest groups have been able to have laws enacted to prevent the illegal hunting of wildlife, and wildlife preserves have been established to protect endangered species. Methods to control the human birth rate have been developed and shared with underdeveloped nations. Conservation measures are currently being applied to conserve arable land.

EFFECTS OF HUMANS ON THE ENVIRONMENT

Effect	Description
Population Growth	The human population of the world has grown so explosively that ecosystems cannot produce food to serve human needs.
Overhunting	Indiscriminate and uncontrolled hunting of wildlife has resulted in the extinction of such species as the great auk, dodo bird, passenger pigeon, Carolina parakeet, and heath hen. A number of other species are threatened with extinction: the blue whale, wild turkey, bison, grizzly bear, and American bald eagle.
Importation of Organisms	By accident or with definite intent, people have transported organisms into ecosystems where they have no natural enemies. Every year the Japanese beetle destroys innumerable species of green plants in the United States, at a cost of millions of dollars. The gypsy moth and the fungus that has destroyed elm trees are other examples of imported organisms that cause damage because of lack of natural enemies.

EFFECTS OF HUMANS ON THE ENVIRONMENT (continued)

Effect	Description
Exploitation	Humans have exploited wildlife for their own profit. Illegal hunting of the African elephant and the Pacific walrus for ivory, capture of the Colombian parrot for the pet trade, and the cutting down of trees in the tropical rain forests for plywood have upset ecosystems.
Thoughtless Land Use Management	The increased use of agricultural lands for the construction of houses, parking lots, huge shopping malls, and the like has destroyed the natural habitats of wildlife. In addition, poor use of agricultural land in the forms of overcropping and overgrazing has resulted in the loss of valuable nutrients from the soil.
Biological Magnification	The process by which toxins, such as pesticides and heavy metals, enter and ascend the food chain. These chemicals first enter rivers and lakes where they are ingested by worms, crustaceans, or small fish, which are then eaten by larger fish. Birds, other animals, and humans eat the contaminated organisms. The toxicity of the original substances becomes more concentrated in tissues or organs as they move up the food chain.

Conservation

The **conservation** of natural resources is of primary importance to humans if the land on which they live is to be preserved for future generations. Natural resources are the air we breathe, the water we drink, the soil that supports plant and animal life, and the minerals under the ground. All of these non-human-made resources are vital to the well being of nations. Ecologists think of natural resources as *renewable* and *nonrenewable*. Forests, grasslands, and the wild species that inhabit them can be restored after their numbers have been somewhat depleted. These are the renewable resources. Once used up, minerals, including the fossil fuels of oil and coal, cannot be restored and therefore are classified as nonrenewable. Desirable conservation practices must be the concern of every human. Understanding principles of ecology and of conservation is paramount to the making of wise decisions.

SOIL CONSERVATION

The soil covering Earth is really a very thin layer measuring about 0.9 meter in depth. Soil is not made quickly. Thousands of years, much erosion and weathering of rock, and the gradual addition of plant and animal remains are needed to make fertile topsoil. Topsoil can be quite fragile and easily destroyed by **sheet erosion**, that is, washed away from sloping hills by rain. Soil may also become exhausted or **leached**, having been overcultivated so that the mineral components are lost.

Conservation measures protect topsoil. Erosion can be prevented by the wise use of ground cover, rooted plants that hold the soil in place. **Terracing**, the planting of trees, bushes, and low-growing plants in steps cut into hillsides, prevents loss of soil by sheet erosion. **Contour plowing**, the circular plowing of hillsides, is another method of preventing the erosion of soil. The leaching of minerals and nitrates from the soil is prevented by **strip cropping** (planting corn or cotton with alternate rows of legumes) and **crop rotation** (planting a field on alternate years).

CONSERVATION OF WATER

Water is rendered useless for drinking, for bathing, for irrigation, and as a habitat for fish when it is polluted by the chemical wastes from industry and by human sewage. Sewage treatment plants clean up sewage before it is dumped into waterways. Special treatment must be given to chemical wastes to detoxify them before disposal.

Water is a renewable resource. However, people are using more water than ever before in industry, refrigeration, agriculture, and the like. Humans are dependent on rainfall to maintain an adequate **water table** (level of groundwater) and to replenish water stores in reservoirs. The wasting of water through careless use can have serious consequences for human life.

CONSERVATION OF FORESTS

Forests are important in the conserving of water, soil, and habitats for wildlife. Forests also supply us with lumber and natural resins that are used in the paint industry and for other industrial processes. Forests are destroyed by over-lumbering through which more trees are cut down than can be replaced by natural processes. Trees grow slowly, requiring about thirty or forty years to reach maturity. We need wood to make paper and to build houses, telephone poles, and fences. However, if forests are lumbered imprudently, trees will not last. Most of the virgin forests in the United States have been destroyed, and now in their places are second growth forests.

REVIEW EXERCISES FOR CHAPTER 18

WORD-STUDY CONNECTION

abiotic	biogeochemical cycle	biotic
autotroph	biosphere	carnivore

climax
commensalism
community
conservation
crop rotation
cycling
deciduous
decomposers
ecological succession
ecology
ecosystem
erosion
food chain
food pyramid
food web
global warming
greenhouse effect
herbivore
heterotroph
limiting factor
mutualism
natural resources
niche
nitrogen fixation
parasitism
photoperiodism
population
predator
prey
primary consumer
producer
scavenger
secondary consumer
strip cropping
succession
symbiosis
taiga
tertiary consumer
tolerance
trophic level
tundra

SELF-TEST CONNECTION

PART A. Completion. *Write in the word that correctly completes each statement.*

1. The living factors in the environment are known as _____ factors.
2. The living community and the nonliving environment are known collectively as the _____.
3. Without the work of _____ the remains of dead organisms would pile up.
4. A nuthatch is a species of _____.
5. The number of species that can successfully occupy a niche is _____.
6. The flow of energy through an ecosystem can be studied by identifying simple _____ (two words).
7. An herbivore is also classified as a _____ consumer.
8. A great deal of water is lost from plants by an evaporation process known as _____.
9. About 80 percent of the volume of air is the elemental gas _____.
10. Among the limiting factors in the environment are light, temperature and _____.
11. Phytochrome is a plant hormone that controls _____.
12. The seeds of most deciduous plants must go through a period of _____ before they can germinate.

13. The compound that all living things need for the dissolving of chemical substances is _____.

14. The orderly change of a biotic community is known as _____.

15. A geographical area having one type of climate is a _____.

16. The general term meaning "living together" is _____.

17. A lichen represents an intricate living-together combination of an alga and a _____.

18. Deer ticks are appropriately classified nutritionally as _____.

19. Excessive carbon dioxide and methane gases continually released into the atmosphere cause the _____ effect.

20. The coldest world biome is the _____.

PART B. Multiple Choice. *Circle the letter of the item that correctly completes each statement.*

1. The nonliving factors in an ecosystem are referred to as
   (a) adiabatic
   (b) abomasum
   (c) abiotic
   (d) abscissant

2. The percent of the air that is made up of oxygen is
   (a) 21
   (b) 50
   (c) 78
   (d) 90

3. The habitat of phytoplankton is the
   (a) deciduous forest
   (b) ocean surface
   (c) Nairobi desert
   (d) Canadian taiga

4. An ecological niche is a (an)
   (a) organism
   (b) geographical region
   (c) habitat
   (d) feeding pattern

5. The lives of organisms in an ecosystem depend upon the
   (a) amount of light in the ecosystem
   (b) quality of environment
   (c) level of pH in the ecosystem
   (d) flow of energy through the ecosystem

6. In a food chain, lettuce is best classified as a
   (a) producer
   (b) deciduous plant
   (c) vegetable
   (d) detritus

7. The direction an arrow points in a food web diagram stands for
   (a) an alternative energy pathway
   (b) increased energy
   (c) "is eaten by"
   (d) "eats up"

8. In a food web, hawks are best classified as
   (a) scavengers
   (b) predators
   (c) prey
   (d) primary consumers

9. In plant cells, phytochrome is associated with
   (a) nuclei
   (b) cytoplasm
   (c) lysosomes
   (d) plasma membranes

10. The washing away of topsoil from the slopes of hills is known as
   (a) weathering
   (b) erosion
   (c) irrigation
   (d) leaching

11. An example of an adaptation that permits the survival of some plant species in the tropical rain forest is
   (a) degenerate stomates
   (b) needlelike leaves
   (c) aerial roots
   (d) cutinized stems

12. Chemical changes on the surfaces of rock may be brought about by
    (a) lichen colonies
    (b) heavy rainfall
    (c) tree roots
    (d) germinating bryophytes

13. Musk ox, caribou, and foxes are part of the mammal population of the
    (a) temperate forest
    (b) taiga
    (c) grasslands
    (d) tundra

14. In desert biomes, the average rainfall is no greater than
    (a) 20 cm
    (b) 6.5 cm
    (c) 15 cm
    (d) 0 cm

15. Examples of primary consumers in the marine biome are
    (a) halibut and cod
    (b) ray and haddock
    (c) sardine and anchovy
    (d) cod and haddock

16. A biome is best defined as
    (a) a feeding pattern of worldwide internal parasites
    (b) major geographic, climatologic regions with consistent inhabiting populations
    (c) the plants and animals in the biosphere
    (d) a flourishing geographical area of a country

17. Aphids live in the intestines of termites and digest cellulose for the termites. The termites provide the aphids with a protective environment. This relationship is best classified as
    (a) symbiosis
    (b) commensalism
    (c) saprophytism
    (d) mutualism

18. One effect of global warming predicted by research geochemists is the
    (a) premature melting of ice and snow
    (b) disappearance of the polar bear
    (c) decrease in the growth of vegetation
    (d) major changes in modes of symbiosis

19. The organism *Treponema pallidum*, which can infect the human genital tract, belongs to the group called
    (a) saprophytes
    (b) viruses
    (c) parasites
    (d) fungi

20. A subarctic biome is designated as
    (a) taiga
    (b) temperate
    (c) tundra
    (d) tropical

PART C. Modified True-False. *If a statement is true, write "true" for your answer. If a statement is incorrect, change the underlined expression to one that will make the statement true.*

1. The cultivation of a field in alternate years is known as <u>crop rotation</u>.

2. Oil, coal and natural gas are <u>renewable</u> resources.

3. <u>Straight row</u> plowing prevents the erosion of hillsides.

4. All of the plant and animal species interacting in a given environment are known as a population.

5. Tertiary consumers are those that feed on smaller <u>herbivores</u>.

6. Vultures are best classified as <u>predators</u>.

7. In ecosystems, fungi and bacteria fill the roles of <u>scavengers</u>.

8. Competition results when two species share the same <u>habitat</u>.

9. Once energy is used, it is converted into <u>fat</u>.

10. A process in which energy is released is <u>photosynthesis</u>.

11. Most of the energy in a food chain is concentrated at the <u>carnivore</u> level.

12. The alternative pathways of energy flow in an ecosystem are best shown in a food <u>pyramid</u>.

13. The relationships between predators and prey help to <u>disrupt</u> an ecological community.

14. The source of hydrogen utilized by green plants for photosynthesis is <u>sugar</u>.

15. A <u>changing</u> community is known as a climax community.

16. In the classification scheme of animals, the plover is appropriately grouped with the <u>mammals</u>.

17. Because of the greenhouse effect, less radiation will be deflected and therefore <u>less</u> radiation will be absorbed by dark earth.

18. In a parasitic relationship, the parasite is always <u>harmed</u> by the relationship.

19. The life cycle of the Chinese liver fluke involves <u>two</u> hosts.

20. Gas pollutants have a <u>warming</u> influence on the atmosphere.

CONNECTING TO CONCEPTS

1. What is an ecosystem?

2. How does an ecological niche differ from an ecosystem?

3. What is the difference between a food chain and a food web?

4. Why is it important to preserve our forests?

5. How might the guinea worm be eradicated?

ANSWERS TO SELF-TEST CONNECTION

PART A

1. biotic
2. ecosystem
3. decomposers
4. bird
5. one
6. food chains
7. primary
8. transpiration
9. nitrogen
10. rainfall
11. flowering
12. cold
13. water
14. succession
15. biome
16. symbiosis
17. fungus
18. parasites
19. greenhouse
20. tundra

PART B

1. **(c)**
2. **(a)**
3. **(b)**
4. **(d)**
5. **(d)**
6. **(a)**
7. **(c)**
8. **(b)**
9. **(d)**
10. **(b)**
11. **(c)**
12. **(a)**
13. **(d)**
14. **(b)**
15. **(c)**
16. **(b)**
17. **(d)**
18. **(a)**
19. **(c)**
20. **(a)**

PART C

1. true
2. nonrenewable
3. Contour
4. community
5. carnivores
6. scavengers
7. decomposers
8. niche
9. heat
10. respiration
11. autotroph
12. web
13. stabilize
14. water
15. stable
16. birds
17. more
18. helped
19. three
20. true

CONNECTING TO LIFE/JOB SKILLS

Science writers combine two special talents: an overwhelming interest in science and the ability to write. Persons in this field are employed by drug companies, newspapers, magazines, wildlife organizations, conservation groups, and insurance companies. In addition to college-level work in biology, science writers take courses in communication and journalism. Facility with the computer is essential.

Chronology of Famous Names in Biology

1840 **Justus von Liebig** (Germany)—developed the ecological concept of the "law of the minimum," in which he showed the limiting effect on growth of environmental factors.

1869 **Ernst Haeckel** (Germany)—coined the word *ecology* to name the science that studies environment.

1920 **W. W. Garner** and **H. A. Allard** (United States)—discovered that daylength affects the flowering of the Biloxi soybean and also the flowering of a mutant form of tobacco called "Maryland Mammoth."

1938 **K. C. Hamner** and **J. Bonner** (United States)—discovered how the cocklebur responds to the photoperiod.

1939 **George Washington Carver** (United States)—discovered that legumes such as peanuts returned nitrates and minerals to the soil. He developed 300 new industrial uses for peanuts and 118 by-products from sweet potatoes.

1942 **Elton Charles** (United States)—made a detailed study of the irruption and death of lemmings.

1948 **Ralph Buchsbaum** (United States)—wrote a very important work on basic ecology.

1952 **Rachel Carson** (United States)—wrote vivid descriptions of the oceans as ecosystems, and organisms that live therein.

1955 **F. W. Went** (United States)—elucidated the various roles of plants in ecosystems.

1955 **Herbert Zim** (United States)—presented important information about the living organisms that inhabit seashores.

1957 **G. S. Avery** (United States)—elucidated reasons for the death of oak trees in U.S. cities.

1958 **Eugene Odum** (United States)—has been a motivator in the field of ecology.

1959 **W. H. Amos** (United States)—researched the living organisms of sand dunes.

1960 **Howard T. Odum** (United States)—developed techniques for measuring the energy flow in forests and marine environments.

1960 **Marston Bates** (United States)—was a leading authority on the ecology of world biomes.

1964 **Rene Catala** (New Caledonia)—was an authority on fluorescent corals.

1967 **George M. Woodwell** (United States)—is a leading authority on the effects of radiation and toxic substances on ecosystems.

1997 **Michael Dombeck** (United States)—is a leading authority on the use of modern ecological principles to manage and preserve public lands.

1999 **Robert H. Mohlenbrock** (United States)—published engaging descriptions of the biological and geological highlights of national forests and parklands in the United States.

2000 **Peter Marchand** (United States)—is a renowned field ecologist who writes about the natural history of animals that occupy a limited ecology range, such as the long-nosed bat of the American Southwest.

Appendix A English/Metric Conversions

METRIC CONVERSION TABLE OF LENGTH

METRIC SYSTEM	ENGLISH SYSTEM
Abbreviations:	in. = inch
mm = millimeter	ft = foot
cm = centimeter	yd = yard
m = meter	mi = mile
METRIC UNITS OF LENGTH	**ENGLISH EQUIVALENTS**
10 mm = 1 cm	2.5 cm = 1 in.
100 cm = 1 m	25 mm = 1 in.
1000 m = 1 km	1 m = 39 in.

HOW TO CONVERT UNITS OF LENGTH FROM THE ENGLISH SYSTEM TO THE METRIC SYSTEM

When you know	inches	*multiply by*	2.5	*to find*	centimeters
	feet		30		centimeters
	yards		0.9		meters
	miles		1.6		kilometers

METRIC CONVERSION TABLE OF VOLUME

METRIC SYSTEM	ENGLISH SYSTEM
Abbreviations:	tsp = teaspoon
mL = milliliter	Tbsp = tablespoon
cc = cubic centimeter	fl oz = fluid ounce
L = liter	pt = pint
	qt = quart
	gal = gallon
METRIC UNITS OF VOLUME	**ENGLISH EQUIVALENTS**
1000 mL = 1.0 L (liter)	29.5 cc = 1 fl oz
1 mL = 1 cc	1.0 L = 1.06 qt
	29.5 mL = 1 fl oz

HOW TO CONVERT UNITS OF VOLUME FROM THE ENGLISH SYSTEM TO THE METRIC SYSTEM

When you know		multiply by		to find	
	teaspoons		5		milliliters
	tablespoons		15		milliliters
	fluid ounces		30		milliliters
	cups		0.24		liters
	pints		0.47		liters
	quarts		0.95		liters
	gallons		3.8		liters

METRIC CONVERSION TABLE OF WEIGHT (MASS)

METRIC SYSTEM
Abbreviations:
mg = milligram
g = gram
kg = kilogram
t = tonne

ENGLISH SYSTEM
oz = ounces
lb = pound

METRIC UNITS OF WEIGHT
1000 mg = 1 g
1000 g = 1 kg

ENGLISH EQUIVALENTS
28.3 g = 1 oz
453.6 g = 1 lb
0.45 kg = 1 lb

HOW TO CONVERT UNITS OF WEIGHT (MASS) FROM THE ENGLISH SYSTEM TO THE METRIC SYSTEM

When you know		multiply by		to find	
	ounces		28		grams
	pounds		0.45		kilograms
	{short tons / 2000 lb}		0.9		tonnes

PREFIXES USED IN THE METRIC SYSTEM

Greek Prefixes Indicating MORE Than the Unit
kilo- = 1000
hecto- = 100
deca- = 10

UNIT
LITER METER GRAM

Latin Prefixes Indicating A PART of the Unit
deci- = 0.1 (one-tenth)
centi- = 0.01 (one-hundredth)
milli- = 0.001 (one-thousandth)

TEMPERATURE CONVERSIONS

When you know		multiply by		to find	
	Fahrenheit		5/9 (after subtracting 32)		Celsius
	Celsius		9/5 (then add 32)		Fahrenheit

Appendix B Terms, Units, and Measurements Used in Microscopy

The Prefix

centi-	means	one-hundreth
milli-	means	one-thousandth
micro-	means	one-millionth
nano-	means	one-billionth

Abbreviations

Å	=	Angstrom unit*
cm	=	centimeter
m	=	meter
mm	=	millimeter
nm	=	nanometer
μ	=	micron† (μ is pronounced "mew")
μm	=	micrometer

** Angstrom units are no longer used by biologists. However, older references contain these units of measurement.*

† The unit "micron" is now called "micrometer" (μm).

Microscopy-Size Units

1 m	=	100 cm	=	39.4 in.		
1 cm	=	10 mm				
1 mm	=	10^{-3} m	=	10^{3} μm		
1 μm	=	10^{-6} m	=	10^{-4} cm	=	103 nm
1 nm	=	10^{-9} m	=	10^{-7} cm	=	Å
1 Å	=	10^{-10} m	=	10^{-8} cm	=	10^{-1} m

Note:	1/10	=	0.1	=	10^{-1}
	1/1000	=	10^{-3}		
	1,000,000	=	10^{6}		

RESOLVING POWERS OF VARIOUS LENS SYSTEMS

Lens System	Resolving Power	
Human eye	0.1 mm	= 100 μm
Light microscope	0.4 μm	
Oil immersion light microscope	0.2 μm	
Ultraviolet microscope	0.1 μm	
Scanning electron microscope	10 nm	= 0.010 μm
Transmission electron microscope	0.2 nm	= 0.0002 μm
Some Average Biological Sizes		
Animal body cell	10–20 μm in diameter	
Plant body cell	30–50 μm in diameter	
Plant chloroplast	5–10 μm long	
Human nerve cell	Up to 2 m (78 in.) long	

Index

MOVE TO THE HEAD OF YOUR CLASS

THE EASY WAY!

Barron's presents **THE E-Z SERIES** (formerly THE EASY WAY SERIES)—specially prepared by top educators, it maximizes effective learning while minimizing the time and effort it takes to raise your grades, brush up on the basics, and build your confidence. Comprehensive and full of clear review examples, **THE E-Z SERIES** is your best bet for better grades, quickly!

ISBN 978-0-7641-1976-7 **Accounting the Easy Way, 4th Ed.**
ISBN 978-0-7641-1972-9 **Algebra the Easy Way, 4th Ed.**
ISBN 978-0-7641-1973-6 **American History the Easy Way, 3rd Ed.**
ISBN 978-0-7641-3428-9 **American Sign Language the Easy Way, 2nd Ed.**
ISBN 978-0-7641-1979-8 **Anatomy and Physiology the Easy Way, 2nd Ed.**
ISBN 978-0-7641-2913-1 **Arithmetic the Easy Way, 4th Ed.**
ISBN 978-0-7641-1079-5 **Bookkeeping the Easy Way, 3rd Ed.**
ISBN 978-0-7641-0314-8 **Business Letters the Easy Way, 3rd Ed.**
ISBN 978-0-7641-1359-8 **Business Math the Easy Way, 3rd Ed.**
ISBN 978-0-7641-2920-9 **Calculus the Easy Way, 4th Ed.**
ISBN 978-0-7641-2579-9 **Creative Writing the Easy Way**
ISBN 978-0-7641-2146-3 **Earth Science the Easy Way**
ISBN 978-0-7641-1981-1 **Electronics the Easy Way, 4th Ed.**
ISBN 978-0-7641-3736-5 **English for Foreign Language Speakers the Easy Way**
ISBN 978-0-7641-1975-0 **English the Easy Way, 4th Ed.**
ISBN 978-0-7641-4134-8 **E-Z Biology, 4th Ed.**
ISBN 978-0-7641-4128-7 **E-Z Chemistry, 5th Ed.**
ISBN 978-0-7641-3918-5 **E-Z Geometry, 4th Ed.**
ISBN 978-0-7641-4261-1 **E-Z Grammar, 2nd Ed.**
ISBN 978-0-7641-4132-4 **E-Z Math, 5th Ed.**
ISBN 978-0-7641-4126-3 **E-Z Physics, 4th Ed.**
ISBN 978-0-7641-4129-4 **E-Z Spanish, 5th Ed.**
ISBN 978-0-7641-3978-9 **E-Z Statistics, 4th Ed.**
ISBN 978-0-7641-3050-2 **Forensics the Easy Way**
ISBN 978-0-7641-3411-1 **French the Easy Way, 4th Ed.**
ISBN 978-0-7641-2435-8 **French Grammar the Easy Way**
ISBN 978-0-7641-3413-5 **Italian the Easy Way, 3rd Ed.**
ISBN 978-0-8120-9627-9 **Japanese the Easy Way**
ISBN 978-0-7641-3237-7 **Macroeconomics the Easy Way**
ISBN 978-0-7641-9369-9 **Mandarin Chinese the Easy Way, 2nd Ed.**
ISBN 978-0-7641-1871-5 **Math Word Problems the Easy Way**
ISBN 978-0-7641-2845-5 **Microbiology the Easy Way**
ISBN 978-0-8120-9601-9 **Microeconomics the Easy Way**
ISBN 978-0-7641-2794-6 **Organic Chemistry the Easy Way**
ISBN 978-0-7641-2892-9 **Precalculus the Easy Way**
ISBN 978-0-7641-2393-1 **Psychology the Easy Way**
ISBN 978-0-7641-2263-7 **Spanish Grammar the Easy Way**
ISBN 978-0-8120-9852-5 **Speed Reading the Easy Way**
ISBN 978-0-7641-3410-4 **Spelling the Easy Way, 4th Ed.**
ISBN 978-0-7641-1360-4 **Trigonometry the Easy Way, 3rd Ed.**
ISBN 978-0-8120-9765-8 **World History the Easy Way, Vol. One**
ISBN 978-0-8120-9766-5 **World History the Easy Way, Vol. Two**
ISBN 978-0-7641-1206-5 **Writing the Easy Way, 3rd Ed.**

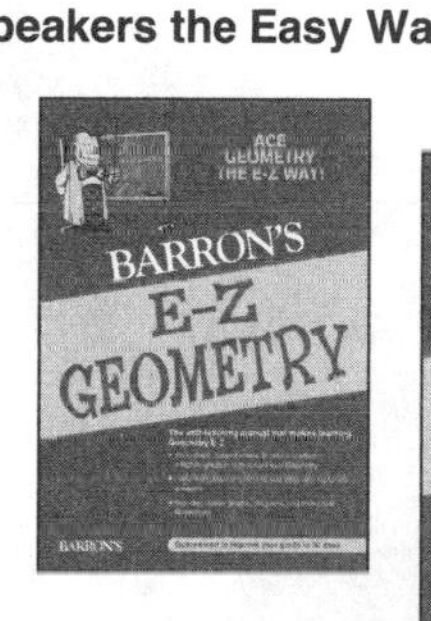

Barron's Educational Series, Inc.
250 Wireless Boulevard
Hauppauge, New York 11788
In Canada: Georgetown Book Warehouse
34 Armstrong Avenue
Georgetown, Ontario L7G 4R9

Please visit
www.barronseduc.com to vie
current prices and to order bo

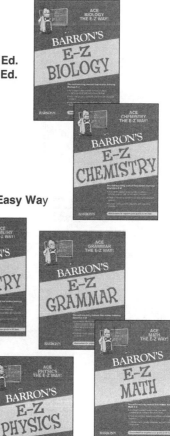

w
ks

45) R11/08